Assessment of Energy–Environment–Economy Interrelations

Assessment of Energy–Environment–Economy Interrelations

Special Issue Editor

George E. Halkos

MDPI • Basel • Beijing • Wuhan • Barcelona • Belgrade • Manchester • Tokyo • Cluj • Tianjin

Special Issue Editor
George E. Halkos
University of Thessaly
Greece

Editorial Office
MDPI
St. Alban-Anlage 66
4052 Basel, Switzerland

This is a reprint of articles from the Special Issue published online in the open access journal *Energies* (ISSN 1996-1073) (available at: https://www.mdpi.com/journal/energies/special_issues/energy_environment_economy_interrelation).

For citation purposes, cite each article independently as indicated on the article page online and as indicated below:

LastName, A.A.; LastName, B.B.; LastName, C.C. Article Title. *Journal Name* **Year**, *Article Number*, Page Range.

ISBN 978-3-03928-809-0 (Pbk)
ISBN 978-3-03928-810-6 (PDF)

Contents

About the Special Issue Editor vii

Preface to "Assessment of Energy–Environment–Economy Interrelations" ix

George E. Halkos and Apostolos S. Tsirivis
Energy Commodities: A Review of Optimal Hedging Strategies
Reprinted from: *Energies* **2019**, *12*, 3979, doi:10.3390/en12203979 1

Juan Uribe-Toril, José Luis Ruiz-Real, Juan Milán-García and Jaime de Pablo Valenciano
Energy, Economy, and Environment: A Worldwide Research Update
Reprinted from: *Energies* **2019**, *12*, 1120, doi:10.3390/en12061120 21

Yu Hao, Zirui Huang and Haitao Wu
Do Carbon Emissions and Economic Growth Decouple in China? An Empirical Analysis Based on Provincial Panel Data
Reprinted from: *Energies* **2019**, *12*, 2411, doi:10.3390/en12122411 41

Abbas Ali Chandio, Yuansheng Jiang, Abdul Rauf, Amir Ali Mirani, Rashid Usman Shar, Fayyaz Ahmad and Khurram Shehzad
Does Energy-Growth and Environment Quality Matter for Agriculture Sector in Pakistan or not? An Application of Cointegration Approach
Reprinted from: *Energies* **2019**, *12*, 1879, doi:10.3390/en12101879 57

Valentijn Stienen and Jacob Engwerda
Measuring Impact of Uncertainty in a Stylized Macroeconomic Climate Model within a Dynamic Game Perspective
Reprinted from: *Energies* **2019**, *12*, 482, doi:10.3390/en13020482 75

Zhi Zhang, Jiaorong Ren, Kaichao Xiao, Zhenzhi Lin, Jiayu Xu, Wei Wang and Chuanxun Pei
Cost Allocation Mechanism Design for Urban Utility Tunnel Construction Based on Cooperative Game and Resource Dependence Theory
Reprinted from: *Energies* **2019**, *12*, 3309, doi:10.3390/en12173309 115

Shining Zhang, Fang Yang, Changyi Liu, Xing Chen, Xin Tan, Yuanbing Zhou, Fei Guo and Weiyi Jiang
Study on Global Industrialization and Industry Emission to Achieve the 2 °C Goal Based on MESSAGE Model and LMDI Approach
Reprinted from: *Energies* **2020**, *13*, 825, doi:10.3390/en13040825 131

Dariusz Mikielewicz, Krzysztof Kosowski, Karol Tucki, Marian Piwowarski, Robert Stępień, Olga Orynycz and Wojciech Włodarski
Influence of Different Biofuels on the Efficiency of Gas Turbine Cycles for Prosumer and Distributed Energy Power Plants
Reprinted from: *Energies* **2019**, *12*, 3173, doi:10.3390/en12163173 153

Paweł Ziemba
Inter-Criteria Dependencies-Based Decision Support in the Sustainable wind Energy Management
Reprinted from: *Energies* **2019**, *12*, 749, doi:10.3390/en12040749 175

Xi Zhang, Zheng Li, Linwei Ma, Chinhao Chong and Weidou Ni
Analyzing Carbon Emissions Embodied in Construction Services: A Dynamic Hybrid Input–Output Model with Structural Decomposition Analysis
Reprinted from: *Energies* **2019**, *12*, 1456, doi:10.3390/en12081456 . . . . **205**

George Halkos and Kleoniki Natalia Petrou
Analysing the Energy Efficiency of EU Member States: The Potential of Energy Recovery from Waste in the Circular Economy
Reprinted from: *Energies* **2019**, *12*, 3718, doi:10.3390/en12193718 . . . . **229**

About the Special Issue Editor

George Halkos (BA, MSc, PhD) is Professor in Economics of Natural Resources. He has worked as a team leader and research fellow in various research and academic institutions for research projects financed by the European Union (PHARE programs), the Hellenic Republic Ministry of Development, the National Statistical Service of Greece, EUROSTAT, and the Ministry of Employment, among others.

He has participated and presented articles at international conferences, and he acts as a referee in various scientific journals. He is Director of the Operations Research Laboratory. His research interests are in the fields of Applied Statistics and Econometrics, Simulations of Economic Modeling, Natural Resource and Environmental Economics, Applied Microeconomics (with an emphasis on Welfare Economics), Air Pollution, Game Theory, and Mathematical Models (Non-Linear Programming).

Preface to "Assessment of Energy–Environment–Economy Interrelations"

The development of new energy and environmental policies, the new climate regime, and the development of new scientific techniques have provided ample opportunity for further research. Projects on energy efficiency have been prioritized in the portfolio of policies in many countries, as these policies are considered no-regret options, meaning that they may even provide gains for the economy. In order to fill a gap in the literature, this book is intended to provide an analysis of the energy–environment–economy interrelations, with special attention paid to the potential role of energy and economic growth in the environment.

Currently, energy is considered a commodity, and continuous access, along with price stability, is of vital importance for every economic agent worldwide. Halkos and Tsirivis (2019) provided a review of energy hedging and discussed the two main hedging strategies related to energy portfolios, namely, the minimum-variance hedge ratio and the expected utility maximization methodology. They showed that when substantiation from the energy market during exceptionally volatile economic climate periods is considered, both hypotheses may be violated, implying that it would be sensible for possible hedgers to take into account both methodologies in constructing a successful and profitable hedging strategy.

Energy consumption and economic growth have been of great interest to researchers and policy-makers. Knowledge of the actual causal relationship between energy and the economy with respect to the environment has important implications for modeling environmental and growth policies. Uribe-Toril et al. (2019) reviewed the international research on interactions between the 3E, that is, Economy, Energy, and Environment, in the 21st century. They used bibliometric and cluster analyses by fractional accounting and relied on the two most extensive databases: Web of Science (WoS) and Scopus. This paper contributed an analysis of keywords from 2001 to 2018, with trends showing that sustainable development and sustainability, together with CO_2 emissions and consumption, were the main common elements. Moreover, Hao et al. (2019) combined the Tapio decoupling model and the environmental Kuznets curve (EKC) framework to explore the relationship between China's carbon emissions and economic growth. Panel data of 29 provinces from 2007 to 2016 were used to estimate the nexus of emissions, development for the nation, and the decoupling status of individual provinces. Similarly, Chandio et al. (2019) considered the LR influence of financial development, economic growth, energy consumption (specifically, electricity consumption in the agriculture sector), foreign direct investments (FDI), and the population on the environment in Pakistan for the period 1980–2016. CO_2 emissions from agriculture were used as an indicator of environmental quality. Their findings showed that increasing financial development and FDI helped environmental quality, whereas higher economic growth and electricity consumption in agriculture damaged the environment in Pakistan.

The game theory set-up may also be helpful, given that uncertainty is a major issue in such analyses. Stienen and Engwerda (2020) considered a stylized dynamic interdependent multi-country energy transition model in an aim to examine the effect of uncertainty in such cases. A simple model based on the standard Solow macroeconomic growth model was developed in a two-country setting using non-cooperative dynamic game perspectives. Total CO_2 emissions were added as a factor that negatively influences growth, and production could be realized with either green or fossil energy. A factor that captures the difficulties in using green energy, such as accessibility in each country, was also added. In general, the model satisfactorily describes energy transitions towards different equilibrium constellations. Additionally, Zhang et al. (2019) considered the high investment cost of utility tunnels and the limitations of common cost-sharing methods (such as spatial proportional, direct-laying cost, and benefit-based proportional methods) in their effort to establish a fair and practical cost-sharing mechanism. For this purpose, an improved Shapley value-based spatial proportional method was suggested, and the resource dependence theory was introduced to enumerate the bargaining power of pipeline companies in negotiating cost allocation. Simulations on the utility data in China demonstrated that the suggested cost allocation mechanism was the most satisfactory, as well as more adequate and more practical, compared with traditional cost allocation methods.

The analysis and quantification of relationships between industrialization, energy systems, and carbon emissions are crucial. Zhang et al. (2020) developed an expanded Kaya identity to explore the main drivers of industrial emissions by employing the logarithmic mean Divisia index method to follow the historical contributions of various sources of emissions, together with forecasts into the future. They found that development and the population were the two main determinants of past industrial CO_2 emissions, while carbon and industry energy intensities were predicted to be the main two factors for the reduction of future industrial CO_2 emissions. Clean supply, electrification, and energy efficiency were suggested for industrial emission reduction

Obviously, energy efficiency and renewable energy sources are essential in addressing environmental degradation. Mikielewicz et al. (2019) estimated the effect of fuel calorific value on turbine performance and analyzed the possibilities for optimizing turbine construction in terms of maximum efficiency for specific fuels. The careful design of such devices attained high efficiency. Thess individually created generation systems that may be used in distributed generation systems to achieve environmental profits.

Along these lines, Ziemba (2019) discussed an approach to solving wind energy-related decision problems that demand many criteria, which are sometimes interrelated and dependent on each other. In these cases, decision systems that rely on multi-criteria decision analysis (MCDA) methods are usually used, but most methods assume independence between criteria, making their use in wind energy decision problems arguable. Therefore, this paper discussed the use of the analytic network process (ANP) method to select the location and design of wind farms. This method captures the complexity of decision problems that involve criteria dependencies. The results of the ANP method were compared with those of the analytic hierarchy process (AHP), which relies on hierarchical dependencies between criteria. They claimed that the rankings extracted from the ANP were of higher quality than those of the AHP.

Furthermore, Zhang et al. (2019) proposed a dynamic hybrid input–output model combined with structural decomposition analysis (DHI/O-SDA model). Taking China as an example, this DHI/O-SDA model was verified with the presentation of bilateral relationships among sectoral responsibilities for energy-related carbon emissions (ERCE) in construction services. Their main finding was that the "Other Tertiary Industry" sector is responsible for ERCE in construction services, which also affect other sectors. In this way, controlling the final demand increase in the service industry was deemed the most effective policy to reduce ERCE generated by construction services.

Finally, Halkos and Petrou (2020) considered energy efficiency in the European Union (EU) member states and reviewed the potential for energy recovery from waste according to derived efficiency scores. These efficiencies were assessed using data envelopment analysis (DEA). According to these efficiency scores, various recommendations were proposed to meet the stated objectives, namely, sufficient and sustainable energy production and effective treatment of municipal solid wastes.

All the above-cited papers aim to help decision-makers further understand relevant issues and adopt appropriate methods to tackle and solve relevant problems.

George E. Halkos
Special Issue Editor

energies

MDPI

Article

Energy Commodities: A Review of Optimal Hedging Strategies

George E. Halkos * and **Apostolos S. Tsirivis**

Laboratory of Operations Research, Department of Economics, University of Thessaly, 33888 Volos, Greece; apostolostsirivis@hotmail.com

* Correspondence: halkos@econ.uth.gr

Received: 11 August 2019; Accepted: 16 October 2019; Published: 18 October 2019

Abstract: Energy is considered as a commodity nowadays and continuous access along with price stability is of vital importance for every economic agent worldwide. The aim of the current review paper is to present in detail the two dominant hedging strategies relative to energy portfolios, the Minimum-Variance hedge ratio and the expected utility maximization methodology. The Minimum-Variance hedge ratio approach is by far the most popular in literature as it is less time consuming and computationally demanding; nevertheless by applying the appropriate multivariate model Garch family volatility model, it can provide a very reliable estimation of the optimal hedge ratio. However, this becomes possible at the cost of a rather restrictive assumption for infinite hedger's risk aversion. Within an uncertain worldwide economic climate and a highly volatile energy market, energy producers, retailers and consumers had to become more adaptive and develop the necessary energy risk management and optimal hedging strategies. The estimation gap of an optimal hedge ratio that would be subject to the investor's risk preferences through time is filled by the relatively more complex and sophisticated expected utility maximization methodology. Nevertheless, if hedgers share infinite risk aversion or if alternatively the expected futures price is approximately zero the two methodologies become equivalent. The current review shows that when evidence from the energy market during periods of extremely volatile economic climate is considered, both hypotheses can be violated, hence it becomes reasonable that especially for extended hedging horizons it would be wise for potential hedgers to take into consideration both methodologies in order to build a successful and profitable hedging strategy.

Keywords: energy commodities; hedging strategies; minimum-variance hedge ratio; expected utility maximization; risk aversion

JEL Classification: O1; P2; Q4; QO2

---

1. Introduction

Continuous access along with price stability of energy commodities is of vital importance for every state or individual economic unit around the globe. As a result, energy risk composes a major risk factor for most firms involved in every key industrial sector in both developed and developing countries [1]. Relatively few studies have been conducted regarding the understanding of energy risk and the measurement of price risk exposure and an even lower number of research has been developed in the field of optimal hedging [2–4]. In those regards optimal hedging strategies are designed under the assumption that managers maximize their expected utility while their income from the firm is increasing with changes in the value of the firm [4].

Apart from diminishing the level of risk exposure regarding energy commodities, it is equally essential for all counterparties in this market to create the optimal strategy or portfolio that will enable them to maximize their expected profit, given their approved level of risk exposure [5]. In order for this goal to be accomplished, a variety of financial derivatives over specific energy commodities is used; the most common and widely accepted ones being the Forwards and Futures contracts, Option rights and Swaps. Derivatives provide the advantage to hedge some or all risk coming from the spot price energy market, ensuring future energy commodity sales (Short position) or purchases (Long position) at a prearranged time, place (some of them do not involve physical delivery or they are cashed out prior to expiry date) and price [5]. Due to the increasing uncertainty in the energy commodity market over the last decade, there is high interest from all participants for more extensive research to be conducted around energy portfolio optimization. Additionally, concerning the ability of estimating and predicting the relative price risk exposure, researchers put their mixed Garch-VaR (Value-at-Risk) (see [6]) type models into another assessment involving revealing the optimal hedging strategy for certain energy products.

Specifically, there are two dominant methodologies for estimating the optimal hedging strategy or optimal hedging ratio, including the Minimum-Variance hedge ratio and the expected utility maximization [5]. The two methodologies vary in one key aspect which involves the expected utility maximization approach incorporating into the estimation of the optimal hedge ratio, a time varying parameter representing the investor's risk aversion [5]. In the expected utility maximization methodology three basic types of utility functions are used, including the quadratic, the logarithmic and the exponential utility function. On the contrary, in the Minimum-Variance hedge ratio methodology infinite risk aversion is assumed, with a number of developed tests showing which model and methodology provides the most appropriate hedging strategy for profit maximization [7].

Within an uncertain worldwide economic climate and a highly volatile energy market, energy producers, retailers and consumers had to become more adaptive and develop the necessary energy risk management and optimal hedging strategies. Subsequently, firms that succeed in securing their access to the required energy sources, while minimizing the relative cost, gain a serious competitive advantage over their rivals and reinforce their established market position and profitability.

Of the first to deal with a nation's competitive advantage was Ricardo [8] who identified that if two countries capable of producing two commodities engage in the free market, then each country will increase its overall consumption by exporting the good for which it has a comparative advantage. Moreover Manoilescu [9] supports the opinion that the engagement of productive forces in industry is always more productive than agriculture and other raw materials.

In those regards Balassa [10] defined the revealed comparative advantage as an index used in international economics for calculating the relative advantage or disadvantage of a certain country in a certain category of goods or services as evidenced by trade flows. Porter [11] shows that a nation's competitiveness depends on its capacity to innovate and upgrade, with the determinants of national competitive advantage being: factor conditions, demand conditions, related and supporting industries and firm strategy and structure.

With regards to competitive advantage in the energy sector, Dogaru [12] shows that the trade flows of elementary and macroeconomic process are explained using comparative and absolute advantage principles (CAAPs). As for renewable energy sources, while competitive advantage appears to remain stable over time for the wind industry, it decreases in the solar PV industry after four or five years [13].

Firms that do not manage to hedge their energy risk effectively, especially in times with intense price fluctuations, increase their overall production cost putting under serious threat their current and future viability. Furthermore, energy producers and retailers who in most cases already possess energy commodities and have a higher motive not to hedge but speculate in the spot market, might suffer huge losses or even go bankrupt within days like in some recent examples coming from the US energy market, unless they develop the necessary mechanisms to constantly measure their risk exposure and build a suitable hedging strategy [14]. In 2019 alone a total of 26 firms with a total debt of $10.96 billion have filed for court restructuring until mid-August [15].

This paper focuses on hedging strategies in the energy sector. More specifically it thoroughly reviews the two dominant hedging methodologies relative to energy portfolios, the Minimum-Variance hedge ratio and the expected utility maximization methodology. In more detail Section 2 explains how hedging the price risk of energy commodities works, while Section 3 presents the main strategies for building the optimal hedging approach, Section 4 demonstrates some studies which have used the approaches described in Section 3. Finally, the last Section (Section 5) concludes the paper.

2. Hedging the Price Risk of Energy Commodities

This section reviews in greater detail the options for hedging the price risk of energy commodities and the relevant tools to be used, which include forward contracts (Section 2.1.1), futures contracts (Section 2.1.2), option contracts (Section 2.1.3) and swap contracts (Section 2.1.4).

The most effective strategy for every business to reduce the danger coming from the various risk types such as price risk, basis risk, credit risk, operational risk to name a few while remaining profitable and solvent is hedging. Especially for businesses belonging in the energy sector or severely depending on certain energy commodities as inputs, their ability to effectively hedge the different risks and particularly the market risk or price risk is crucial, due to the extreme price volatility of energy commodities.

Reference [16] reports that based on a sample of 100 oil and gas producers, the companies using hedging strategies increased significantly within a three-year period in which they were investigated, where a noteworthy number of them ended up hedging more than 28% of their total production. Furthermore, in an attempt to reduce their overall risk, it was noted that firms with higher debt, tended to also hedge a higher percentage of their total production, while firms possessing a larger number of assets were more likely to develop a hedging strategy.

Additionally, according to [17] firms hedge either because the management has a risk averse mentality in general, or they want to lower the probability of falling in financial distress and be unable to fund any profitable projects. Moreover, based on [18] firms decide to hedge to diminish any risks deriving from business processes that do not have any insight or control, enabling them to concentrate on their core competences improving the firm's effectiveness and efficiency. References [19,20] report, that especially energy consumer firms which hedge against the price risk of the energy products they use as basic inputs, can obtain significant benefits and grow the overall firm value.

Ceczy et al. [21], found evidence that companies using commodity derivatives for hedging purposes appear to have significantly less volatile stock prices, with the companies having a lower bond rating hedging substantially more than those with higher credit reliability. Additionally, reference [22] using a large sample of oil and gas producers in the United States, also found strong indication of significant sensitivity reduction in the stock prices of the most active companies in terms of hedging practices.

Therefore, as it comes forward, energy prices are by far the most volatile of all other commodities with the volatility difference increasing considerably during the last decade. Therefore, hedging of energy risk can add value to firms in many ways, enabling the firm to have a greater debt capacity, low cost of capital, capital availability for optimal investing even through periods of unexpectedly low cash flows and of course avoid the cost of financial distress.

2.1. Hedging Tools for Managing Energy Commodities' Price Risk

Financial derivatives are the key hedging instrument that enables firms to manage the risks coming from the persistent high volatility and uncertainty in energy commodities' prices. Derivatives are secondary market contracts that instead of directly depicting certain ownership rights about an asset, they derive their current value from an underlying commodity or asset [23]. The wise use of derivatives for hedging purposes allows for an effective reduction of price risk exposure, as in this way derivative investors accomplish to transfer part of their overall risk exposure to a third party in exchange for potential profit.

The party transferring risk ensures price certainty for a given time period mentioned in the contract, though sacrificing the potential of making extra profit from a price movement towards its favorable side [23]. In contrast, the party accepting the risk will realize a loss if the price movement confirms the initial fears of the other party. In general, derivatives are becoming more and more important for everyone involved in financial markets and play an even more crucial role in hedging the extreme price risk, that have to deal with both energy producers as well as industries heavily dependent on high demand energy commodities [5].

The reason behind this, is that the characteristic features of energy commodities provide additional flexibility when facing the extreme energy price risk, while they can provide enterprises with the necessary security and certainty about any future expected cash flows [24]. The largest number of available derivative contracts in the market consist of forward contracts, futures contracts, options and swaps.

2.1.1. Forward Contracts

Forward contracts are more commonly used in the electricity market among individual producers and industrial firms and can mainly be described as a contractual agreement, which specifies the buying or selling of a particular commodity or asset of given quality and quantity at a pre-specified price and delivery place in a future time [24]. Forward contracts are basically a step further of the traditional cash and carry exchanges with the delivery taking place in the future.

In the oil market forward contracts are commonly used by industrial firms to make sure that they will maintain the necessary oil reserve that is needed to guarantee their future operational ability, while avoiding the extra costs required for the storage until the time of use [25]. The inability of electricity to be directly stored except from the excessively expensive possibility of saving remaining production capacity of power plants, as well as the flexibility to adjust in the exact needs of both the producers or retailers and the large consumer firms made this type of derivative contracts extremely useful and popular among electricity market participants [26].

Nevertheless, because of the aforementioned unique characteristics of forward contracts several issues may arise, as it can prove to be rather difficult to find a suitable counterparty that will match the exact needs of the producer or the consumer. A common problem that is mostly present in the electricity market refers to the difficulty of delivering the purchased electricity when making a forward contract with a producer that is far away from the supplying network of the consumer's region [27].

Additionally, both counterparties face the default and credit risk exposure of the other party, with the risk significantly rising for forward contracts with long future time delivery or when the contract value is moving too much in favor of one of the two parties, in a way strangling the other party and thus making inevitable to fail meeting its contractual obligations [27]. For that reason, investors who intend to get involved in a forward contract, need to first thoroughly investigate their potential counterparties' reliability and credit ratings, or set collateral requirements prior to final agreement to secure the viability and validity of the contract. Finally, there is a chance that the needs or the operational conditions of one of the two involved parties change during the time of the contract; this usually leads to renegotiation of certain contractual clauses facing very strict penalties [25].

2.1.2. Futures Contracts

Futures are very similar to forward contracts in terms that they also represent the obligation to sell or buy a specific quantity of a certain commodity for a pre-specified price and delivery place, however counterparties in futures contracts avoid several risks and problems [28]. Specifically, involved parties in futures contracts avoid are much easier to find the most appropriate counterparty which will be able to cover their exact needs, as it is not necessary to search for the other party on their own.

Instead there are futures exchanges which take on the role of getting together the potential futures investors and specialized dealers who are responsible to represent these investors to the exchange, while at the same time they are in charge of fulfilling the clauses of the contractual agreement [29]. Additionally, futures investors avoid counterparty risk, as when they enter a futures agreement they are obliged to disburse an initial amount which is used to cover any losses from the daily 'marking to market' examination of the party's position. Nevertheless, in case this amount is not sufficient to cover the losses then it is the broker's obligation to cover his client's or defaulting party's losses and close his position [29]. If the broker cannot fulfill this either, then the exchange will bear the responsibility to compensate the other party.

Especially, in the oil market, futures contracts can prove to be very useful as they allow for additional hedging strategies and contracts such as the crack spread contracts [30]. For example, refiners mainly fear the price difference between their basic input and the price of their produced output products, rather than the actual level of prices. Being able to safely estimate all other costs except oil, refiners are primarily concerned about this price spread and as a result they are interested in strategies to ensure this spread, which their potential profits are largely depended on [30].

The most common hedging strategy in this case is to buy futures on oil and simultaneously sell futures on their oil refined products. In order to cover these multiple transactions, the crack spread contract was created, which incorporates all the above necessary hedging actions to ensure the price spread in just one trade [30]. A rather popular crack spread contract is the one including initially buying three crude oil futures, while selling two gasoline futures and one heating oil future one calendar month later. However because this 3-2-1 ratio based crack spread contract cannot meet the needs of all refiners in an over-the-counter market outside exchanges that have been developed [31].

On the other hand, as in the case of forwards, futures contracts are also accompanied by some risks. It is most common for counterparties in futures agreements to close their position prior to maturity, hence physical delivery rarely is taking place with both parties exploiting their chances for making profit until the settlement date [31]. However, it is possible for someone to sell futures on a specific energy commodity without having the obligation to actually possess the underlying commodity in the first place. This fact is widely taken advantage by speculators, who are willing to get transferred the producers' price risk and gamble on the price movements of the energy commodities by selling or buying futures contracts and close their position prior to the delivery date [29].

Furthermore, another characteristic of futures contracts that may arise further questions concerning the risks in that particular secondary commodity market, is that the necessary initial margin required by the participants in a futures contract is significantly smaller relative to their overall commitment to buy or sell a specific energy commodity [32]. This allows investors to leverage their position realizing enormous profits or losses for only small price changes. Finally, because futures contracts are only available for a few specific energy commodities, very limited number of delivery locations and a shorter to a decade ahead time horizon, there is a fast growing over-the-counter market outside exchanges that covers the gap between the investors' needs and what is offered in futures exchanges [31].

2.1.3. Option Contracts

The purchaser of an option contract for an underlying energy commodity, is basically buying the right to sell to (put option) or buy (call option) from the contract issuer a specific amount of the energy commodity for a pre-determined price over a specific future time period [31]. There are two main types of options contracts, where the first one, the American type, provides the contract owner the ability to exercise the described in the contract right at any point until the maturity date [33]. The second type, referred as the European type of option contracts, can be exercised only on the pre-defined maturity date [33]. Nevertheless, both types of option contracts regardless of being purchased on an exchange or over-the-counter, they are paid in advance [33].

Additionally, option contracts offer an alternative to employing a hedging strategy relying on futures like hedging using crack spread contracts. This alternative strategy involves buying call options on an energy commodity which is used as a basic input such as oil, while simultaneously selling put options on the refined products [33]. Moreover, in the electricity market it is most common for suppliers to purchase electricity options in order to eliminate the risk of their clients consuming more electricity than the relative amount corresponding in the futures contracts that are in the supplier's possession. Finally, similarly to crack spread contracts in the oil market, in the case of electricity there are the spark spread contracts aiming to diminish price risks and specifically price fluctuations between the electricity's selling price and the price of the necessary input fuels for its production [34].

2.1.4. Swap Contracts

Swap contracts is the latest development in the financial derivatives market and were created in an attempt to provide price security at a lower cost than option contracts. Swaps contrary to other derivatives do not involve actual physical delivery of the underlying commodity, but instead works as an agreement between two parties to exchange a number of cash flows based on the price changes of the underlying asset or commodity [35]. This type of contracts mainly concerns external agreements which do not take place in a traditional exchange or other established trading facility, hence they are considered as over-the-counter derivatives [35].

As a result of no physical delivery taking place and no amounts being initially paid as a security, a principal base or notional amount is being determined upon which the various future cash flow transfers will take place [36]. This notional amount can represent the current market value of the two assets that are about to be swapped between counterparties or the quantity of a specific underlying commodity for which the cash flow settlements of each month will be arranged based on its price fluctuations.

In general, swaps share a large number of similar characteristics with futures and option contracts, allowing hedging price risk exposure without obliging the counterparties involved in the agreement to possess the actual underlying commodity or asset [31]. However, the fact that swaps are individually negotiated contracts allows counterparties to be more flexible and customize their swap agreements, enabling them to better manage the risks that arise from their business activities and are vital to be hedged in order to ensure the financial stability and future viability of their firms [36].

Nevertheless, the lack of security that accompanies swap contracts as they are not guaranteed by an established clearinghouse and hence the high counterparty and credit risk exposure, often lead to less liquid swap contracts as it very often is the case that counterparties renegotiate very much in detail all the relative contractual terms before they decide to offset or terminate a swap agreement prior to its expiry date [27].

3. Building the Optimal Hedging Strategy

Following the review of the most commonly used hedging strategies, this Section focuses on the main strategies for building the optimal hedging approach, namely the Minimum-Variance hedging strategy (Section 3.1) and the expected utility maximization methodology for hedging (Section 3.2) and a few alternative hedging strategies (Section 3.3).

Managing the energy price risk is becoming more and more crucial for all businesses and investors that are involved in a direct or less direct way with that particular market, as the volatility in energy product prices and their derivative contracts is by far the highest than any other asset [37]. Nevertheless, not all interested parties deciding to deal with energy price risk and develop a relative hedging strategy belong in the same group. Participants in the energy derivatives market are often driven by various and most often opposite incentives [37].

In the existing literature, hedgers are typically separated in two basic groups including short hedgers that are most likely to represent the position of an energy commodity producer, greatly worrying about potential price decreases and long hedgers, which are mainly heavily dependent firms using energy commodities as one of their basic inputs and are deeply worrying about potential price increases [38]. As a result, it is clear that the two groups are concerned about the exactly opposite side of the return distribution, as [39] found evidence deriving from the oil futures market that the vast majority of short hedgers (producers) merely hedge the difference of their present production to the minimum economic production level and the extreme correlation between oil producers' profits and actual prices strongly indicate that producers hedge only a small portion of their overall production.

During the last two decades a series of academic studies have been developed trying to explore what should be considered as an optimal hedging strategy. There were two fundamental approaches in the field in which researchers based their studies to provide estimations of the optimal hedge ratio.

The first, and perhaps most popular approach refers to minimizing the return volatility as an optimal hedging strategy and is well known as the Minimum-Variance hedge ratio. This approach may be by far less time consuming and computationally demanding than others, however it may lead to rather unrealistic and even false conclusions if the limitations of the particular method are not taken into serious consideration by the researcher [5]. These limitations refer to the assumption for zero expected return on the futures contract or for constant, infinite risk aversion regarding the general risk attitude of the hedgers, which is a fairly important hedging parameter considering that there can be significant alternations of this factor between the two subsequent groups of hedgers [40].

The second more popular hedging strategy is the expected utility maximization methodology and is used in both financial and energy risk management, as it takes into serious account the aspect of risk aversion and relies on the utility maximization framework to estimate the optimal hedge ratio [5].

In general, investors aim to secure their portfolio's position in the spot market by using financial derivatives and especially futures contracts, hence the optimal hedging ratio represents the exact combination of spot market investments and futures that would eliminate or minimize to the lowest possible degree the volatility of the overall portfolio value [5]. As a result, considering a portfolio of A_s assets in the spot market (long spot position) and A_F assets in the futures market (short futures position), P_{S_t} and P_{F_t} denoting the spot and futures prices at a specific time t and r_{S_t} and r_{F_t} the net returns for a single period from $t-1$ to t, then the total return of the hedged portfolio r_h is estimated as follows:

$$r_h = \frac{A_S P_S r_{S_t} - A_F P_F r_{F_t}}{A_S P_s} = r_{S_t} - \delta_{h_{t-1}} r_{F_t} \tag{1}$$

where $\delta_{h_{t-1}}$ represents the hedge ratio and is basically defined as the ratio of the futures position value to the value of the spot position at time $t-1$ showing how many currency units are invested in the futures market for each unit invested in the spot market.

Nevertheless, due to the fact that the optimal hedge ratio plays a key role in every successful hedging strategy, it is of vital importance to mention that its estimation is always subject to the specific objective function that needs to be optimized based on the chosen hedging methodology [6]. Therefore, the optimal hedge ratio which according to existing literature can be both static and dynamic, may either represent the investment strategy that minimizes the variance of the total portfolio value or maximizes a particular utility function or is in line with the limitations set by a prespecified VaR level [41].

3.1. *The Minimum-Variance Hedging Strategy*

The vast majority of academic research relies on [42] variance minimization concept for building the most effective hedging strategy regarding a single or a portfolio of energy commodities. This fundamental methodology which was further developed by [43–45] relies on decreasing the variance of the proposed hedged portfolio to the lowest possible degree.

Under the minimum variance approach, the optimal hedging strategy is the one that simply offers the higher price risk reduction. This particular framework is less computationally demanding, while it also allows for easier interpretation, however it emphasizes solely on the risk reduction and completely ignores the risk aversion and the expected return parameters for the optimal hedging planning [5]. Therefore, on a minimum variance hedging model it is arbitrarily been assumed that all investor groups in energy market share infinite risk aversion, a hypothesis which is rather unrealistic even for the most conservative and modest economic organizations (i.e., public companies, pension funds etc.), as infinite risk aversion means that investors would reject investment opportunities which could offer significant potential returns for even the slightest amount of additional risk [46].

The fact that the minimum variance approach fails to distinguish hedgers both based on their interests (e.g., refiners, producers, consumers etc.), as well as on their individual investor characteristics (e.g., investors, speculators) and hence their attitude towards risk is a very important factor that needs to be taken into account, when estimating the optimal hedging ratio since hedgers may vary from the point of being reluctant to take any risk, up to the point where hedgers are found to adopt unexpectedly risky hedging strategies [47]. Figure 1 shows the graphical representation of the optimal hedging ratio.

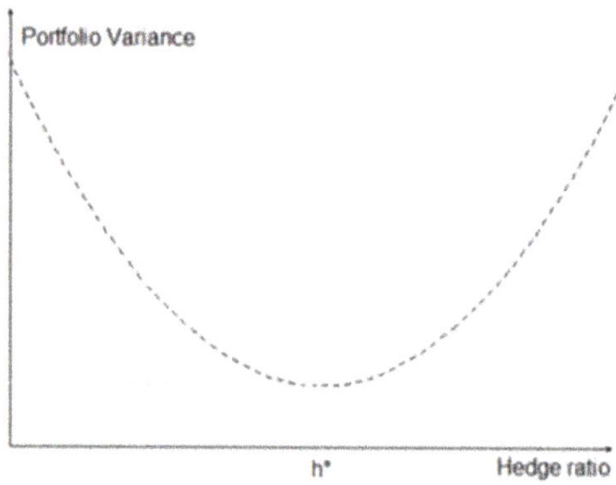

Figure 1. Hedge position and the shape of portfolio variance [48].

Proper assessment of energy risk relies on models that reflect a number of important properties of the underlying assets which affect the performance of the participants' portfolios such as time-dependent volatility and heavy tails [49]. Some of the main factors affecting hedging behavior are: profit opportunities and shareholder values which may solve conflicts over different contract preferences between companies in commodity-marketing channels [50].

3.1.1. Estimation of the Minimum-Variance hedge ratio based on the OLS methodology

The most simplistic method to estimate the Minimum-Variance hedge ratio by taking into account any potential price volatility differentials between the spot and futures prices, involves the use of the OLS regression technique between spot returns and futures returns of the examined energy portfolio [46]. Nevertheless, it is important to mention that the resulting Minimum-Variance hedge ratio of this basic analysis is static and not dynamic. Based on Equation (1) the variance of the portfolio return can be mathematically estimated as follows:

$$\sigma_{h_t}^2 = \sigma_{S_t}^2 + \delta_h^2\sigma_{F_t}^2 - 2\delta_h cov(r_{S_t}, r_{F_t}) \tag{2}$$

where $\sigma_{h_t}^2$ symbolizes the portfolio's conditional variance and the $\sigma_{S_t}^2$ and $\sigma_{F_t}^2$ the conditional variances of the spot and futures positions respectively, while $cov(r_{S_t}, r_{F_t})$ indicates the conditional covariance.

Hence, the Minimum-Variance hedge ratio can be estimated by minimizing the portfolio's conditional variance $\left(\sigma_{h_t}^2\right)$ with respect to δ_h:

$$\delta_h = \frac{cov(r_{S_t}, r_{F_t})}{\sigma_{F_t}^2} = \rho\frac{\sigma_{S_t}}{\sigma_{F_t}} \tag{3}$$

where ρ denotes the correlation between spot and futures returns, while σ_{S_t} and σ_{F_t} represent the standard deviations.

Assuming that the variance-covariance matrix is constant and not time variant, the optimal hedge ratio can be calculated by performing an OLS regression of the spot returns on the futures returns. In this regression the slope parameter will refer to the optimal δ_h^{OLS}. Nevertheless, since most energy commodities are characterized by excess price volatility and thus prices are not reasonable to be considered as stationary for any reason, a dynamic analysis is required instead that will allow for a time variant optimal hedge ratio [5]. Finally, [51] point that the resulting optimal δ_h^{OLS} can be trustworthy only when all OLS methodology criteria are met. However, this is relatively rare to happen mostly due to the heteroskedasticity problem that econometric tests find to be present in the vast majority of energy commodity price data sets.

3.1.2. Estimation of the Minimum-Variance Hedge Ratio based on Nonlinear Multivariate Garch Models

Researchers trying to deal with the problematic and unrealistic hypothesis of the OLS approach for constant variance-covariance matrix of returns that leads to an estimation of a static optimal hedge ratio, started to implement the Arch-Garch methodology in their studies [52]. With the implementation of the appropriate Garch type volatility model (see reference [53] for a complete analysis of the econometric procedures and tests that lead to the choice of the appropriate Garch type volatility model in an energy portfolio), researchers in their estimates for the optimal hedge ratio use the conditional sample variance and covariance resulting from the chosen model.

This particular econometric technique allows for time varying variances and covariances supporting updates of the optimal hedge ratio during the hedging period. Furthermore, the Arch-Garch methodology overcomes another limitation of the OLS approach, which has to do with the presence of heteroskedasticity in the energy commodities' data sample of price returns, as OLS regression provides unreliable and less efficient results in case of heteroskedasticity in the error term. Reference [44] suggest that the use of conditional variance and covariance in the estimation of the optimal hedge ratio for a portfolio of commodities with highly volatile returns provides significantly more accurate estimates. Additionally, [45] using a data set for six different commodities they conclude that the implementation of a static optimal hedge ratio as a hedging strategy can prove to be rather costly.

The estimation of the optimal hedge ratio even for a single energy commodity, which includes the spot and futures returns, requires the application of sophisticated nonlinear multivariate Garch models [54]. The type of volatility models are more commonly used in risk management for energy portfolios, as they are found to be superior in terms of incorporating and revealing the dynamics of variances and covariances, as well as allowing for dynamic interactions between spot and futures returns [55].

The most simplistic version of these type of models that is widely used for the estimation of the optimal hedge ratio is the VECH model, which was initially introduced by [56] and can be considered as a straightforward extension of the basic univariate GARCH model.

The VECH model is estimated as follows:

$$\mathrm{VECH}(H_t) = C + \sum\nolimits_{i=1}^{q} A_i\, vech\left(\varepsilon_{t-1}\varepsilon'_{t-i}\right) + \sum\nolimits_{j=1}^{p} B_j\, vech\left(H_{t-j}\right), \tag{4}$$

All conditional variances and covariances are functions of their own lagged values, along with lagged squared returns and cross-products of returns. Vech(.) denotes an operator, stacking the columns of the lower triangular elements of its suggested square matrix, while H_t represents the resulting conditional covariance matrix, C is an $[N(N+1)/2 \times N(N+1)/2]$ vector and A_i, B_J are $[N(N+1)/2 \times N(N+1)/2]$ parameter matrices.

The model has the advantage of being rather simple and flexible: however it is accompanied by some serious drawbacks and limitations. That is because, firstly, as H_t necessarily remains positive for all ε_t, in order to reasonably estimate all the parameters that are specified by the model, this can be diffic ult to investigate. Secondly, the large number of required parameters, as well as the demanding computational time, critically limit the model given the difficulty to consider more than two basic factors. As a result it is limited to a bivariate model [53].

According to [45] the Minimum-Variance hedge ratio in a bivariate VECH model is estimated as follows:

Let $y_t = (P_s, P_F)'$ denote a (2×1) vector containing cash and futures prices:

$$\begin{bmatrix} \Delta P_S \\ \Delta P_F \end{bmatrix} = \begin{bmatrix} \mu_1 \\ \mu_2 \end{bmatrix} + \begin{bmatrix} \varepsilon_{1t} \\ \varepsilon_{2t} \end{bmatrix} \Leftrightarrow \Delta y_t = \mu + \varepsilon_t \tag{5}$$

With:

$$\varepsilon_t|\Omega_{t-1} \sim N(0, H_t) \text{ and } H_t = \begin{bmatrix} H_{11,t} & H_{12,t} \\ H_{21,t} & H_{22,t} \end{bmatrix} \tag{6}$$

Considering Equation (1) the optimal hedge ratio at time t is given by:

$$\delta_{t-1} = \frac{\sigma_{21,t}}{\sigma_{22,t}} \tag{7}$$

where $\sigma_{ij,t}$ corresponds to the value of the exact same position in the H_t conditional covariance matrix.

Despite its drawbacks and limitations the VECH model remains popular in estimating the Minimum-Variance hedge ratio in portfolios with a very small number of assets as it provides a time variant optimal hedge ratio rather than a single static one for the total hedging period [55].

However, in the vast majority of studies related to risk management and hedging in energy portfolios more sophisticated and complex multivariate Garch models are being used, such as the constant correlations (CCC) model and the dynamic conditional correlation model (DCC) model [53]. This is merely happening due to the advantages that these models offer to researchers compared to VECH model. Nevertheless, the more popular of the two models is [57] DCC model, which allows for a more realistic time-varying correlation structure enabling the model to capture any interactions between portfolio's assets. As a result, the DCC model has been used recently in a larger series of studies such as [58–62].

The Minimum-Variance hedge ratio based on the dynamic conditional correlation model (DCC) is estimated as follows:

$$\text{Consider}: \; y_t\big|\Omega_{t-1} \sim N(0, Q_t), t = 1, 2, \ldots, \text{n} \tag{8}$$

and:

$$Q_t = D_t \rho_t D_t \tag{9}$$

where Ω_t is the existing data set at time t, $D_t = \left(h_1^{1/2}, \ldots, h_m^{1/2}\right)$ represents the diagonal matrix of conditional volatility with h_{it} denoting the conditional variances that can be calculated using the basic Garch model and ρ_t the dynamic conditional correlation.Let $r_{S_t} = \mu + \varepsilon_{F_t}$ and $r_{F_t} = \mu + \varepsilon_{F_t}$ the returns on the spot and futures position respectively, the following two equation provide the relative conditional variances:

$$h_{S_t} = \omega_S + \alpha_S \varepsilon_{S_t}^2 + \beta_S \, h_{S_{t-1}} \tag{10}$$

$$h_{F_t} = \omega_F + a_t \varepsilon_{F_t}^2 + \beta_F \, h_{F_{t-1}} \tag{11}$$

While (9) with respect to ρ_t becomes:

$$\rho_t = \left\{ \left(diag(Q_t)^{-\frac{1}{2}} \right) Q_t \left(diag(Q_t)^{-\frac{1}{2}} \right) \right\} \tag{12}$$

where the conditional covariance matrix Q_t is estimated as follows:

$$\begin{aligned} & Q_t = \overline{Q}_t + \gamma\left(\varepsilon_{S_{t-1}}\varepsilon_{F_{t-1}} - \overline{Q}_t\right) + \delta\left(Q_{t-1} - \overline{Q}_t\right) \\ \Leftrightarrow \;\; & Q_t = (1 - \gamma - \delta)\overline{Q}_t + \gamma \varepsilon_{S_{t-1}}\varepsilon_{F_{t-1}} + \delta Q_{t-1} \end{aligned} \tag{13}$$

$\overline{Q}_t$ symbolizes the unconditional correlation coefficient and $\varepsilon_{S_{t-1}}, \varepsilon_{F_{t-1}}$ the standardized residuals of the spot and futures returns, respectively.

Hence the time variant Minimum-Variance hedge ratio is given by:

$$\delta_t^* \big| \Omega_{t-1} = \frac{h_{SF_t}}{h_{F_t}} \tag{14}$$

where h_{SF_t} denotes the conditional covariance between the spot and futures returns, and h_{F_t} the conditional variance of futures returns.

3.2. Hedging via the Expected Utility Maximization Methodology

The maximization of the expected utility constitutes the alternative hedging approach to Minimum-Variance hedge ratio. In expected utility maximization methodology the hedger's attitude towards risk is explicitly taken into consideration instead of assuming infinite risk aversion, by which it is implied that investors would reject investments that offer significantly high potential returns for even a relatively small additional risk [63]. This hypothesis is reasonably considered as irrational for the vast majority of hedgers, constituting risk aversion a critical factor for every risk management analysis and for the estimation of the optimal hedge ratio.

Furthermore, in contrast with Minimum-Variance hedge ratio the expected utility maximization methodology also examines the parameter of the expected return, combining elements of both risk and expected return in its estimates for the optimal hedge ratio. Nevertheless, the implementation of the risk aversion aspect requires the use of the appropriate utility function that would ideally match with the hedger's risk preference [5]. Reference [64] analyzed data for a number of energy commodities and reported that the presence of excess skewness and kurtosis in the return distribution lead to important differentiations in the resulting optimal hedge ratios that where estimated based on specific applied utility functions.

As a result, specifying the appropriate utility function becomes a critical matter considering that these statistical characteristics are found to be present in almost every risk management analysis regarding energy commodities, while they appear to be more intense during periods of economic turmoil [64]. Specifically, large movements of certain commodities are noticed during severe crisis such as for the prices of oil [65]. Finally, another parameter of great importance that needs to be taken into account is the time variance in the hedger's risk attitude. Table 1 shows the volatility under normal and crisis market conditions for a few commodities.

Similarly to evidence coming from financial markets, which is an even less volatile market compared to energy market, investors tend to adjust their risk preferences over time. Perhaps the most characteristic example is the 2007 economic crisis, during which investors' perception towards risk was found to have changed dramatically. Hence, applying arbitrary risk aversion values to a hedging strategy analysis for an energy portfolio is a practice that can lead to suboptimal hedge ratios [61].

The utility concept was first introduced by Georgescu-Roegen [66,67] to explain economic values. Though since then, it has now become an obsolete concept since nobody has been able to provide a specific measurement [68]. One business model in the evolving energy sector is the energy service utility model that, unlike conventional investor-owned energy utilities, provides services such as hot water, clean electricity, or sustainable materials rather than commodities like kilowatt-hours, therms, and so on [69].

Additionally, the current business model of the utility industry is based on increasing sales and needs to be revised as electricity consumption continues to decline, at the same time energy efficiency should be a main function of the utility business model in order to reduce carbon emissions while maintaining the long-term stability of the industry [70]. More efficient distribution utility models can be designed taking into account forward looking strategies, regulatory tools, financial incentives, performance incentives and incentives for long term innovation [71].

Table 1. Risk analysis data—volatility under normal and crisis market conditions and sensitivity factors (adapted from [65]).

Commodity Name	Monthly Volatility (Normal Market) (%)	Monthly Volatility (Crisis Market) (%)	Annual Volatility (Normal Market) (%)	Annual Volatility (Crisis Market) (%)	Sensitivity Factors
Petroleum: average crude price	8.1	24.6	28.1	85.2	1.72
Gasoline	10.4	25.4	30.4	87.2	1.88
Natural gas	5.8	20.6	20.0	71.2	0.14
Coal	4.0	13.5	13.9	46.9	0.26
Gold	3.3	12.5	11.3	43.2	0.18
Silver	5.4	21.6	18.7	75.0	0.18
Copper	6.2	20.0	21.5	69.2	0.48
Zinc	6.1	24.9	21.3	86.4	0.34
Lead	6.3	23.8	21.9	82.3	0.15
Aluminum	5.8	32.6	20.0	133.1	0.31
Nickel	8.9	22.2	30.7	76.9	0.54
Iron ore	4.4	12.9	15.2	44.7	0.18
Phosphate rock	2.3	21.7	8.1	75.2	0.01
Wheat	5.1	15.1	17.7	52.3	0.08
Cotton	4.9	12.6	17.0	43.5	0.14.9
Sugar	2.1	11.0	7.3	38.2	−0.05
Maize	5.3	25.2	18.4	87.2	−0.08
Tobacco	1.8	4.9	6.2	16.8	0.01
Coffee	8.0	37.1	27.6	128.6	0.04
Tea	7.7	23.6	26.8	81.8	0.11
Rubber	6.0	18.1	20.8	62.7	0.37
Wool	4.7	16.5	16.4	57.3	−0.02
All commodities	3.6	12.3	12.5	42.5	1.00

3.2.1. Measuring Risk Aversion

Determining the degree of risk aversion has always been a challenge for researchers, nevertheless there are two measures that are more commonly used by the vast majority of researchers in the field of hedging and energy economics, consisting of the coefficients of absolute and relative risk aversion [61]. In general, the term risk aversion is basically defined as the investor's assessment regarding the tradeoff between in taking risk that needs to be accepted for potential future return coming from the particular investment. This relationship is depicted by the investor's utility function and the relative risk aversion is approximated by the slope change that is observed between each individual point in the function [64].

The coefficient of absolute risk aversion (CARA) examines the percentage changes of the investor's portfolio that is invested in the risky and the risk free asset respectively regardless of the investor's wealth level and it is mathematically described as follows:

$$\text{CARA} = -\frac{U''(\text{Wealth})}{U'(\text{Wealth})} \tag{15}$$

From the above equation it is evident that an investor with CARA in absolute terms will invest a smaller part of the total portfolio value in the risky asset as wealth (W) increases.

On the contrary, the coefficient of relative risk aversion (CRRA) investigates percentage changes in the part of the investor's portfolio that is invested in risky and risk free asset respectively, given specific changes in wealth and it is mathematically described as follows:

$$\text{CRRA} = -W * \frac{U''(\text{Wealth})}{U'(\text{Wealth})} \tag{16}$$

The above equation allows for the investor's risk aversion to be expressed numerically, while a scale factor is used to represent the investor's present wealth level. Nevertheless, the whole concept of the CRRA relies on the market risk premium, showing the demanded excess return by the investor's side in order to be compensated for the additionally accepted systematic risk.

3.2.2. Optimal Hedge Ratio Estimation based on Expected Utility Theory

Assuming that $\left(r_{\delta_h}\right)$ and $\sigma^2_{\delta_h}$ represent the expected return and variance of the hedged portfolio then the expected utility function is mathematically described as follows:

$$\mathrm{EU}\left(r_{\delta_h}\right) = E\left(r_{\delta_h}\right) - \lambda\sigma^2_{\delta_h}, \text{ for } \lambda > 0 \tag{17}$$

where λ denotes the risk aversion parameter. As a result, the hedger's expected utility maximization is given by:

$$\mathrm{EU} = \max_{\delta_h} \mathrm{EU}\left(r_{\delta_h}\right) = \max_{\delta_h}\left[E(r_s) - \delta_h E\left(r_f\right) - \lambda\left(\sigma_s^2 + \delta_h^2\sigma_f^2 - 2\delta_h cov\left(r_s, r_f\right)\right)\right] \tag{18}$$

Hence the optimal hedge ratio based on the investor's expected utility maximization can be estimated as follows:

$$\delta_h = \frac{cov\left(r_s, r_f\right)}{\sigma_f^2} - \frac{E\left(r_f\right)}{2\lambda\sigma_f^2} \tag{19}$$

From the above equation it is evident that in case of absolute risk aversion or the futures expected return is zero (i.e., futures prices follow a martingale), the speculative term of the equation becomes zero and as a result the estimated optimal hedge ratio becomes equivalent to the Minimum-Variance hedge ratio [6].

3.3. *Alternative Hedging Strategies*

Although the vast majority of academic researches incorporate primarily the Minimum-Variance methodology to estimate the optimal hedge ratio and ultimately build the most effective hedging strategy for a particular energy portfolio, there are also other approaches that aim to solve the same problem through a different perspective. The most characteristic example is the mean-risk hedge ratios, in which the optimal hedge ratios are estimated by maximizing the utility function or a specific objective function of the expected return [6]. In this alternative methodology there are three main derivations regarding risk measurement including the Sharp hedge ratio, the mean-extended-Gini (MEG) coefficient hedge ratio and the generalized semi-variance (GSV) hedge ratio.

The Sharp hedge ratio introduced by [72] actually comprises a tradeoff between risk and return, including the element of portfolio return into the hedging strategy. The Sharp hedge ratio is derived by maximizing the portfolio's excess return relative to the portfolio's volatility and can be calculated using the following function:

$$\max_{\delta_h}\theta = \frac{E(r_h) - r_{free}}{\sigma_h} \tag{20}$$

where, r_{free} denotes the risk-free rate and σ_h^2 is equal to the portfolio variance $\mathrm{Var}(r_h)$ The Sharp hedge ratio in case the futures contracts return follows a pure martingale, it becomes equal to the optimal hedge ratio estimated by the Minimum-Variance hedge ratio.

Similarly, the MEG coefficient $\Gamma_u(r_h)$ as proposed by [73–75] is estimated by minimizing the following equation:

$$\Gamma_u(r_h) = -\lambda Cov(r_h, (1 - G(r_h)))^{\lambda-1} \tag{21}$$

where G represents the cumulative probability distribution.

While alternatively [76] suggest maximizing the utility function below:

$$\mathrm{U}(r_h) = E(r_h) - \Gamma_\lambda(r_h) \tag{22}$$

In this case the estimated hedge ratio is called M-MEG and it differs from MEG hedge ratio, as it incorporates the expected return parameter into the developing hedging strategy. The two hedge ratios become equivalent when the expected return is zero (i.e., the futures returns follow a martingale). Additionally, [77] proved that the MEG hedge ratio reduces to the Minimum-Variance hedge ratio whenever the assumption that both spot and futures returns are normally distributed is confirmed. Furthermore, [73,76] show that for low values of the risk aversion parameter λ the MEG hedge ratio converges to Minimum-Variance hedge ratio, while significantly differentiating for increased risk aversion. Contrary, in case of high levels of risk aversion the M-MEG hedge ratio was found to become equivalent to the Minimum-Variance hedge ratio.

The third alternative mean-risk ratio is the GSV hedge ratio, which was developed by [78–80] and further extended by [81]. The optimal hedge ratio is estimated by minimizing the following GSV equation:

$$U(r_h) = E(r_h) - \Gamma_\lambda(r_h) \tag{23}$$

where $G(r_h)$ represents the probability distribution of the hedged portfolio's return while γ denotes the portfolio's target return. In equation (23) it is assumed that investors consider lower than the targeted returns to be riskier, meaning that risk is measured based on a lower part of the hedged portfolio's distribution. [79] verify that the GSV hedge ratio becomes equivalent to the Minimum-Variance hedge ratio, provided that both assumptions for joint normality in the return distribution and pure martingale price process are met.

Extending the abovementioned hedge ratio methodology, [81] alter the GSV hedge ratio by including the mean return parameter in the derivation of the optimal hedge ratio. In this case, the produced mean-GSV hedge ratio is estimated by maximizing the below utility function:

$$U(r_h) = E(r_h) - V_{\gamma,\lambda}(r_h) \tag{24}$$

Reference [81] show that the M-GSV hedge ratio would become equivalent to the Minimum-Variance hedge ratio if both the pure martingale and joint normality hypotheses hold. Reference [80] suggest that adopting a conventional Minimum-Variance hedge strategy is unsuitable for hedgers that are mostly worried about downside risk. As a result, because of its conceptual simplicity for measuring the downside risk of a hedged portfolio the VaR methodology is being adopted by researchers as an alternative approach to build the optimal hedging strategy. Such case is the VaR constraint approach, which involves estimating the optimal hedge ratio based on a certain acceptable amount of risk or expected profit. Reference [82] where the first to build a VaR constraint hedging optimization model motivated by the high level of risk in the US electricity, with [83] following.

4. State of the Art—Relevant Studies Using Hedging Strategies

Although the specific research field in general became popular only recently, there is a number of very interesting research papers trying to explore the most important aspects for employing a successful hedging strategy regarding energy commodities, which are further presented in this Section.

4.1. Optimal Hedging Strategies based on the Minimum Variance Methodology

The Minimum-Variance hedge ratio methodology, despite its demerits, is by far the most widely used in academic literature offering useful advice about the mixture of the hedging strategy that should be employed by a firm that is exposed to energy price risk [5]. Therefore, many researchers use the Minimum-Variance hedge ratio to provide guidance to risk managers relative to the most appropriate hedging policy that would lead to reduced stock price volatility and increased firm value.

Reference [84] are some of the first that focused on estimating the most suitable hedging ratios regarding the crude oil market. The authors were using a four-year weekly spot price data sample for crude oil futures and the basic ARCH and GARCH models, and they concluded that the optimal hedging ratio is time-varying and that is positively affected by the duration of the contracts. Next, [61] using a much larger sample of daily spot prices for both Brent and WTI oil from a period from 1997 to 2009, tested several multivariate volatility models for their ability to estimate the most effective optimal hedge ratio, determining that the diagonal BEKK model is the most appropriate. Furthermore, they suggested a hedging strategy which involves an increased proportion of shorting in crude oil futures.

Reference [85] while emphasizing on the oil refining industry, they examine the weekly spot and futures prices for a 15-year period (1994–2009), regarding three of the most important and characteristic energy products in that specific industry, including the WTI crude oil, heating oil and gasoline. Their empirical results reveal that a combination of a dynamic conditional correlation (DCC) model and error correction GARCH model is better compared to simplified GARCH based models to capture the risk for refiners stemming both from the crude oil market itself, as well as the oil product market.

Finally, [86] were the first to examine the natural gas market and build an optimal hedging strategy using a single Henry-Hub futures contract. Specifically, they reveal that an adjusted error correction GARCH model is by far more capable of estimating the most effective time-varying hedging ratio relative to the conventional OLS methodology and several basic GARCH models. Moreover, based on the results of their research it is also supported that taking into consideration the elements of co-integration and time varying volatility doesn't have a significant impact on the hedging effectiveness of a specific strategy.

4.2. Incorporating the Elements of Risk Aversion and Expected Return into the Optimal Hedging Strategy

A rather interesting factor, when investigating the hedging policies of particular industries, is the tendency as well as the willingness of the market participants to take risks. The risk attitude of an industrial firm, which is exposed to the energy price risk may seriously affect the overall hedging policy of the firm regarding energy products considering their increased price volatility [37].

Although the above issue constitutes a quite interesting topic for further research, the fact that incorporating a risk aversion factor in such a study can prove to be tricky, as well as computationally demanding and time consuming, probably discouraged most academics to deal with the aforementioned topic. Specifically, [64] are the first who try to address the problem of risk aversion incorporation in an optimal hedging strategy regarding energy products.

The researchers while applying a GARCH in Mean model, estimate the optimal hedging ratio relying on a sample for gasoline futures prices for a 16-year period between 1992 and 2008. However, the innovative element of their study, is that in their model's estimates a time varying risk aversion factor is taken into consideration, enabling them to forecast risk aversion and thus better adjust the hedging strategy to the hedger's future needs. Finally, [87] further extend the previous study by incorporating the factor of risk aversion to the most popular and frequently applied utility functions and through that they end up to the most appropriate and efficient hedging strategy.

5. Conclusions

Energy is considered as a commodity nowadays and continuous access along with price stability is of vital importance for every economic agent worldwide. The current study comprehensively reviews the advantages and disadvantages of the main hedging methodologies regarding risk management in energy portfolios. Additionally, it enables the reader to explain, analyze and interpret the variations in the results for the proposed optimal hedging strategies of each methodology, while advising when and why choosing a specific methodology over the others and if more than one methodology is required to build a more reliable hedging strategy due to special economic characteristics of a certain time period.

Based on the conducted review, it is clear that the Minimum-Variance hedge ratio methodology if the appropriate nonlinear multivariate model is used, the Garch family volatility model can provide a reliable optimal hedge ratio with relatively low computational effort. However, that becomes possible by applying a rather restrictive assumption for infinite risk aversion on behalf of the investor into the analysis. This estimation gap of an optimal hedge ratio that would be subject to the investor's risk preferences through time is filled by the relatively more complex and sophisticated expected utility maximization methodology. Nevertheless, if hedgers share infinite risk aversion or if alternatively the expected futures price variation is approximately zero, the two methodologies become equivalent.

In general, considering evidence from the energy market the assumption that energy futures prices follow a martingale is confirmed for extended periods of time, yet during periods with extremely volatile economic climate and financial crisis this may change until the market returns to normality. Finally, it is important to note that during periods of extreme uncertainty and high risk it is common also for the investors' risk attitude to show significant variations. Hence, it becomes reasonable that especially for extended hedging horizons it would be wise for potential hedgers to take into consideration both methodologies in order to build a successful and profitable hedging strategy.

A numerical analysis is proposed as a further extension of the present paper, in which the resulting hedging strategies from the different methodologies would be tested regarding their effectiveness and profitability through multiple time horizons and for several energy commodities. An important limitation of this research has to do with resources decoupling and keeping under control the increasing of the value added and decreasing energy consumption.

Author Contributions: G.E.H. conceived and designed the analysis. Both G.E.H. and A.S.T. wrote the manuscript and contributed to the final version of the manuscript. G.E.H. supervised the paper, provided critical feedback and helped shape the research.

Funding: This research received no external funding.

Acknowledgments: We would like to thank the Editor and the three anonymous reviewers for the helpful and constructive comments on earlier drafts of our paper. Any remaining errors are solely the authors' responsibility.

Conflicts of Interest: The authors declare no conflicts of interest

References

1. European Commission. *Study on Energy Efficiency and Energy Saving Potential in Industry and on Possible Policy Mechanism*; EC: Brussels, Belgium, 2015.
2. Boroumand, R.H.; Goutte, S.; Porcher, S.; Porcher, T. Hedging strategies in energy markets: The case of electricity retailers. *Energy Econ.* **2015**, *51*, 503–509. [CrossRef]
3. Zhang, J.; Tan, K.S.; Weng, C. Optimal hedging with basis risk under mean–variance criterion. *Insur. Math. Econ.* **2017**, *75*, 1–15. [CrossRef]
4. Stulz, R.M. Optimal Hedging Policies. *J. Financ. Quant. Anal.* **1984**, *19*, 127–140. [CrossRef]
5. Dewally, M.; Marriott, L. Effective Basemetal Hedging: The Optimal Hedge Ratio and Hedging Horizon. *J. Risk Financ. Manag.* **2008**, *1*, 41–76. [CrossRef]
6. Hung, J.-C.; Chiu, C.-L.; Lee, M.-C. Hedging with zero-value at risk hedge ratio. *Appl. Financ. Econ.* **2006**, *16*, 259–269. [CrossRef]
7. Halkos, G.E.; Tsirivis, A.S. Value-at-risk methodologies for effective energy portfolio risk management. *Econ. Anal. Policy* **2019**, *62*, 197–212. [CrossRef]
8. Ricardo, D. *On the Principles of Political Economy and Taxation*; John Murray: London, UK, 1817.
9. Manoilescu, M. *Die Nationalen Producktivkräfte und der Aussenhandel. Theorie des Internationalen Warenaustausches, Junker und Dünnhaupt, Name?* Juncker & Dünnhaupt: Berlin, Germany, 1937.
10. Balassa, B. Trade Liberalisation and "Revealed" Comparative Advantage. *Manch. Sch.* **1965**, *33*, 99–123. [CrossRef]
11. Porter, M.E. The Comparative Advantage of Nations. *Harv. Bus. Rev.* **1990**, 73–91. [CrossRef]
12. Dogaru, V. The expanding of constructal law in economics—A justification for crossed flows of similar macro goods. *Int. J. Heat Technol.* **2016**, *34*, 59–74. [CrossRef]
13. Kuik, O.; Branger, F.; Quirion, P. Competitive advantage in the renewable energy industry: Evidence from a gravity model. *Renew. Energy* **2019**, *131*, 472–481. [CrossRef]
14. Blum, J. *Energy Bankruptcies Back on the Rise in 2019*; Houston Chronicle: Houston, TA, USA, 2019.
15. Seba, E. *Bankruptcy Filings by U.S. Energy Producers Pick up Speed: Law Firm Analysis*; Reuters: London, UK, 2019.
16. Haushalter, G.D. Financing Policy, Basis Risk, and Corporate Hedging: Evidence from Oil and Gas Producers. *J. Financ.* **2000**, *55*, 107–152. [CrossRef]

17. Haushalter, G.D. Why Hedge? Some Evidence from Oil and Gas Producers. *J. Appl. Corp. Financ.* **2001**, *13*, 87–92. [CrossRef]
18. Chew, D. Energy Derivatives and the Transformation of the U.S. Corporate Energy Sector. *J. Appl. Corp. Finance* **2001**, *13*, 50–75.
19. Smithson, C.; Simkins, B.J. Does Risk Management Add Value? A Survey of the Evidence. *J. Appl. Corp. Finance* **2005**, *17*, 8–17. [CrossRef]
20. Carter, D.A.; Rogers, D.A.; Simkins, B.J. Does Hedging Affect Firm Value? Evidence from the US Airline Industry. *Financ. Manag.* **2006**, *35*, 53–86. [CrossRef]
21. Geczy, C.; Minton, B.; Schrand, C. *Choices among Alternative Risk Management Strategies: Evidence from the Natural Gas Industry*; Working Paper Wharton School of Economics; The Rodney L. White Center for Financial Research: Pennsylvania, PA, USA, 2002.
22. Jin, Y.; Jorion, P. Firm Value and Hedging: Evidence from U.S. Oil and Gas Producers. *J. Financ.* **2006**, *61*, 893–919. [CrossRef]
23. Islam, M.; Chakraborti, J. Futures and forward contract as a route of hedging the risk. *Risk Gov. Control. Financ. Mark. Inst.* **2015**, *5*, 68–78. [CrossRef]
24. Trafigura. Commodities Demystified: A Guide to Trading and the Global Supply Chain 2018. Available online: www.trafigura.com (accessed on 10 July 2019).
25. Difiglio, C. Oil, economic growth and strategic petroleum stocks. *Energy Strat. Rev.* **2014**, *5*, 48–58. [CrossRef]
26. Economic Consulting Associates Limited. European Electricity: Forward Markets and Hedging Products—State of Play and Elements for Monitoring 2015. Available online: www.acer.europa.eu (accessed on 10 July 2019).
27. Malik, F. Risk Management: Understanding Credit Risk. Available online: www.medium.com (accessed on 10 July 2019).
28. Mas, I.; Saa-Requejo, J. *Using Financial Futures in Trading and Risk Management*; Policy Research Working Paper; World Bank: Washington, DC, USA, 1995.
29. IOSCO. Report on the International Regulation of Derivative Markets, Products and Financial Intermediaries 2012. Available online: www.trafigura.com (accessed on 15 July 2019).
30. Carmona, R.; Durrleman, V. Pricing and Hedging Spread Options. *SIAM Rev.* **2003**, *45*, 627–685. [CrossRef]
31. New York Mercantile Exchange. A guide to Energy Hedging. Available online: www.renesource.com (accessed on 5 July 2019).
32. Coffman, D.; Lockley, A. Carbon dioxide removal and the futures market. *Environ. Res. Lett.* **2017**, *12*, 015003. [CrossRef]
33. Keytrade Bank. *Options Manual of Keytrade Bank*; Keytrade Bank: Brussels, Belgium, 2017.
34. Mercatus. An Introduction to Crack Spread (Refiner) Hedging 2019. Available online: www.mercatusenergy.com (accessed on 30 July 2019).
35. Vitale, L. Interest Rate Swaps under the Commodity Exchange Act. Case West. *Reserve Law Rev.* **2001**, *51*, 539–591.
36. Ernst & Young. Financial Reporting Developments A Comprehensive Guide Derivatives and Hedging. Available online: www.ey.com (accessed on 25 July 2019).
37. Deloitte. Commodity Price Risk Management: A manual of Hedging Commodity Price Risk for Corporates 2018. Available online: www2.deloitte.com (accessed on 16 July 2019).
38. Stoft, S.; Belden, T.; Goldman, C.; Pickle, S. *Primer on Electricity Futures and Other Derivatives*; Environmental Energy Technologies Division, Ernest Orlando Lawrence Berkeley National Laboratory, University of California Berkeley: Berkeley, CA, USA, 1998.
39. Devlin, J.; Titman, S. Managing Oil Price Risk in Developing Countries. *World Bank Res. Obs.* **2004**, *19*, 119–139. [CrossRef]
40. Ritchken, P. Hedging with Futures Contracts. Case West-ern Reserve University—Class Notes 1999. Available online: http://faculty.weatherhead.case.edu (accessed on 5 July 2019).
41. Castellino, M.G. Hedge effectiveness: Basis risk and minimum-variance hedging. *J. Futures Mark.* **2000**, *20*, 89–103. [CrossRef]
42. Johnson, L.L. The Theory of Hedging and Speculation in Commodity Futures. *Rev. Econ. Stud.* **1960**, *27*, 139. [CrossRef]

43. Ederington, L.H. The Hedging Performance of the New Futures Markets. *J. Financ.* **1979**, *34*, 157–170. [CrossRef]
44. Myers, R.J.; Thompson, S.R. Generalized Optimal Hedge Ratio Estimation. *Am. J. Agric. Econ.* **1989**, *71*, 858–868. [CrossRef]
45. Baillie, R.T.; Myers, R.J. Bivariate garch estimation of the optimal commodity futures Hedge. *J. Appl. Econ.* **1991**, *6*, 109–124. [CrossRef]
46. Yu, H. The Implication of Currency Hedging Strategies for Pension Funds. Master's Thesis, MSc Finance—Pension Track Tilburg University, Tilburg, The Netherlands, 2014.
47. Arthur, J.N.; Williams, R.J.; Delfabbro, P.H. The conceptual and empirical relationship between gambling, investing, and speculation. *J. Behav. Addict.* **2016**, *5*, 580–591. [CrossRef]
48. Benada, L. Hedging of Energy Commodities. Ph.D. Thesis, Masaryk University, Brno, Czech Republic, 2017.
49. Pouliasis, P. Essays on the Empirical Analysis of Energy Risk. Ph.D. Thesis, Cass Business School, London, UK, 2011.
50. Pennings, J.M.E. What Drives Actual Hedging Behaviour? Developing Risk Management Instruments. In *Agribusiness and Commodity Risk: Strategies and Management*; Risk Books: London, UK, 2003; pp. 63–74.
51. Chen, S.S.; Lee, C.; Shrestha, K. Futures hedge ratios: A review. *Quart. Rev. Econ. Financ.* **2003**, *43*, 433–465. [CrossRef]
52. Harris, R.D.F.; Stoja, E.; Tucker, J. A Simplified Approach to Modelling the Comovement of Asset Returns. *J. Futures Mark.* **2007**, *27*, 575–598. [CrossRef]
53. Halkos, G.E.; Tsirivis, A.S. Effective energy commodity risk management: Econometric modeling of price volatility. *Econ. Anal. Pol.* **2019**, *63*, 234–250. [CrossRef]
54. Alizadeh, A.H.; Nomikos, N.K.; Pouliasis, P.K. A Markov regime switching approach for hedging energy commodities. *J. Bank. Financ.* **2008**, *32*, 1970–1983. [CrossRef]
55. Casillo, A. Model Specification for the Estimation of the Optimal Hedge Ratio with Stock Index Futures: An Application to the Italian Derivatives Market. Available online: www.luiss.it (accessed on 18 October 2019).
56. Bollerslev, T.; Engle, R.F.; Wooldridge, J.M. A Capital Asset Pricing Model with Time-Varying Covariances. *J. Polit. Econ.* **1988**, *96*, 116–131. [CrossRef]
57. Engle, R.F. Dynamic Conditional Correlation: A Simple Class of Multivariate Generalized Autoregressive Conditional Heteroskedasticity Models. *J. Bus. Econ. Stat.* **2002**, *20*, 339–350. [CrossRef]
58. Manera, M.; McAleer, M.; Grasso, M. Modelling time-varying conditional correlations in the volatility of Tapis oil spot and forward returns. *Appl. Financ. Econ.* **2006**, *16*, 525–533. [CrossRef]
59. Liu, S.D.; Jian, J.B.; Wang, Y.Y. Optimal dynamic hedging of electricity futures based on copula-GARCH models. In Proceedings of the 2010 IEEE International Conference on Industrial Engineering and Engineering Management, Macao, China, 7–10 December 2010; pp. 2498–2502.
60. Zanotti, G.; Gabbi, G.; Geranio, M. Hedging with futures: Efficacy of GARCH correlation models to European electricity markets. *J. Int. Financ. Mark. Inst. Money* **2010**, *20*, 135–148. [CrossRef]
61. Chang, C.-L.; McAleer, M.; Tansuchat, R. Crude oil hedging strategies using dynamic multivariate GARCH. *Energy Econ.* **2011**, *33*, 912–923. [CrossRef]
62. Basher, S.A.; Sadorsky, P. Hedging emerging market stock prices with oil, gold, VIX, and bonds: A comparison between DCC, ADCC and GO-GARCH. *Energy Econ.* **2016**, *54*, 235–247. [CrossRef]
63. Heisler, A. 7 Critical Risks Impacting the Energy Industry. Risk & Insurance. Available online: www.riskandinsurance.com (accessed on 5 July 2019).
64. Cotter, J.; Hanly, J. Time-varying risk aversion: An application to energy hedging. *Energy Econ.* **2010**, *32*, 432–441. [CrossRef]
65. Al Janabi, M.A.M. Commodity price risk management: Valuation of large trading portfolios under adverse and illiquid market settings. *J. Deriv. Hedge Funds* **2009**, *15*, 15–50. [CrossRef]
66. Georgescu-Roegen, N. Choice, Expectations and Measurability. *Quart. J. Econ.* **1954**, *68*, 503–534. [CrossRef]
67. Georgescu-Roegen, N. Utility. *Int. Encycl. Soc. Sci.* **1968**, *16*, 236–267.
68. Georgescu-Roegen, N. The Entropy Law and the Economic Process. *East. Econ. J.* **1971**, *12*.
69. Byrne, J.; Taminiau, J. A review of sustainable energy utility and energy service utility concepts and applications: Realizing ecological and social sustainability with a community utility. *WIREs Energy Environ.* **2016**, *5*, 136–154. [CrossRef]

70. Fox-Penner, P. The Smart Grid–Enabled Energy Services Utility: How Utilities Can Become Sustainable by Selling Less. *Solutions* **2010**, *1*, 42–48.
71. MIT. Utility of the Future. *2016, An MIT Energy Initiative Response to an Industry in Transition.* Available online: energy.mit.edu (accessed on 9 July 2019).
72. Howard, C.T.; D'Antonio, L.J. A Risk-Return Measure of Hedging Effectiveness. *J. Financ. Quant. Anal.* **1984**, *19*, 101. [CrossRef]
73. Kolb, R.W.; Okunev, J. An empirical evaluation of the extended mean-gini coefficient for futures hedging. *J. Futures Mark.* **1992**, *12*, 177–186. [CrossRef]
74. Lien, D.; Luo, X. Estimating the extended mean-gini coefficient for futures hedging. *J. Futures Mark.* **1993**, *13*, 665–676. [CrossRef]
75. Gregory-Allen, R.B.; Shalit, H. The Estimation of Systematic Risk under Differentiated Risk Aversion: A Mean-Extended Gini Approach. *Rev. Quant. Financ. Account.* **1999**, *12*, 135–158. [CrossRef]
76. Kolb, R.W.; Okunev, J. Utility maximizing hedge ratios in the extended mean gini framework. *J. Futures Mark.* **1993**, *13*, 597–609. [CrossRef]
77. Shalit, H. Mean-Gini hedging in futures markets. *J. Futures Mark.* **1995**, *15*, 617–635. [CrossRef]
78. De Jong, A.; De Roon, F.; Veld, C. Out-of-sample hedging effectiveness of currency futures for alternative models and hedging strategies. *J. Futures Mark.* **1997**, *17*, 817–837. [CrossRef]
79. Lien, D.; Tse, Y.K. Hedging time-varying downside risk. *J. Futures Mark.* **1998**, *18*, 705–722. [CrossRef]
80. Lien, D.; Tse, Y.K. Hedging downside risk with futures contracts. *Appl. Financ. Econ.* **2000**, *10*, 163–170. [CrossRef]
81. Chen, S.-S.; Lee, C.-F.; Shrestha, K. On a Mean—Generalized Semivariance Approach to Determining the Hedge Ratio. *J. Futures Mark.* **2001**, *21*, 581–598. [CrossRef]
82. Kleindorfer, P.R.; Li, L. Multi-Period VaR-Constrained Portfolio Optimization with Applications to the Electric Power Sector. *Energy J.* **2005**, *26*, 1–26. [CrossRef]
83. Oum, Y.; Oren, S. VaR constrained hedging of fixed price load-following obligations in competitive electricity markets. *Risk Dec. Anal.* **2009**, *1*, 43–56.
84. Jalali-Naini, A.R.; Manesh, M.K. Price volatility, hedging and variable risk premium in the crude oil market. *OPEC Rev.* **2006**, *30*, 55–70. [CrossRef]
85. Ji, Q.; Fan, Y. A dynamic hedging approach for refineries in multiproduct oil markets. *Energy* **2011**, *36*, 881–887. [CrossRef]
86. Ghoddusi, H.; Emamzadehfard, S. Optimal hedging in the US natural gas market: The effect of maturity and cointegration. *Energy Econ.* **2017**, *63*, 92–105. [CrossRef]
87. Cotter, J.; Hanly, J. Performance of utility based hedges. *Energy Econ.* **2015**, *49*, 718–726. [CrossRef]

energies

MDPI

Article

Energy, Economy, and Environment: A Worldwide Research Update

Juan Uribe-Toril *, José Luis Ruiz-Real, Juan Milán-García and Jaime de Pablo Valenciano

Faculty of Economics and Business, University of Almería, Ctra. de Sacramento, s/n, 04120 Almería, Spain; jlruizreal@ual.es (J.L.R.-R.); jmg483@ual.es (J.M.-G.); jdepablo@ual.es (J.d.P.V.)

* Correspondence: juribe@ual.es

Received: 23 February 2019; Accepted: 19 March 2019; Published: 22 March 2019

Abstract: This paper has reviewed the international research on the interactions between the Economy, Energy, and Environment (3E) in the 21st century. For this purpose, a bibliometric and cluster analysis by fractional accounting has been carried out based on the two most important databases: Web of Science (WoS) and Scopus. The research found and studied 2230 documents from the WoS Core Collection and 3,149 from Scopus. The results show a continuous increase in the number of articles that were published and citations during the whole period. They also showed that China and the United States (U.S.) were the most productive countries and there was a predominance of Asian organizations supporting and fostering researches. The main contribution of this article is the analysis of keywords from 2001 to 2018. The trends show that the main common elements are sustainable development and sustainability and they also include CO_2 emissions and consumption. Future research in this field should address the energy transition issue in the area of sustainable development by adapting it to the restrictions of this economic model.

Keywords: energy; economy; environment; 3E; sustainable development

1. Introduction

Throughout history, the concept of economy has evolved in parallel with society: from a perspective that is solely focused on obtaining wealth, to a holistic and integrated vision in which a growing number of factors are interrelated with it.

The scientific literature includes numerous articles in which the interrelation between Energy, Economy, and Environment is identified with the nomenclature "3E" [1–3]. In this sense, the eight Millennium Development Goals that were proposed by the United Nations Development Organization [4] at the beginning of the 21st century show the global importance of this triple helix in the global economic scenario. Ensuring environmental sustainability is the seventh of these goals, while energy appears as an indicator of this objective: carbon dioxide emissions or use of water resources, among others. The evolution of these objectives in the Sustainable Development Goals [5] expands the importance of energy and the environment in the form of the following: affordable and non-polluting energy, sustainable cities and communities, or climate action. In light of this, 3E is more present today than ever before.

In recent years, the problems that are related with 3E have been studied and evaluated in a deeper way than ever before in history [6–8]. Recently, the academic community has linked these three elements in the form of diverse currents or lines of research [9–11].

One of these research lines emphasizes the impact of the energy management of productive units on the economic growth of the regions. On the one hand, the impact of human activities, such as mining or tourism, on the state of natural resources (even human capital) is analysed [12–14]. In contrast, some authors apply a long-term approach and analyse the viability of cleaner energy in developing economies [15,16].

There are also research studies in which the efficiency and effectiveness of the energy sources used are related to the environment and the restrictions that it poses. For example, the solid waste recycling strategy in Brazil to solve the problem of the growing amount of electronic waste because of an increase in the use of new technologies and electrical energy [17], or the impact of efficient coal waste management [18] while using environmental indicators that characterize the combustion of different ranges of coal (gas, flame, coke, or uncooked coal).

There is a factor used by the research community that manages to relate the three elements in an integral way: the concept of renewable energy [19,20]. Through the inclusion of alternative energies, the productive system of an economy adapts the intervening elements in the value chain to develop an "environment-centred" strategy [21,22]. For example, an integrated sustainability model that is used to understand how changes in the bioenergy system influence environmental measures, economic development, and society, showing that an increase in the share of bioenergy in total electricity generation will stimulate the electricity market [23].

Other authors follow this trend regarding energy from different perspectives: a reorientation of productive energy distribution towards others, such as biogas, biodiesel, or bioethanol to close the carbon cycle in nature [24]; the achievement of a given objective of carbon dioxide elimination through the framework of Modelling and Optimisation of Negative Emission Technologies (MONET) [25], or an analysis of the impact of traditional and alternative energy resources on economic growth, the transport sector, and the carbon dioxide emissions [26].

Furthermore, another highlighted line of investigation focuses more on the environmental aspect: from a legal and educational approach of the issue [27], the application of the concept of eco-efficiency to assess the suitability of renewable energies [28–30] to multi-target models or a qualitative comparative analysis to study the relationship between economic growth and the environment [31–33].

The so-called carbon footprint, which can be defined as a measure of the greenhouse gas emissions of human activity, is another element of growing research attention that associates the three concepts [34,35]. This line of research applies the method of accumulated energy demand to develop and validate indicators of urban environmental sustainability, using the five urban systems in Italy as case studies to analyse their ecological footprint. Other researchers [36] use the life cycle assessment method to propose a strategy to maximise the benefits of the cold chain of table grapes by integrating its carbon footprint. In the same way, the concept of the seasonal footprint avoided by imports has been used to analyse whether proximity and seasonal consumption are consumption patterns that buyers can use to improve the sustainability of the economy [37]. An approach that is focused on the Internet of Things has also been used to analyse the management of "Smart" cities and how households can reduce their carbon footprint [38].

However, the most relevant expression of the union of these concepts is found in the concept of sustainability, which made its first appearance on the international stage at the United Nations Conference on Environment and Development [39] that was held in Stockholm in 1972. Since then, the question of how to improve and stabilize the economic situation of countries is linked to the restrictions that are imposed by the natural environment [40] and it has materialized in numerous research articles that interrelate the economy with the environment. In fact, sustainability is identified as the perfect conciliation between the environment and the economy [41].

In accordance with this approach, several authors focus on the following methodologies to assess the impact of human activities on the sustainability of the environment: an analysis of the social and environmental impact of these activities by using the life cycle assessment method [42], agency theory [43], or the development of different indicators [44,45]. On the other hand, another stream of authors study how the different agents of the economy of a country influence the sustainability: multinational companies [46] or final consumers [47].

Nevertheless, the researchers also point to examples of economic growth using environmentally damaging energies [48], recommending that authorities should take the path of sustainable development with low resource consumption, less pollution, and high ecological security.

There are similar articles that analyse the economy, energy, and environment from a bibliometric analysis, either independently or in pairs [49–51]. However, there is no article in the existing literature analysing the latest trends in 3E that are based on a bibliometric analysis. From this analysis, it can be seen that the junction of the three terms results in new research areas, such as sustainability, and a focus on CO_2 emissions.

2. Materials and Methods

The methodology that is used to analyse the concept of 3E is bibliometric analysis, a scientific method that is widely accepted and used by institutions, such as the European Commission or the National Science Foundation [52]. Bibliometric is a scientific method that uses mathematical techniques and statistics to evaluate a given scientific output [53]. The principle on which it is based is the citation network, from which the sub-methods of citation analysis [54] and scientific cartography, which are essential for the evaluation of research performance, are derived. In order to understand the performance or production of a researcher, this research has also applied the index h, as developed by Hirsch [55] and defined as the number of articles with citation number ≥h. To this end, the indexes of publications in the core collection of the WoS and Scopus online databases are considered.

All types of documents (articles, books, proceedings and so on) were included for the general analysis, but the impact analysis was filtered to only include articles (Table 1). The reason why this filter was applied is that this type of scientific document has undergone a rigorous review process to guarantee its quality and will, thus also guarantee the quality of our conclusions. Finally, information that is related to 3E was also filtered, coding the recovered material, and analysing the results.

Table 1. Distribution of publications by type of document.

Type of Document	WoS	Scopus
Article	2230	3149
Proceedings Paper	1462	1894
Review	268	437
Book and Book Chapter	110	518

The cluster analyses were built while using VOSviewer software tool for constructing and visualizing bibliometric networks [56]. For this analysis, a fractional counting method was chosen. The basic idea of the fractional counting approach is that each action, such as co-authoring or citing of a publication, should have equal weight, regardless of, for instance, the number of authors, citations, or references of a publication [57].

This bibliometric analysis followed the following steps (Figure 1). First, the search criteria, keywords, and period were defined. In this work, we have chosen the words "energy", "environment", and "economy". The reason why these words have been chosen lies with the scientific community's continued use of 3E terminology to name the development and growth models in which these elements are integrated [1,2,58]. The study period corresponds to the 21st century: from 2001 to 2018 so that new trends can be better defined. Subsequently, Scopus and WoS were the chosen databases in which the analysis was conducted, since they are the two largest databases that follow a rigorous protocol for the inclusion of research work in order to ensure scientific quality [59].

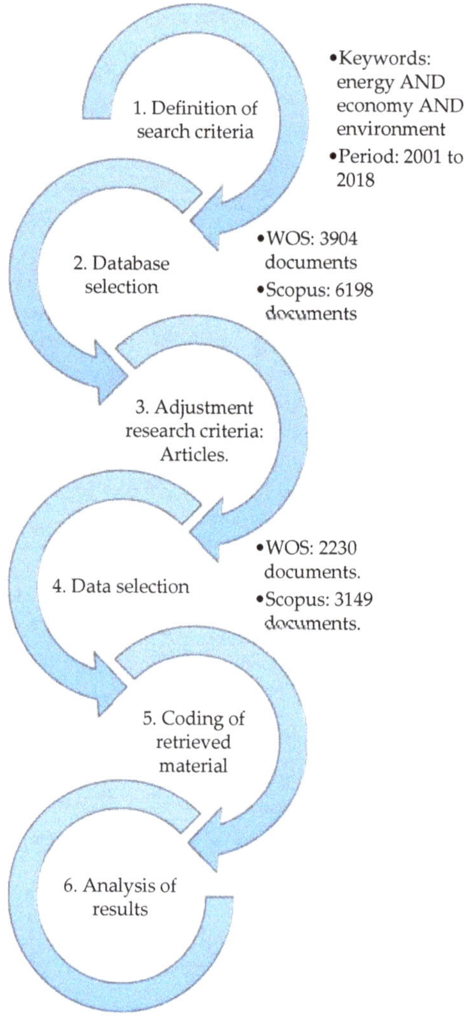

Figure 1. Bibliometric analysis steps followed.

In order to identify new trends involving the three elements, 3904 relevant research studies have been identified from 2001 to 2018 in the core collection of Web of Science (WoS). The list was then filtered down to 2230 publications that link Energy, Economy, and Environment as keywords in the documents recorded. The process was then repeated for the Scopus database. This time, 6198 documents were founded and the results were filtered down to 3149 research articles that were published in impact journals.

3. Results and Discussion

3.1. Number of Publications per Year

Below, a series of data is displayed, which shows the status of the research activity about 3E with reference to the results of the WoS and Scopus databases in the 21st century.

WoS opens the new century with the article entitled Energy relations of gas estimated from flare radiation in Nigeria [60], in which the economic and environmental impact of oil extraction in Nigeria is studied. Scopus, on the other hand, includes, as its first article for the period, the work entitled Food security, agricultural subsidies, energy, and the environment. A process of 'glocalization' in Sri Lanka [61], in which the interaction of the political dilemma in the areas of food security, agricultural subsidies, energy consumption, and the environment in the process of 'glocalization' in Sri Lanka are analysed.

Throughout the study period, it is observed that the scientific contribution that was collected in Scopus is higher than that of WoS in terms of the number of articles and citations, with the only exception of 2016, in which the latter is slightly higher. The evolution of the h-index follows a similar pattern, with WoS being higher in 2007. On the other hand, the ratio of scientific production (represented by the number of citations per article) does not follow such a clearly defined path. However, there is convergence in the number of articles that are indexed in each of the databases by the end of the period (Table 2).

Table 2. Annual distribution of the publication of Economy, Energy, and Environment (3E) scientific articles.

	WOS				SCOPUS			
Year	**Articles**	**H-Index**	**Citations**	**TC/Art**	**Articles**	**H-Index**	**Citations**	**TC/Art**
2018	356	9	553	1.55	401	11	687	1.71
2017	327	16	1366	4.18	387	17	1767	4.57
2016	227	20	2179	9.60	269	20	1868	6.94
2015	187	20	2480	13.26	277	24	2656	9.59
2014	166	26	2220	13.37	279	28	3254	11.66
2013	147	24	2195	14.93	195	28	3289	16.87
2012	125	24	1971	15.77	203	29	3260	16.06
2011	121	27	2502	20.68	185	30	4665	25.22
2010	121	32	3134	25.90	177	35	4047	22.86
2009	108	25	2656	24.59	165	31	4673	28.32
2008	69	23	2033	29.46	120	29	3376	28.13
2007	65	25	2145	33.00	99	24	2329	23.53
2006	57	21	2564	44.98	79	21	2939	37.20
2005	42	16	1170	27.86	68	16	1217	17.90
2004	23	13	792	34.43	48	14	906	18.88
2003	35	18	1238	35.37	68	23	2567	37.75
2002	24	15	1959	81.63	49	21	2424	49.47
2001	20	9	343	17.15	63	14	1433	22.75

TC/Art: Total Citations/Article.

With consideration of the total number of articles in both databases, the trend is positive, exponentially growing in recent years, and even surpassing 400 articles published in 2018 on the Scopus database. The dynamics of WoS with respect to this issue is positive throughout the period, only decreasing in 2004. The Scopus trend, on the other hand, shows several moments of decreasing scientific contribution: in 2004, 2013, 2014, 2015, and 2016. However, the overall positive evolution of this factor indicates that research into the interrelationship between the economy, energy, and the environment is a safe bet, and currently at a high point in terms of the number of studies being published on this issue (Figures 2 and 3).

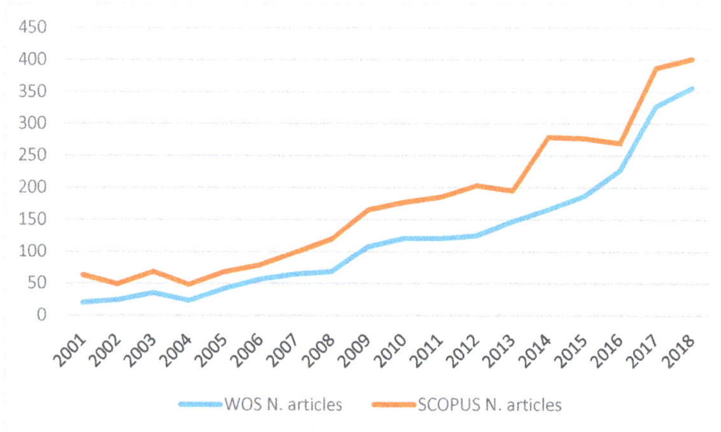

Figure 2. Evolution in number of articles.

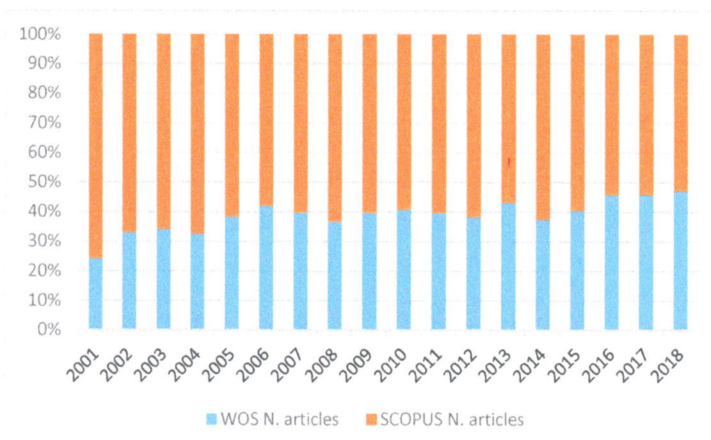

Figure 3. Average number of articles on WOS vs Scopus.

The evolution of the number of citations does not present as stable a path as in the previous variable, drawing an inverted U shape. The highest peak was in 2009 and 2011 for Scopus and 2010 for WoS, decreasing in recent years. The most quoted article throughout this period in WoS and Scopus is the work, A class of non-precious metal composite catalysts for fuel cells [62], with 1477 and 1531 citations, respectively (Figure 4).

Figure 4. Evolution of total cites.

3.2. *Language and Most Influential Countries*

The main countries in terms of publication on 3E for both databases are represented below. The ranking by country is practically the same for both databases: China leads the economy, energy, and environment research field, followed by the United States of America (USA) and the United Kingdom (UK), although the difference between the Asian countries and Anglo-Saxon ones is higher in Scopus than in WoS. The only difference that was observed in both geographical distributions is the last place: Netherlands for WoS and Russia for Scopus, with a similar production of articles, but with much greater capacity for dissemination in the case of the Netherlands as compared to Russia.

The distribution by language shows the complete predominance of English over other languages. With regards to other languages that are used in this field of research, the most commonly used in both databases are Russian, Spanish, and German. Chinese and Portuguese are the divergent languages, especially the latter if the results of WoS and Scopus are compared: three versus 311 articles (Tables 3 and 4).

Table 3. Distribution of articles per country.

WOS					SCOPUS				
Country	**Articles**	**H-Index**	**Citations**	**TC/Art.**	**Country**	**Articles**	**H-Index**	**Citations**	**TC/Art.**
China	582	43	7191	12.36	China	933	49	9472	10.15
USA	365	51	11538	31.61	USA	471	57	15599	33.12
UK	202	34	4789	23.71	UK	205	34	7252	35.38
Italy	98	17	1174	11.98	India	132	19	1218	9.23
Germany	92	24	2802	30.46	Germany	128	26	2156	16.84
Canada	89	18	1477	16.60	Italy	115	22	1904	16.56
Australia	85	23	2469	29.05	Australia	101	27	3734	36.97
India	83	14	662	7.98	Canada	99	19	2804	28.32
Japan	74	19	2588	34.97	Japan	90	17	1211	13.46
Turkey	72	13	954	13.25	Turkey	74	19	2259	30.53
Spain	66	15	1595	24.17	Spain	73	19	1560	21.37
France	61	20	2646	43.38	France	65	18	1370	21.08
Netherlands	57	21	2291	40.19	Rusia	61	6	143	2.34

TC/Art: Total Citations/Article.

Table 4. Distribution by language.

WoS		SCOPUS	
Languages	**Articles**	**Languages**	**Articles**
English	2083	English	2640
Russian	33	Chinese	311
Spanish	23	German	38
Portuguese	19	Russian	26
German	15	Spanish	24
Chinese	3	Portuguese	10

3.3. *Journals and Authors*

The distribution of scientific production by authors with respect to 3E shows a situation in which the vast majority of authors are of Asian origin, especially from the perspective of WoS. However, Professor Terry Barker (Department of Applied Economics at the University of Cambridge (UK)) is the most prolific author in both datasets, with 13 and 10 articles in WoS and Scopus, respectively (Table 5).

Table 5. Distribution of articles by author (Web of Science (WoS)/Scopus).

Author	ID	Ranking (W/S)	Articles (W/S)	H-Index (W/S)	Citations (W/S)	TC/A (W/S)
Barker, T.	7103052504	1/1	13/10	9/8	375/465	28.85/46.5
Chen, B.	55503929500	2/3	11/8	8/5	363/200	33/27.25
Zhang, Y.	57203830670	3/-	11/-	6/-	173/-	15.73/-
Ulgiati, S.	6701799759	-/2	-/10	-/7	-/200	-/20
Wang, L.	NA	4/-	10/-	5/-	143/-	14.30/-
Lin, BQ.	35098935000	5/4	9/8	4/6	151/178	16.78/30.14
Zhang, J.	57193255205	6/-	9/-	4/-	95/-	10.56/-
Liu, Y.	57200105972	7/-	9/-	4/-	82/-	9.11/-
Chen, GQ.	7406541589	8/-	8/-	7/-	351/-	43.88/-
Huang, GH.	55489745300	9/-	8/-	6/-	193/-	24.13/-
Yang, L.	57203351492	10/-	8/-	6/-	133/-	16.63/-
Zhu, L.	56701286100	11/-	8/-	6/-	95/-	11.88/-
Song, ML.	NA	12/-	8/-	4/-	75/-	9.38/-
Fan, Y.	7403491920	-/5	-/7	-/6	-/211	-/30.14
Yuan, X.	15066382000	-/6	-/7	-/5	-/75	-/10.71
Antunes, C.H.	57191244701	-/7	-/6	-/6	-/113	-/18.83
Krausmann, F.	6602183651	-/8	-/6	-/6	-/412	-/68.67
Lutz, C.	7103325863	-/9	-/6	-/5	-/108	-/18
Pollitt, H.	22954406100	-/10	-/6	-/5	-/134	-/22.33
Zuo, J.	23020460400	-/11	-/6	-/4	-/70	-/11.67
Edenhofer, E.	55868364000	-/12	-/5	-/5	-/430	-/86

ID: Identification author number on Scopus database; W/S: WoS/Scopus values; TC/Art: Total Citations/Article.

In both databases, the journals Energy Policy, Journal of Cleaner Production, Energy, and Sustainability are the most influential journals on 3E-related issues. In fact, half of the most influential journals are the same in WoS and Scopus. The main difference is the inclusion in Scopus of Asian journals: Shengtai Xuebao Acta Ecologica Sinica and Nongye Gongcheng Xuebao Transactions of the Chinese Society of Agricultural Engineering (Table 6).

Table 6. Distribution of articles by journal (WoS/Scopus).

Journal	Ranking (W/S)	Articles (W/S)	H-Index (W/S)	Citations (W/S)	TC/A (W/S)
Energy Policy	1/1	116/124	36/38	4128/4681	35.58/37.75
Journal of Cleaner Production	2/2	107/87	18/19	1252/1323	11.70/15.21
Sustainability	3/4	60/56	8/8	219/276	3.65/4.93
Energy	4/3	51/63	17/23	1243/1654	24.37/26.25
Applied Energy	5/5	44/45	18/20	1361/1840	30.93/40.89
Shengtai Xuebao Acta Ecologica Sinica	-/6	-/43	-/5	-/96	2.23
Ecological Economics	6/-	29/-	15/-	1139/-	39.27/-
Resources conservation and recycling	8/7	28/31	10/12	311/437	11.10/14.10
Nongye Gongcheng Xuebao Transactions of the Chinese Society of Agricultural Engineering	-/8	-/28	-/7	-/204	-/7.29
Energies	7/9	28/26	6/6	114/111	4.07/4.27
International Journal of Hydrogen Energy	9/-	27/-	12/-	938/-	34.74/-
Ecological Indicators	10/-	24/-	11/-	342/-	14.25/-
Energy Economics	-/10	-/26	-/14	-/792	-/30.46

W/S: WOS/Scopus values.

3.4. Areas of Knowledge

The analysis of knowledge areas has been carried out with an initial homogenization of the existing categories in WoS and Scopus, in order to extract more conclusive results. The adaptation of the categories in WoS has been conducted with the inclusion of the Environmental, Chemical, Civil, Industrial, and Agricultural Engineering subsections, while Computer Science includes Artificial Intelligence, Interdisciplinary Applications, Software Engineering, and Information Systems. The rest of the thematic areas correspond to the distribution that is presented in Table 7. The revision of categories in Scopus has required the inclusion of Chemical Engineering in the Engineering area, while the rest of categories correspond to those existing in this database.

Table 7. Distribution of articles by knowledge areas.

	WOS				SCOPUS			
Subject Area	**Articles**	**H-Index**	**Citations**	**TC/A**	**Articles**	**H-Index**	**Citations**	**TC/A**
Environmental Sciences	620	52	10779	17.4	1300	67	21212	16.32
Engineering, Chemical	452	42	7400	16.4	1076	54	12413	11.54
Energy Fuels	483	55	11417	23.6	966	63	17199	17.80
Business Economics	345	48	7762	22.5	519	45	7770	14.97
Science Technology	299	25	3203	10.7	296	27	2951	9.97
Computer Science	54	14	705	13.1	178	25	2495	14.02
Social Science	98	15	736	7.51	608	34	5995	9.86
Agriculture	32	10	224	7	309	27	2834	9.17

TC/Art: Total Citations/Article.

The areas of Environmental Sciences, Engineering, and Energy Fuels are the most predominant in terms of the number of articles published, especially in the case of Scopus, where this trio is separated from the rest with a gap of almost double the number of works. However, if the influence or productivity, as indicated by the number of citations per article, is observed, the category of Business

Economics has the greatest impact in the scientific field, occupying second and third place in WoS and Scopus, respectively (Table 7).

3.5. Institutions

The distribution of the 3E's scientific contribution with respect to the institutions shows a predominance of Asian organizations. In fact, nine of the 13 institutions that were analysed are located in China, which is consistent with the results that were obtained in relation to geographical distribution and the most productive and influential authors.

Both of the databases show that the top three positions are held by the Chinese Academy of Science, Tsinghua University, and North China Electric Power. The first of these is one of the most relevant research centres in the world with around 60,000 researchers, standing out in the field of chemistry. Tsinghua University is dedicated to academic excellence, the benefit of Chinese society, and global development. It is considered to be one of the best academic institutions in China and Asia, ranking in the top 20 of the Times Higher Education World Reputation Rankings. Finally, North China Electric Power has been fostering talent in the areas of engineering technology, management, economics, and the social sciences.

The exceptions to the Asian institutions are the University of Cambridge, the U.S. Department of Energy, the University of London, and the University of California. In other words, the main organizations researching the relationship between the economy, energy, and environment are of Chinese and Anglo-Saxon origin (Table 8).

Table 8. Distribution of articles by institution (WOS/Scopus).

Institution	Ranking (W/S)	Articles (W/S)	H-Index (W/S)	Citations (W/S)	TC/A (W/S)
Chinese Academy of Science	1/1	76/113	20/21	1412/1614	18.58/14.28
Tsinghua University	2/2	44/61	17/18	850/1064	19.32/17.44
North China Electric Power University	3/3	40/46	12/12	542/522	13.55/11.34
Beijing Normal University	4/5	31/35	13/11	568/567	18.32/16.2
Peking University	5/8	27/22	13/12	654/488	24.22/22.18
University of Cambridge	6/6	27/29	14/17	828/1432	30.67/49.38
U.S. Department of Energy	7/-	26/-	14/-	3509/-	134.96/-
University of London	8/-	25/-	10/-	330/-	13.2/-
University of Chinese Academy of Science	9/4	24/36	10/8	433/355	18.04/9.86
Ministry of Education China	-/7	-/23	-/8	-/392	-/17.04
University of California System	10/-	21/-	15/-	1840/-	87.62/-
Zheijang University	-/9	-/21	-/8	-/243	-/11.57
Beijing Institute of Technology	-/10	-/21	-/5	169	-/8.05

W/S: WoS/Scopus values; TC/Art: Total Citations/Article.

3.6. Linked Areas: Clustering 3E

In order to have a better understanding of the evolution of the literature from 2001 to 2018, a fractional counting cluster analysis of keywords throughout the study period has been carried out. The different configurations of the clusters can be observed in Figures 5 and 6, and they also show how the main and central topics have changed.

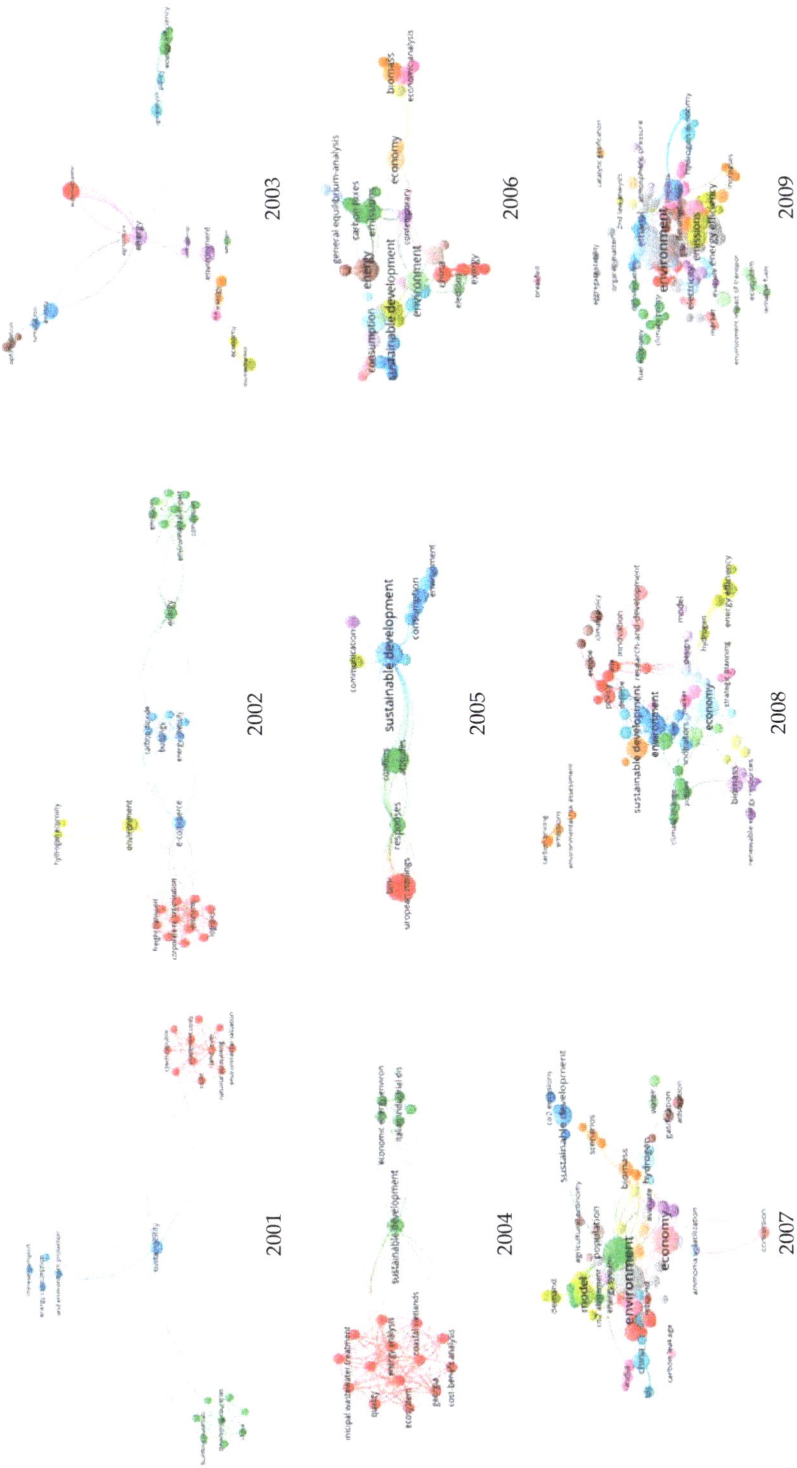

Figure 5. Evolution of keywords from 2001 to 2009.

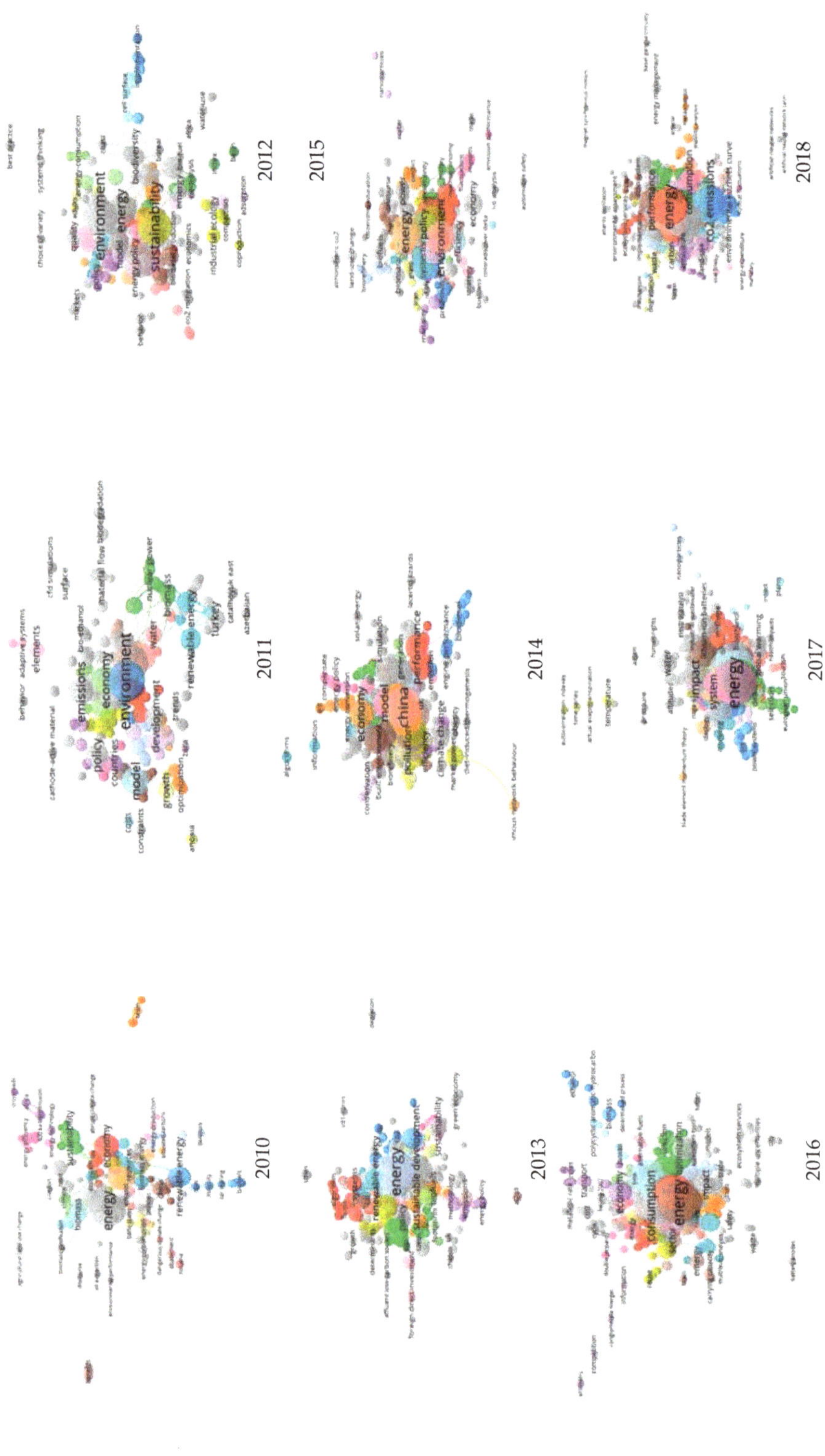

Figure 6. Evolution of keywords from 2010 to 2018.

The articles published during 2001 are distributed into three clusters, with the term *Sustainability* linking them (Figure 5). The first cluster includes: cost effectiveness, Czech Republic, environmental assessment, water, and national accounting; the second cluster: construction materials, governance, developing countries, public policy, and technology transfer; and, the third one: environmental protection, energy consumption, and transport policy, all being within the geographical scope of the Czech Republic and China.

The following two years follow a similar group structure, with a greater distribution in 2003: four groups in 2002 as compared to 11 in 2003. The 2002 groups show the incorporation of new elements in the 3E research field, such as logistics, structural change, land use, environmental impact, energy, sustainable development, carbon dioxide, and hydrogen economy. In 2003, energy was the central axis of the documents, being closely related to the environment in models of bottom-up approach. The same cluster also includes keywords, such as integrated econometric models, welfare, or trade reforms, all being framed within the geographical scope of China. The economy, on the other hand, is found in another cluster, along with terms, such as biomechanics or energy efficiency.

There was a variation in the distribution of the research in 2004, with the predominance of two large clusters that are united by the concept of sustainable development. The first cluster includes elements, such as cost-benefit analysis, energy analysis, and quality. The second cluster focuses on the value chain, the input-output technique, the explicit integration of economy, energy, and environment and industrial district, focusing the issue in countries, such as Italy and the USA. In 2005, sustainable development continued to be the central trend in the relationship between economy, environment, and environment. Six clusters were identified, in which biomass energy is related to environmental conservation, energy policies, renewable and rural energies, as well as environmental management or the analysis of life cycle assessment in the territories of China and the European Union.

In the following year, the importance of the economy was greater within the dimension of 3E, while sustainable development continued to be the central concept. Keywords, such as energy efficiency, taxes on coal and emissions, as well as biomass, renewable energy sources, climate change, Jevons paradox, or eco-efficiency within New Zealand and Turkey revolve around it. In 2007, there was a convergence between the three concepts of 3E, surrounded by elements, such as the agricultural economy, biomass, hydrogen, energy efficiency, gasification, carbon dioxide emissions, production of biohydrogen, or energy in the geographical areas of China and India.

The scenario drawn in 2008 and 2009 is very similar to that of 2007, although the importance of energy is lower when compared to the presence of the environment and the economy in those years' articles. New concepts that were incorporated in those years include wind energy, research and development policies, strategic planning, exergy analysis, uncertainty, and the price of carbon emissions in the Balkans, Europe, Asia, and India. Similar to 2008, 2009 saw a waning of the importance of the economy with respect to energy and the environment, as well as a lesser presence of sustainable development. Newer elements in 3E include the analysis of the environmental impact of transport, critical discursive analysis of ecological modernization, deforestation, biodiesel, biofuel, and ecological footprint, all being framed in countries such as China, Japan, or the African continent.

In 2010 (Figure 6), there was an increase in the importance of the elements of 3E, together with the concept of sustainable development. In this period, new factors appear, such as the change in the use of agricultural land, the dangers of climate change, the energy footprint, ecological modernisation, greenhouse gas, environmental strategy, and the responsibility of consumers and pressure groups on the state of the environment. The main territories at this time were China, Japan, Denmark, Europe, and Africa. In 2011, the environment was the main element of 3E. Sustainability and climate change are at the same level in terms of presence, while aspects such as renewable energies or emissions management appear in the research to a greater extent than in previous periods. In addition, the geographical scope of the studies broadened to include territories, such as Azerbaijan and the United Arab Emirates. In 2012, the concept of sustainability once more gained space, along with the terms biomass, biodiversity, efficiency, and performance. Economy, environment, and energy are at the same level of presence, but behind sustainability. The number of researches that were carried out in China increased, as well as those that were focused on developing countries.

The following year, 2013, shows a similar structure to that of 2009, with a recovery of the concept of sustainable development. It is in this year that the green economy had greater presence in the research scenario, including terms, such as recycling or energy efficiency. 2014 stands out for the high prominence that China acquires in the investigation on 3E. The concept of environment is at the same level, to the detriment of energy and economy. Sustainability and energy efficiency also have a high presence, and it is at this time that solar energy is analysed to a greater extent. In that year, there was a distancing of the concept of economy in relation to energy and the environment, which had a greater presence in the research landscape than the first. In fact, the greatest division takes place between economy and energy, with the environment being the connecting element. China continues to have a high presence, as well as a growing number of articles analysing the uncertainty, performance, and management of energy efficiency.

Something similar happened the following year, 2015, where energy and the environment are more closely related than the economy, which is less relevant. Along with the first two elements, sustainable development, sustainability, consumption, economic growth, and CO2 emissions also stand out, with China and the USA being the main countries under investigation. In 2016 and 2017, the economy was once again linked to the other two components of 3E, with a concentration of keywords around 3E. Finally, energy and sustainable development predominate in 2018. Factors, such as the environmental curve of Kuznets, the carbon footprint, the circular economy, and the optimization of the efficiency of greenhouse gas emissions are also present. These coincide with the latest trends that were observed in the cluster analysis for the whole period (Figure 7).

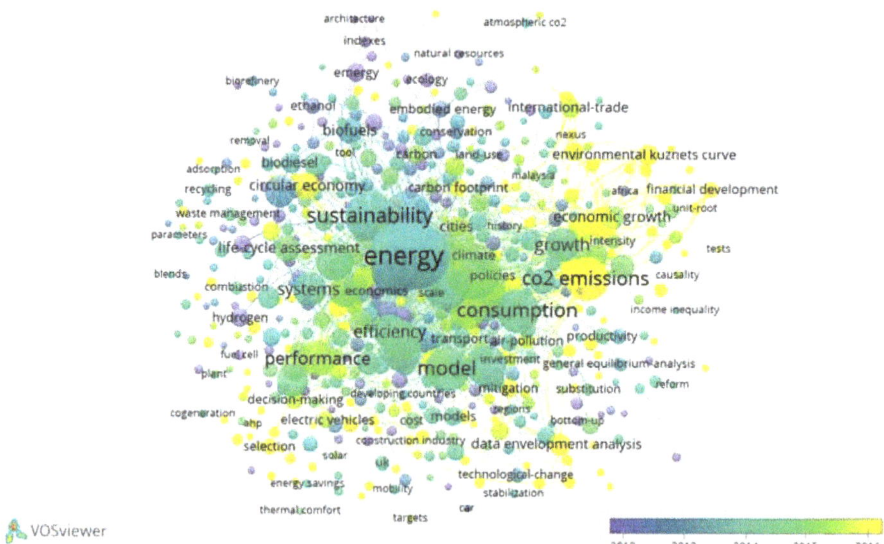

Figure 7. Keyword analysis from 2001 to 2018.

The analysis of the 3E keyword trend shows that the most prominent common elements were sustainable development and sustainability. The inclusion of this in the analysis shows the different clusters that link them and the latest trends.

With regard to the clusters, these are distributed in six groups. The first of them relates to economic growth in the circular economy, the efficient use and consumption of energy, and the management of emissions, all being framed in the geographical scope of China and in the methodology of surrounding data analysis. The second group focuses more on the field of economic development, climate change, and greenhouse gas emissions, with its methodological element being the analysis of the environmental curve of Kuznets. The following two groups follow the line of climate change, identifying new energy sources, such as biomass or biofuel, and including elements, such as green economy, development, and sustainable energy, as well as the territory of Turkey. The fifth encompasses the concept of sustainability and energy, while the last cluster integrates innovation with development policies and dynamic systems.

Figure 8 shows the latest trends in research with the integration of energy, economy, and environment terms in a single cluster analysis. An initial reflection in these trends indicates that the Kuznets environmental curve is used under the sustainable development approach. Currently, the enveloping analysis of data is used to study the impact of variables, such as carbon dioxide emissions or energy consumption on economic growth. On the other hand, due to the absence of some terms that were commonly used in previous years, such as petrol, pollution, or even some that are lagging behind, such as climate change, which is in accordance with this cluster analysis, future research on 3E may emphasise on Circular Economy and Green Economy as the main solution for achieving sustainable development.

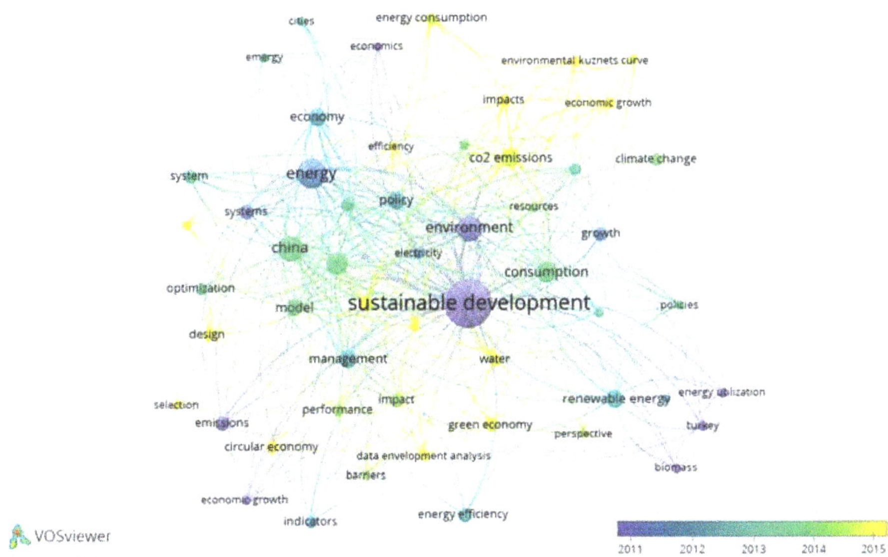

Figure 8. Keywords of sustaniable development from 2001 to 2018.

4. Conclusions

The interrelation between Energy, Economy, and Environment has been studied in depth by academia and the number of publications increases year on year. The Millennium Development Goals that was proposed by the United Nations Development Organization has contributed to highlight the importance of this triple helix in the global economic scenario.

The bibliometric and cluster analysis has shown that the main thematic categories that are linked to the integrated concept of 3E correspond to environmental sciences, engineering, energy fuels, and business and economics. However, the diversity of topics with which this concept is related is so great that it demonstrates its multidisciplinarity and transversal character.

The study of the most prolific countries shows the hegemony of China, followed by the USA and the United Kingdom. This leadership of China as a research country in economy, energy, and environment is more evident after analysing the distribution by authors and institutions, wherein most of them are from said country. However, English continues to be the main language used by researchers due to the fact that it is the preferred language for the publication of articles in the principal high impact journals.

The analysis of keywords shows that the evolution of the interrelation between 3E from 2001 to 2018 has been marked by a process of progressive integration of the three concepts. From 2001 to 2005, there was a clear differentiation in a few groups between energy, economy, and environment. However, from 2009 onwards, a progressive change can be observed in this relationship, integrating itself more and more until 2018, when the concept of 3E culminates in the term sustainable development, being linked to the environmental curve of Kuznets, the carbon footprint, the circular economy, the green economy, and the optimisation of the efficiency of greenhouse gas emissions. These latest trends are framed within the enveloping data analysis methodology. The main common element is sustainable development and sustainability. It can be observed that topics regarding renewable power, such as solar energy, have a relevant role from 2010. The inclusion of this in the analysis shows the different clusters that link them and the latest trends.

According to the results that were obtained, the future of 3E studies revolves around the concept of sustainable development, in which China, with the Chinese Academy of Science at the forefront, is positioned as the driving country of this trend, and journals, such as Energy Policy, are the main drivers to concentrate the research effort of the scientific community of institutions with greater research capacity in the field of environmental sciences, energy fuels, engineering, business, and economics.

Specifically, in relation with energy, one of the most important topic are energy saving, energy efficiency, recycling, and renewable energy sources, highlighting the importance of green energy. In this line, concepts, such as eco-efficiency and energy production, have a greater presence in the academia.

This work is placed as an identification of the latest trends that relate to Energy, Economy, and Environment, at a time when the transition to energy from less polluting sources is being considered in view of the imminent arrival of climate change. Therefore, it marks new lines of research that is related to the concept of sustainable development and other complementary terms, such as the circular economy or carbon footprint. In other words, new research in the field of 3E should address the energy transition issue in the area of sustainable development, by adapting it to the restrictions of this economic model.

With regards to the limitations of this research, firstly, the field of study has focused on the most influential academic databases (WoS and Scopus). Secondly, only articles have been analysed and therefore it would be interesting to open a broader line of research that includes other databases and other types of publications, such as books or conference proceedings.

Author Contributions: The authors contributed equally to this work.

Funding: This research received no external funding.

Conflicts of Interest: The authors declare no conflict of interest.

References

1. Roques, F.; Sassi, O.; Hourcade, J.C.; Guivarch, C.; Waisman, H.; Crassous, R. The impact of China and India's economic growth on energy use and CO_2 emissions-integrated modelling of economic-energy-environment scenarios. *IOP Conf. Ser. Earth Environ. Sci.* **2009**, *6*, 212003. [CrossRef]
2. Capros, P. Integrated economy-energy-environment models. In Proceedings of the International Symposium on Electricity, Health and the Environment: Comparative Assessment in Support of Decision Making, Vienna, Austria, 16–19 October 1995.
3. Liu, D.; Tian, X.; Wu, R.; Wang, L. Study on integrated simulation model of economic, energy and environment safety system under the low-carbon policy in Beijing. *Procedia Environ. Sci.* **2011**, *5*, 120–130. [CrossRef]
4. United Nations. *United Nations Millennium Declaration*; United Nations: New York, NY, USA, 2000.

5. United Nations. *Sustainable Development Goals Report*; United Nations: New York, NY, USA, 2016.
6. Yi, Q.; Feng, J.; Wu, Y.; Li, W. 3E (energy, environmental, and economy) evaluation and assessment to an innovative dual-gas polygeneration system. *Energy* **2014**, *66*, 285–294. [CrossRef]
7. Uno, K. Economy-energy-environment: The 3E compass model. In *Integrated Global Models of Sustainable Development*; Onishi, A., Ed.; EOLSS Publishers Co Ltd.: Tokio, Japan, 2009; Volume 2, pp. 131–153. ISBN 978-1-905839-18-6.
8. Jorgenson, D.W.; Goettle, R.J.; Ho, M.S.; Wilcoxen, P.J. Energy, the environment and US economic growth. In *Handbook of Computable General Equilibrium Modeling*; Dixon, P., Jorgenson, D., Eds.; Elsevier: Amsterdam, The Netherlands, 2013; Volume 1, pp. 477–552. ISBN 9780444595560.
9. Nakata, T. Energy-economic models and the environment. *Prog. Energy Combust.* **2004**, *30*, 417–475. [CrossRef]
10. Yanqing, X.; Mingsheng, X. A 3E Model on Energy Consumption, Environment Pollution and Economic Growth—An Empirical Research Based on Panel Data. *Energy Procedia* **2012**, *16*, 2011–2018. [CrossRef]
11. Besstremyannaya, G.; Dasher, R.; Golovan, S. Technological change, energy, environment and economic growth in Japan. In *Energy, Environment and Economic Growth in Japan*; USAEE Working Paper; Center for Economic and Financial Research (CEFIR): Moscow, Russia, 2018; pp. 18–377.
12. Zaman, K.; Moemen, M.A.E.; Islam, T. Dynamic linkages between tourism transportation expenditures, carbon dioxide emission, energy consumption and growth factors: Evidence from the transition economies. *Curr. Issues Tour.* **2017**, *20*, 1720–1735. [CrossRef]
13. Ping, J.; Yan, S.; Gu, P.; Wu, Z.; Hu, C. Application of MIKE SHE to study the impact of coal mining on river runoff in Gujiao mining area, Shanxi, China. *PLoS ONE* **2017**, *12*, e0188949. [CrossRef]
14. Dzikuć, M. Problems associated with the low emission limitation in Zielona Góra (Poland): Prospects and challenges. *J. Clean. Prod.* **2017**, *166*, 81–87. [CrossRef]
15. Darko, A.; Chan, A.P.C.; Gyamfi, S.; Olanipekun, A.O.; He, B.J.; Yu, Y. Driving forces for green building technologies adoption in the construction industry: Ghanaian perspective. *Build. Environ.* **2017**, *125*, 206–215. [CrossRef]
16. Bashir, S.; Ahmad, I.; Rashid Ahmad, S. Low-Emission Modeling for Energy Demand in the Household Sector: A Study of Pakistan as a Developing Economy. *Sustainability* **2018**, *10*, 3971. [CrossRef]
17. Campolina, J.M.; Sigrist, C.S.L.; de Paiva, J.M.F.; Nunes, A.O.; da Silva Moris, V.A. A study on the environmental aspects of WEEE plastic recycling in a Brazilian company. *Int. J. LCA* **2017**, *22*, 1957–1968. [CrossRef]
18. Dmitrienko, M.A.; Strizhak, P.A. Environmentally and economically efficient utilization of coal processing waste. *Sci. Total Environ.* **2017**, *598*, 21–27. [CrossRef]
19. Prakash, R.; Bhat, I.K. Energy, economics and environmental impacts of renewable energy systems. *Renew. Sustain. Energy Rev.* **2009**, *13*, 2716–2721. [CrossRef]
20. Grover, S. *Energy, Economic, and Environmental Benefits of the Solar America Initiative*; Report by the National Renewable Energy Laboratory; National Renewable Energy Laboratory: Golden, CO, USA, 2007. [CrossRef]
21. Cavalcanti, C. Conceptions of ecological economics: Its relationship with mainstream and environmental economics. *Estudos Avançados* **2010**, *24*, 53–67. [CrossRef]
22. Gowdy, J.; Erickson, J.D. The approach of ecological economics. *Camb. J. Econ.* **2007**, *29*, 207–222. [CrossRef]
23. Jin, E.; Sutherland, J.W. An integrated sustainability model for a bioenergy system: Forest residues for electricity generation. *Biomass Bioenergy* **2018**, *119*, 10–21. [CrossRef]
24. Beschkov, V. Biogas, Biodiesel and Bioethanol as Multifunctional Renewable Fuels and Raw Materials. In *Frontiers in Bioenergy and Biofuels*; JacobLopes, E., Zepka, L.Q., Eds.; InTech: Vienna, Austria, 2017; pp. 185–205.
25. Fajardy, M.; Chiquier, S.; Mac Dowell, N. Investigating the BECCS resource nexus: Delivering sustainable negative emissions. *Energy Environ. Sci.* **2018**, *11*, 3408–3430. [CrossRef]
26. Neves, S.A.; Marques, A.C.; Fuinhas, J.A. Could alternative energy sources in the transport sector decarbonise the economy without compromising economic growth? *Environ. Dev. Sustain.* **2018**, *20*, 1–18. [CrossRef]
27. Yousefi, H.; Roumi, S.; Tabasi, S.; Hamlehdar, M. Economic and air pollution effects of city council legislations on renewable energy utilisation in Tehran. *Int. J. Ambient. Energy* **2018**, *39*, 626–631. [CrossRef]
28. Middleton, P. Sustainable living education: Techniques to help advance the renewable energy transformation. *Sol. Energy* **2018**, *174*, 1016–1018. [CrossRef]

29. Muradin, M.; Joachimiak-Lechman, K.; Foltynowicz, Z. Evaluation of Eco-Efficiency of Two Alternative Agricultural Biogas Plants. *Appl. Sci.* **2018**, *8*, 2083. [CrossRef]
30. Cicea, C.; Marinescu, C.; Popa, I.; Dobrin, C. Environmental efficiency of investments in renewable energy: Comparative analysis at macroeconomic level. *Renew. Sustain. Energy Rev.* **2014**, *30*, 555–564. [CrossRef]
31. Wang, Y.; Xiao, W.; Wang, Y.; Zhao, Y.; Wang, J.; Hou, B.; Song, X.Y.; Zhang, X. Impact of China's Urbanization on Water Use and Energy Consumption: An Econometric Method and Spatiotemporal Analysis. *Water* **2018**, *10*, 1323. [CrossRef]
32. Andreas, J.J.; Burns, C.; Touza, J. Renewable Energy as a Luxury? A Qualitative Comparative Analysis of the role of the Economy in the EU's Renewable Energy Transitions during the 'Double Crisis'. *Ecol. Econ.* **2017**, *142*, 81–90. [CrossRef]
33. Armeanu, D.Ş.; Vintilă, G.; Gherghina, Ş.C. Does renewable energy drive sustainable economic growth? multivariate panel data evidence for EU-28 countries. *Energies* **2017**, *10*, 381. [CrossRef]
34. Goodier, C. Carbon footprint. In *Green Cities*; Cohen, N., Robbins, P., Eds.; SAGE Publications: London, UK, 2010; pp. 49–53.
35. Viglia, S.; Civitillo, D.F.; Cacciapuoti, G.; Ulgiati, S. Indicators of environmental loading and sustainability of urban systems. An emergy-based environmental footprint. *Ecol. Indic.* **2018**, *94*, 82–99. [CrossRef]
36. Xiao, X.; Zhu, Z.; Fu, Z.; Mu, W.; Zhang, X. Carbon Footprint Constrained Profit Maximization of Table Grapes Cold Chain. *Agronomy* **2018**, *8*, 125. [CrossRef]
37. Tobarra, M.A.; López, L.A.; Cadarso, M.A.; Gómez, N.; Cazcarro, I. Is Seasonal Households' Consumption Good for the Nexus Carbon/Water Footprint? The Spanish Fruits and Vegetables Case. *Environ. Sci. Technol.* **2018**, *52*, 12066–12077. [CrossRef]
38. Mahapatra, C.; Moharana, A.K.; Leung, V. Energy management in smart cities based on Internet of Things: Peak demand reduction and energy savings. *Sensors* **2017**, *17*, 2812. [CrossRef]
39. United Nations. *Report of the United Nations Conference on the Human Environment*; United Nations Publications: New York, NY, USA, 1972.
40. Millimet, D.L.; Roy, S.; Sengupta, A. Environmental regulations and economic activity: Influence on market structure. *Annu. Rev. Resour. Econ.* **2009**, *1*, 99–118. [CrossRef]
41. Mahmood, F.; Belhouchette, H.; Nasim, W.; Shahzada, T.; Hussain, A.; Therond, O.; Fahad, E.; Refat, S.S.; Wéry, J. Economic and environmental impacts of introducing grain legumes in farming systems of Midi-Pyrenees region (France): A simulation approach. *Int. J. Plant. Prod.* **2017**, *11*, 65–88.
42. Corona, B.; Bozhilova-Kisheva, K.P.; Olsen, S.I.; San Miguel, G. Social life cycle assessment of a concentrated solar power plant in Spain: A methodological proposal. *J. Ind. Ecol.* **2017**, *21*, 1566–1577. [CrossRef]
43. Li, D.; Cao, C.; Zhang, L.; Chen, X.; Ren, S.; Zhao, Y. Effects of corporate environmental responsibility on financial performance: The moderating role of government regulation and organizational slack. *J. Clean. Prod.* **2017**, *166*, 1323–1334. [CrossRef]
44. Ordouei, M.H.; Elkamel, A. New composite sustainability indices for Cradle-to-Cradle process design: Case study on thinner recovery from waste paint in auto industries. *J. Clean. Prod.* **2017**, *166*, 253–262. [CrossRef]
45. Qi, Y.; Zhang, X.; Yang, X.; Lv, Y.; Wu, J.; Lin, L.; Xiao, Y.; Qi, H.; Yu, X.; Zhang, Y. The environmental sustainability evaluation of an urban tap water treatment plant based on emergy. *Ecol. Indic.* **2018**, *94*, 28–38. [CrossRef]
46. Ishak, M.I.S.; Ishak, N.F.A.; Hassan, M.S.; Amran, A.; Jaafar, M.H.; Samsurijan, M.S. The role of multinational companies for world sustainable development agenda. *J. Sustain. Sci. Manag.* **2017**, *12*, 228–252.
47. Marques, A.C.; Fuinhas, J.A.; Pais, D. Economic growth, sustainable development and food consumption: Evidence across different income groups of countries. *J. Clean. Prod.* **2018**, *20*, 245–258. [CrossRef]
48. Wu, Z.; Wang, R.; Xu, S. Strategic choice and practice of low carbon urbanization in China. *Model. Meas. Control C* **2017**, *78*, 455–466. [CrossRef]
49. Archambault, É.; Caruso, J.; Côté, G.; Larivière, V. Bibliometric Analysis of Leading Countries in Energy Research. In Proceedings of the 12th International Conference of the International Society for Scientometrics and Informetrics (ISSI), Rio de Janeiro, Brazil, 14–17 July 2009; pp. 80–91.
50. Ruiz-Real, J.L.; Uribe-Toril, J.; De Pablo, J.; Gázquez-Abad, J.C. Worldwide Research on Circular Economy and Environment: A Bibliometric Analysis. *Int. J. Environ. Res. Public Health* **2018**, *15*, 2699. [CrossRef] [PubMed]

51. Zheng, T.; Li, P.; Shi, Z.; Liu, J. Benchmarking the scientific research on wastewater-energy nexus by using bibliometric analysis. *Environ. Sci. Pollut. Res. Int.* **2017**, *24*, 27613–27630. [CrossRef]
52. Reuters, T. A Guide to Evaluating Research Performance with Citation Data. Available online: http://ip-science.thomsonreuters.com/m/pdfs/325133_thomson.pdf (accessed on 15 January 2019).
53. Pritchard, A. Statistical bibliography or bibliometrics. *J. Doc.* **1969**, *25*, 348–349.
54. Osareh, F. Bibliometrics, citation analysis and co-citation analysis: A review of literature I. *Libri* **1996**, *46*, 149–158. [CrossRef]
55. Hirsch, J.E. An index to quantify an individual's scientific research output. *Proc. Natl. Acad. Sci. USA* **2005**, *102*, 16569–16572. [CrossRef] [PubMed]
56. Van Eck, N.J.; Waltman, L. Citation-based clustering of publications using CitNetExplorer and VOSviewer. *Scientometrics* **2017**, *111*, 1053–1070. [CrossRef]
57. Perianes-Rodriguez, A.; Waltman, L.; Van Eck, N. Constructing bibliometric networks: A comparison between full and fractional counting. *J. Informetr.* **2016**, *10*, 1178–1195. [CrossRef]
58. Waltman, L.; Van Eck, N.J. A new methodology for constructing a publication-level classification system of science. *J. Assoc. Inf. Sci. Technol.* **2012**, *63*, 2378–2392. [CrossRef]
59. Orduña-Malea, E.; Ayllón, J.M.; Martín-Martín, A.; López-Cózar, E.D. Methods for estimating the size of Google Scholar. *Scientometrics* **2015**, *104*, 931–949. [CrossRef]
60. Ede, P.N.; Johnson, G.A. Energy relations of gas estimated from flare radiation in Nigeria. *Int. J. Energy Res.* **2001**, *25*, 85–91. [CrossRef]
61. Mendis, P. Food Security, Agricultural Subsidies, Energy, and the Environment: A Process of 'Glocalization' in Sri Lanka. *Energy Environ.* **2001**, *12*, 55–71. [CrossRef]
62. Bashyam, R.; Zelenay, P. A class of non-precious metal composite catalysts for fuel cells. In *Materials for Sustainable Energy: A Collection of Peer-Reviewed Research and Review Articles from Nature Publishing Group*; Dusastre, V., Ed.; Nature Publishing Group: London, UK, 2011; pp. 247–250. ISBN 978-981-4317-66-5.

energies

MDPI

Article

Do Carbon Emissions and Economic Growth Decouple in China? An Empirical Analysis Based on Provincial Panel Data

Yu Hao [1,2,3,4,5,*,†], **Zirui Huang** [6,†] **and Haitao Wu** [7,†]

1 Center for Energy and Environmental Policy Research, Beijing Institute of Technology, Beijing 100081, China
2 School of Management and Economics, Beijing Institute of Technology, Beijing 100081, China
3 Sustainable Development Research Institute for Economy and Society of Beijing, Beijing 100081, China
4 Collaborative Innovation Center of Electric Vehicles in Beijing, Beijing 100081, China
5 Beijing Key Lab of Energy Economics and Environmental Management, Beijing 100081, China
6 School of Humanities and Social Sciences, Beijing Institute of Technology, Beijing 100081, China; 1120163193@bit.edu.cn
7 College of Economics and Management, Xinjiang University, Urumqi 830047, China; haitao_wu4015@126.com
* Correspondence: haoyuking@bit.edu.cn; Tel.: +86-10-68914938
† These authors contributed equally to this study and share first authorship.

Received: 16 May 2019; Accepted: 20 June 2019; Published: 23 June 2019

Abstract: Global warming has emerged as a serious threat to humans and sustainable development. China is under increasing pressure to curb its carbon emissions as the world's largest emitter of carbon dioxide. By combining the Tapio decoupling model and the environmental Kuznets curve (EKC) framework, this paper explores the relationship between China's carbon emissions and economic growth. Based on panel data of 29 provinces from 2007 to 2016, this paper quantitatively estimates the nexus of carbon emissions and economic development for the whole nation and the decoupling status of individual provinces. There is empirical evidence for the conventional EKC hypothesis, showing that the relationship between carbon emissions and per capita gross domestic product (GDP) is an inverted U shape and that the inflection point will not be attained soon. Moreover, following the estimation results of the Tapio decoupling model, there were significant differences between individual provinces in decoupling status. As a result, differentiated and targeted environmental regulations and policies regarding energy consumption and carbon emissions should be reasonably formulated for different provinces and regions based on the corresponding level of economic development and decoupling status.

Keywords: environmental Kuznets curve (EKC); decoupling theory; panel data; differential GMM estimation; Tapio decoupling model

1. Introduction

Climate change has emerged as an important issue that threatens humans and sustainable development. In response to global warming, in 2016, 196 parties signed the Paris Agreement, a long-term agreement to control climate change, whose main objective is to limit the increase in global average temperature to 2° Celsius within this century [1]. Among the factors affecting global warming, the impact of carbon dioxide is crucial, and it has been confirmed that carbon dioxide is responsible for about 60% of the world's greenhouse effect [2]. Meanwhile, the short-term outlook for climate governance is not optimistic. Carbon dioxide emissions from fossil fuels and industries were expected to increase by 2% in 2017, the first increase following three consecutive years of decline since 2014 [3]. According to the World Resources Institute's 2017 report [4], the three largest greenhouse gas emitters in the world are China, the European Union, and the United States, accounting for more than half of total emissions. According to the estimations of the Carbon Dioxide Information Analysis Center (CDIAC), since 2007, China has overtaken the United States to become the world's largest carbon emitter. As a large country with international responsibility, China has enacted a number of low-carbon development policies in order to curb the growth of carbon emissions. China's 13th Five-Year Plan highlights low-carbon development as a major strategy for economic and social development, and it is also an important way to construct an ecological civilization [5].

However, partly due to the remarkable gap in economic and social development, the scales and growth rates of CO_2 emissions between different provinces in China differ significantly [6]. Formulating a unified national low-carbon development policy may not effectively achieve carbon emission reduction, and may even have the opposite effect on low-carbon development. Furthermore, economic growth in Chinese provinces varies quite a bit. The per capita gross domestic product (GDP) of some provinces is close to that of developed countries, while some provinces are relatively lagging behind. Therefore, the government's CO_2 emission reduction policies for different provinces should be connected to regional economic development levels, and the chronic trend characteristics of regional CO_2 emissions need to be considered. Decoupling indicators and the environmental Kuznets curve (EKC) hypothesis are two important methods for measuring the nexus of economic growth and carbon emissions. If we can effectively measure the decoupling status of each province, we can figure out the CO_2 emissions status. If the EKC framework could be effectively utilized to analyze the nexus of economic growth and carbon emissions, it would be possible to predict carbon emissions in different provinces in China. Therefore, by combining the two methods, we can develop a deeper understanding of the carbon emission conditions and development trends in each province, and then plan a low-carbon development policy that is suitable for each individual province.

Previous studies have conducted extensive research on the decoupling between economic growth and carbon emissions. The decoupling theory originally came from physics [7]. The Organisation for Economic Co-operation and Development (OECD) pioneered the concept of decoupling in 2002 to explore how to block the relevance of environmental damage and economic growth [8]. Based on the OECD decoupling index, a comprehensive decoupling index of arithmetic mean sum was developed. To eliminate error caused by the selection of the base period, Tapio uses the elastic coefficient decomposition method [9]. This method makes it possible to analyze the internal causes of and responses to decoupling within environmental pressure and economic growth, and opens up a new research path for decoupling theory. Domestic and foreign scholars divide the degree of decoupling into four categories: dichotomy [10], trichotomy [8], sextant, and octave [11]. The Tapio [12] decoupling model is one of the major research methods to explore the nexus of carbon emissions and economic growth. Generalized decoupling in this context refers to the nexus of environmental pollution and economic growth changing from a positive correlation to a negative correlation. Some scholars have summarized the existing methods of decoupling measurement as "speed decoupling" and "quantity decoupling".

Considering the EKC hypothesis, according to the research of Simon Kuznets [13], the degree of income inequality decreases with the expansion of the economy, and the relationship between the two shows the characteristics of an inverted U-shaped curve. Following the study by Kuznets [13], Grossman and Krueger [14] defined the relationship between environment and per capita income as the "environmental Kuznets curve (EKC)", which is also known as the EKC hypothesis. In early empirical studies of the EKC hypothesis with carbon dioxide as a pollutant, panel data were often used. Some scholars have verified the inverted U-type nexus of per capita GDP and per capita carbon emissions, that is, the existence of the EKC curve. Some scholars have determined an N-type relationship through empirical studies [7,15,16], and per capita income at the inflection point of the EKC curve varies greatly [17]. Some scholars have concluded that per capita carbon dioxide emissions are monotonous with per capita income, and an inflection point does not exist [18]. Missing variable errors, integral variables, false regressions, and identification of time effects all affect the final results of

EKC estimates [19]. As Stern [18] stressed, the fact that different data and models may yield different results reflects that the controversy over EKC has not yet been resolved and needs further exploration. It is worth noting that the Kuznets curve model is classified as "quantity decoupling" by some scholars, and the Tapio elastic coefficient method is classified as "speed decoupling." The EKC hypothesis uses the cross-sectional data of per capita GDP and pollution emissions to describe the inverse U relationship between the two in absolute quantity, which is called "quantity decoupling" [15,20,21].

Although more studies have investigated the nexus of CO_2 emissions and economic development, some issues are still being ignored and deserve further investigation. First, the potential endogeneity problem that may be caused by bilateral causality has been largely neglected. Second, the literature on the decoupling effect is usually based on national or aggregate data, while the decoupling situations of individual provinces over time have not been fully investigated. Third, some scholars from China have measured the decoupling status of carbon emissions with similar measurements; however, they have concentrated on a certain region or economic field, and have never made a comprehensive calculation of the decoupling status of various provinces in China [22–26]. Xia [27] measured the economic growth threshold of SO_2 emissions, but the causes of CO_2 and SO_2 are quite different. The research results have less significance for policies aimed at reducing CO_2 emissions.

In order to fill the research gap, this paper used the Tapio decoupling index to measure the decoupling status of economic growth and CO_2 emissions, and used the EKC hypothesis to describe the non-linear nexus of carbon emissions and economic growth to explore their absolute and relative changes. To sum up, the main contributions of this study are fourfold. First, this study quantitatively investigated the nexus of CO_2 emissions and economic development in China by employing the generalized method of moments (GMMs), which can deal with potential endogeneity. Second, the decoupling effects of CO_2 emissions and economic growth in China's provinces were estimated separately, so that the empirical results can serve as an important reference for policy-makers to enact differentiated and targeted regional and provincial carbon reduction policies. Third, the decoupling coefficients for each province were calculated and are discussed. Fourth, in the policy recommendations section, targeted emission reduction policies can be developed based on calculations for the individual provinces with different decoupling types.

2. Methods and Data

2.1. Carbon Emissions Calculation

In general, CO_2 emissions are mainly generated from fossil fuel combustion and the production of cement and lime. According to the estimations of CDIAC and some researchers, CO_2 emissions from fossil fuels account for approximately 90% of the total amount, while the other 10% comes from cement and lime production in China [28,29]. In academia, there is no uniform standard method for measuring carbon emissions. Actual measurement, system simulation, and carbon emission coefficient are the three main research methods. Specifically, the carbon emission coefficient method was employed to calculate CO_2 emissions in this paper. Following the ideas and procedures of the Intergovernmental Panel on Climate Change (IPCC) (2006) [12], the carbon dioxide emissions of 29 provinces (there are currently 23 provinces, four centrally administered municipalities, and five autonomous regions in mainland China (excluding Taiwan). Because they are administratively equal, the term "provincial" is used to refer to them throughout this paper. To ensure comparability of the data, Chongqing was combined with Sichuan Province. Tibet was excluded due to the unavailability of data) in China from 2007 to 2016 were estimated and summed to obtain the total amount of CO_2 emissions in the country. At the same time, referring to Cheng et al. [30], this paper measured CO_2 emissions from fossil fuel combustion and estimated CO_2 emissions from cement production. In addition, fossil fuel consumption data come from the China Energy Statistics Yearbook. Data from cement production were collected from the China Stock Market and Accounting Research (CSMAR) Economic and Financial Database. The emission coefficients were taken directly from the IPCC [12].

According to a study by Hao et al. [28], the formula to calculate CO_2 emissions caused by fossil fuel combustion is as follows:

$$FC = \sum_{s=1}^{n} FC_s = \sum_{s=1}^{n} (F_s \times CF_s \times CC_s - SC_s) \times OF_s \tag{1}$$

Equation (1) represents the calculation result of CO_2 emitted from fossil fuel. The *FC* denotes CO_2 emissions, *s* represents the specific type of energy consumption, *n* is the number of energy types, F_s is the fuel consumption, CF_s is the calorific factor, CC_s is the potential carbon emission factor, SC_s is the carbon sequestration, and OF_S is the oxidation factor. For each fuel type, the *CF*, *CC*, and *OF* values recommended by the IPCC [12] were used directly. To ensure the accuracy of the calculation, data from energy balance sheets in China Energy Statistical Yearbooks (2008–2017) [30] were utilized. Specifically, fossil fuel was further subdivided into different types (i.e., coal, oil, natural gas, other energy sources). Coal included raw coal, cleaned coal, other washed coal, briquettes, coke, and other coking products. Petroleum included crude oil, gasoline, kerosene, diesel, fuel oil, naphtha, lubrication, paraffin, white spirit, bitumen asphalt, petroleum coke, liquefied petroleum gas, and other petroleum products. Natural gas mainly comprised liquefied natural gas. Other energy sources were converted into units of tons of standard coal equivalent (SCE). Because CO_2 emissions generated from fossil fuels are mainly produced from combustion processes, the energy input of transformation was not included. This study followed Hao's et al. [28] calculation method because they strictly followed the IPCC's ideas and obtained relatively reasonable and accurate estimation results for provincial CO_2 emissions. It is noteworthy that they pointed out that the data of provincial carbon sequestration products were not available, and directly ignored the calculation of SC_S. So far, there is still no unified provincial carbon sequestration product measurement method, and data are still unavailable. Since the calculation of provincial carbon sequestration products is not the focus of this study, this paper followed Hao's idea and ignored the *SCs*.

The formula for calculating CO_2 emissions during cement production is as follows:

$$KC = T \times K_{cement} \tag{2}$$

Equation (2) represents the results of CO_2 emissions during cement production. *KC* stands for CO_2 emissions generated from cement production, as the cement industry also contributes significantly to CO_2 emissions in China; *T* is total cement production; and K_{cement} indicates the coefficient of cement production.

The formula to calculate the total amount of CO_2 emissions is as follows:

$$C_{it} = FC_{it} + KC_{it} \tag{3}$$

Equation (3) indicates the results of the total amount of CO_2 emissions. C_{it} represents the first *t* year's CO_2 emissions in city *i*, FC_{it} represents the first *t* year's CO_2 emissions in city *i* generated by burning fossil fuel, and KC_{it} represents the first *t* year's CO_2 emissions in city *i* from cement production.

2.2. Decoupling Theory

Many previous studies found evidence that energy consumption and CO_2 emissions are highly correlated with economic growth in China [15,31]. "Decoupling" is a term that refers to the status when the relationship between energy consumption/environmental deterioration and economic growth begins to break up. In other words, the growth rate of energy consumption or environmental pollution becomes slower than the growth rate of economic development [23]. In this regard, the decoupling function can be expressed as Equation (4):

$$e = \frac{\Delta C/C}{\Delta Y/Y} \quad (4)$$

Equation (4) calculates the decoupling elastic coefficient between economic growth and carbon emissions. e is the decoupling elastic coefficient, C indicates CO_2 emissions, ΔC is current CO_2 emissions minus previous CO_2 emissions (indicating the change in carbon emissions), Y is the GDP of the region, and ΔY is the difference between two adjacent periods. Following Wang et al. [31], decoupling status can be divided into eight types in Tapio research, as shown in Table 1.

Table 1. Eight types of relationships for the Tapio decoupling model. GDP, gross domestic product.

Classification	Status	Carbon Emissions Change (ΔC)	GDP Change (ΔY)	Elastic Coefficient
Decoupling	Weak decoupling	>0	>0	0 ≤ e < 0.8
	Strong decoupling	<0	>0	e < 0
	Recessive decoupling	<0	<0	e > 1.2
Negative decoupling	Expansive negative decoupling	>0	>0	e > 1.2
	Strong negative decoupling	>0	<0	e < 0
	Weak negative decoupling	<0	<0	0 ≤ e < 0.8
Coupling	Expansive coupling	>0	>0	0.8 ≤ e < 1.2
	Recessive coupling	<0	<0	0.8 ≤ e < 1.2

2.3. EKC Framework

The environmental Kuznets hypothesis indicates an inverted U-type nexus of economic growth and environmental pollution, which means that along with economic development, CO_2 emissions will first increase and then decrease after the peak level is achieved. The corresponding regression equation is as follows:

$$C_{it} = \alpha_1 y_{it} + \alpha_2 y_{it}^2 + \alpha_0 \quad (5)$$

Equation (5) describes the EKC relationship between economic growth and carbon emissions. C_{it} represents the CO_2 emissions of city i in year t; y_{it} represents per capita GDP of city i in year t; assuming population growth is 0 and the population of city i is n_i, then we have $Y_{it} = n_i y_{it}$, where Y is GDP; α_1 and α_2 are the quadratic and primary coefficients of per capita GDP; and α_0 represents the factors affecting carbon emissions in other areas, including urbanization, regional openness, proportion of thermal power generation, research and development (R&D) intensity, industrial structure, etc., and is regarded as a constant term.

However, in subsequent studies, scholars found that a simple quadratic curve may not fully describe the complex relationship between CO_2 emissions and economic growth, as the actual relationship may be N-, inverted N-, or even bell-shaped, except for the inverted U shape as described by conventional EKC studies [15,32].

In order to carry out in-depth research on the two phenomena of existing research, we introduced a cubic curve to extend and supplement the quadratic curve in the EKC model, trying to analyze the long-term changes in CO_2 emissions and economic growth:

$$C_{it} = \alpha_1 y_{it} + \alpha_2 y_{it}^2 + \alpha_3 y_{it}^3 + \alpha_0 \quad (6)$$

Equation (6) reflects the long-term relationship between CO_2 emissions and economic growth. C_{it}, y_{it}, and α_0 remain unchanged, while α_1, α_2, and α_3 are the cubic, quadratic, and primary term coefficients of per capita GDP, respectively.

To some extent, the presence of EKC could be treated as the result of decoupling. The presence of an inverted U-shaped EKC actually implies the existence of the strong decoupling, as the emissions would eventually decrease as GDP per capita continuously increases over time. However, decoupling does not necessarily lead to conventional inverted U-shaped EKC, because even if the pollutant emissions do not decline along with economic development, there might still be weak decoupling as long as the emissions do not grow as fast as GDP per capita (i.e., the slope of the projection of emissions gradually decreases) [32].

3. Estimation Method and Data

3.1. Model Relationship Derivation

According to previous assumptions, the non-linear measurement model in economic growth and carbon emissions can be set as follows:

$$C_{it} = u_i + \alpha_1 y_{it} + \alpha_2 y_{it}^2 + \alpha_3 X_{it} + \varepsilon_{it} \tag{7}$$

$$C_{it} = u_i + \alpha_1 y_{it} + \alpha_2 y_{it}^2 + \alpha_3 y_{it}^3 + \alpha_4 X_{it} + \varepsilon_{it} \tag{8}$$

Equations (7) and (8) are the extended forms of Equations (5) and (6) by adding control variables, respectively. y_{it} indicates per capita GDP; to test the EKC hypothesis, y_{it}^2 and y_{it}^3 were added to the model. To reduce the influence of heteroscedasticity and gauge the elasticity of carbon emissions with respect to corresponding explanatory variables, the data of C_{it} was logarithmized, that is, $\ln C_{it}$.

During data processing, 2000 was taken as the base year, and the real GDP for the period of 2007–2016 were computed using the deflator of the base year (i.e., the prices of all commodities were fixed at the levels of the year 2000). This study also utilized the number of permanent residents in China's provinces from 2007 to 2016 to eliminate the influence of population factors on GDP. In this way, the real per capita GDP of China's provinces in 2007–2016 were obtained. The empirical study used the logarithmic per capita GDP, so that the estimated coefficient reflected the elasticity. Since carbon emissions have an inertial effect, the indicators were susceptible to the previous year's data. Therefore, after taking the logarithm, the first-order lag of carbon emissions ($\ln C_{i,t-1}$) was added to the model. The dynamic panel data model was used to examine the EKC hypothesis to deal with potential endogeneity and allow for dynamics. Generally speaking, fixed-effects (hereafter FE) estimators are prone to endogeneity bias, because regressors may be correlated with the unobserved fixed effects and the possible bilateral causality between the dependent and independent variables. Using the instrumental variable estimators and specifically the generalized method of moments (GMM) approach, the potential endogeneity problem could be well addressed. Therefore, GMM was utilized for the empirical analysis in this study.

In Equations (7) and (8), X_{it} represents a collection of control variables, including urbanization rate (the larger the size of a town, the greater the energy consumption, which increases carbon emissions); secondary industry added value as a share of GDP (the secondary industry consumes a lot of fossil fuels); the actual use of foreign capital (the degree of openness affects the quality and level of the town), and carries on the logarithm, expressed as $\ln fdi$; the proportion of thermal power generation (for which coal is the dominant fossil energy source, which directly affects the amount of CO_2 emissions); and R&D intensity (more efficient use of fossil fuels in technologically developed areas means lower carbon emissions), and carries on the logarithm, expressed as $\ln tec$.

In summary, the final forms of Equations (7) and (8) are Equations (9) and (10):

$$\begin{aligned}\ln C_{it} = u_i + \lambda \ln C_{i,t-1} + \alpha_1 \ln gdp_{it} + \alpha_2 \ln^2 gdp_{it} + \alpha_3 \ln fdi + \alpha_4 urban \\ + \alpha_5 \sec + \alpha_6 hermal + \alpha_7 \ln tec + \varepsilon_{it}\end{aligned} \tag{9}$$

$$\begin{aligned}\ln C_{it} = u_i + \lambda \ln C_{i,t-1} + \alpha_1 \ln gdp_{it} + \alpha_2 \ln^2 gdp_{it} + \alpha_3 \ln^3 gdp_{it} + \alpha_4 \ln fdi \\ + \alpha_5 urban + \alpha_6 \sec + \alpha_7 hermal + \alpha_8 \ln tec + \varepsilon_{it}\end{aligned} \tag{10}$$

Equations (9) and (10) are the concrete regression equations used in the empirical study, and they are the specific forms of Equations (7) and (8), respectively. u_i is the ith city intercept item and ε_{it} is the error term.

3.2. Data Sources and Descriptions

Among the explanatory variables, per capita GDP and its index were taken from the China Statistical Yearbook, and the data for the other control variables come from the provincial statistical yearbooks. CO_2 emission (*C*) was measured by fossil fuel consumption and cement production. Fossil fuel consumption data come from the China Energy Statistics Yearbook. Data from cement production come from the China Stock Market and Accounting Research (CSMAR) Economic and Financial Database. Emission coefficients were taken directly from the IPCC (2006) [12]. The decoupling elasticity coefficient (e_c) was calculated by CO_2 emission (*C*) and GDP (*Y*) after reduction.

The descriptive statistics for the selected variables utilized in this study are summarized in Table 2.

Table 2. Descriptive statistics of the variables.

Variable Name	Unit	Mean	Maximum	Minimum Value	Standard Deviation	Observation Value
Total per capita CO_2 emissions	Kilogram/person	8222.72	26,287.15	2588.47	4565.15	290
Real per capita GDP	Yuan/person	28,644.14	92,400	5800	16,533.48	290
Actual use of foreign capital	Yuan/person	227.63	2415.01	0	408.84	290
Urbanization rate	%	53.52	89.60	28.24	13.68	290
The added value of the secondary industry Accounts for the proportion of GDP	%	47.06	61.50	19.26	8.19	290
Proportion of thermal power generation	%	0.78	1	0.09	0.23	290
R&D intensity (R&D)	—	1.45	6.01	0.21	1.08	290
Carbon emission decoupling coefficient	—	0.55	4.35	−2.78	0.84	290

To show the changing trends in individual provinces, the per capita CO_2 emissions in 2007 and 2016 (beginning and ending years of the sample period) and the corresponding average annual growth rates of all provinces are depicted in Figure 1. As can be seen from the figure, the more developed provinces had lower average CO_2 emissions growth. For example, in Beijing, Shanghai, Zhejiang, and Guangdong provinces, the average annual growth rate of CO_2 remained around 2%, and Beijing even showed a negative value of −4.3%. The less developed provinces showed higher average CO_2 emissions growth, including some central provinces (e.g., Anhui, Jiangxi, Shanxi) and most western provinces (e.g., Xinjiang, Qinghai, Ningxia). From Figure 1, we can make a preliminary conclusion that CO_2 emissions were connected to economic level, and thus, closely related to economic growth.

Figure 1. Average CO_2 emissions and annual growth rates for the provinces in 2007 and 2016.

4. Empirical Results and Discussions

4.1. Environmental Kuznets Curve in China

Based on panel data of 29 provinces in China (excluding Hong Kong, Macao, Taiwan, and Tibet) during 2008–2017, Stata15 software was used for the empirical analysis. The first-order differential GMMs estimation was used to process the panel data to ensure reasonable and accurate regression results. To deal with potential endogeneity from possible bilateral causality between CO_2 emissions and economic growth, the lag period instrumental variable method was adopted in this study. The lag phases of the logarithmic number of per capita GDP and the logarithmic quadratic term of per capita GDP were regarded as the current instrumental variables of the quadratic function. The current instrumental variable of the cubic function was based on the instrumental variable of the quadratic function, adding a lag period of the logarithmic cubic term of per capita GDP.

In this study, the validity of the model was tested using the Autocorrelation and Hansen tests. The Autocorrelation test was used to ensure that the model had first-order but not second-order differential autocorrelation. The Hansen test was used to ensure that there was no overidentification in the instrumental variables. Table 2 shows the benchmark regression and robustness test results of the EKC quadratic function and cubic function models.

The rationality of the instrumental variables selected in this study can be confirmed by the results of the Hansen [31], AR(1), and AR(2) tests, as shown in Table 3. In Equation (1) of Table 3, the cube of per capita GDP was not introduced, and the regression results indicated that there existed an inverted U-shaped EKC relationship between CO_2 emissions and economic growth, as the quadratic term (lngdp2) of the real per capita GDP was estimated to be a significantly negative coefficient, while the original level of logarithmic real per capita GDP (lngdp) was estimated to be significantly positive.

Table 3. Generalized method of moments (GMMs) estimation results of environmental Kuznets curve (EKC) relationship for per capita CO_2 emissions with different specifications and estimation methods.

	Benchmark Regression		Robustness Test					
Variable	Quadratic Model	Cubic Model	Quadratic Model			Cubic Model		
	(1)	(2)	(3)	(4)	(5)	(6)	(7)	(8)
L.lnC	0.716 ***	0.714 ***	0.815 ***	0.673 ***	0.747 ***	1.074 ***	0.662 ***	0.765 ***
	(0.027)	(0.036)	(0.023)	(0.028)	(0.029)	(0.077)	(0.038)	(0.029)
lngdp	0.207 ***	−0.101 ***	0.096 ***	0.102 **	0.187 ***	−1.011 ***	−0.229 **	−0.073 *
	(0.062)	(0.039)	(0.031)	(0.040)	(0.056)	(0.318)	(0.094)	(0.045)
lngdp2	−0.043 ***	0.215 ***	−0.024 ***	−0.023 *	−0.041 ***	0.882 ***	0.339 ***	0.078 ***
	(0.016)	(0.039)	(0.009)	(0.015)	(0.014)	(0.289)	(0.084)	(0.034)
lngdp3	—	−0.060 ***	—	—	—	−0.246 ***	−0.087 ***	−0.016 ***
		(0.012)				(0.076)	(0.021)	(0.009)
lnfdi	0.022 ***	0.016 ***	—	0.030 ***	0.019 ***	—	0.022 ***	0.015 ***
	(0.003)	(0.003)		(0.004)	(0.004)		(0.004)	(0.003)
urban	−0.006 ***	−0.003 ***	—	0.005 ***	−0.006 ***	—	—	0.000 ***
	(0.002)	(0.001)		(0.005)	(0.002)			(0.001)
sec	0.001	0.006 ***	—	—	0.002	—	0.008 ***	0.006 ***
	(0.001)	(0.001)			(0.001)		(0.001)	(0.001)
hermal	0.459 ***	0.232 ***	—	—	0.462 ***	—	—	0.165 ***
	(0.089)	(0.053)			(0.092)			(0.046)
lntec	−0.112 ***	−0.123 ***	−0.061 ***	−0.255 ***	−0.092 ***	—	−0.214 ***	−0.112 ***
	(0.027)	(0.016)	(0.020)	(0.045)	(0.024)		(0.039)	(0.013)
_cons	2.275 ***	2.212 ***	1.637 ***	2.570 ***	2.001 ***	−0.28	2.598 ***	1.693 ***
	(0.201)	(0.284)	(0.183)	(0.206)	(0.221)	9(0.649)	(0.300)	(0.222)
AR(1)	0.002	0.001	0.002	0.002	0.002	0.013	0.003	0.002
AR(2)	0.755	0.594	0.644	0.929	0.720	0.937	0.950	0.731
Hansen	0.586	0.723	0.586	0.592	0.540	0.063	0.768	0.695
Curve shape	Invert U	Invert N	Invert U	Invert U	Invert U	Invert N	Invert N	Invert N
Turning point	111,178	86,476	76,990	88,891	98,581	41,719	88,822	148,476

Numbers in parentheses are standard errors. ***, **, and * indicate significance at the 0.01, 0.05, and 0.1 levels, respectively. *_cons* represents a constant term.

Equation (2) was a regression result of the inverted N-type curve relationship of the EKC curve, and its shape was determined by the sign of the discriminant and the sign of α_1 in Equation (10) after one derivation. By multiplying the primary term (*lngdp*) coefficient 0.060, the quadratic term (*lngdp2*) coefficient 0.215, and the cubic term (*lngdp3*) coefficient 0.101 of the real per capita GDP, the discriminant could be constructed. The calculated discriminant value was more than 0.113, and the cubic term coefficient was negative. It proved that the shape of the EKC relationship was an inverted N. At the same time, the levels of real per capita GDP of the two inflection points in the inverted N-type curve can be calculated based on the estimation results.

To test the robustness of the regression results, this study attempted to obtain different regression equations by reducing the explanatory variables and using orthogonal differential GMM estimation. Equations (3), (4), (6), and (7) were used to test the benchmark regression by using the first-order differential GMM estimation and reducing the explanatory variables. Equations (5) and (8) were used to test the benchmark regression using the orthogonal differential GMM method. As can be seen from Table 3, the Hansen, AR(1), and AR(2) tests for all six robustness check equations suggested that the chosen instrumental variables were reasonable, and the shape of the curve was the same as the results of the benchmark regression. The inflection points of the curves were also close to the reference regression inflection point, which proved the robustness of the benchmark regression.

According to the primary (*lngdp*) and quadratic (*lngdp2*) coefficients of the quadratic curve, the real per capita GDP of the inflection point in each inverted U-type quadratic curve was calculated, and the median was about 83,000 yuan. The estimation results of the cubic curve proved that the shape of the EKC relationship was an inverted N. However, because the level of per capita GDP corresponding to the first inflection point in the inverted N curve was relatively low (approximately 15,000 yuan) and most of the observations were higher than this level, the actual EKC relationship for CO_2 emissions after considering the cubic term of (*lngdp*) was still a conventional inverted U shape [33]. It should also be noted that the levels of per capita GDP corresponding to the inflection points for most regressions were similar, and the median value was about 85,000 yuan (at 2000 prices).

These estimations and findings were mostly consistent with previous findings [34–36]. Moreover, given that the average national per capita GDP in 2016 was 17,522 yuan, which is considerably lower than the estimated inflection points of the EKC curve, it is reasonable to expect that China's CO_2 emissions will keep growing in the foreseeable future, and more efforts must be made to achieve the ambitious goal of having carbon emissions peak before 2030.

In summary, this study investigated the non-linear nexus of CO_2 emissions and per capita GDP by employing the first-order differential GMM estimation method under the control of urbanization rate, openness, thermal power generation ratio, and R&D intensity. The results include two curves: quadratic and cubic. The quadratic curve conformed to the inverted U shape, the cubic curve presented the inverted N shape, and the positions and shapes of the two curves coincided. In addition, this study tested the robustness of the benchmark regression using the orthogonal difference GMM estimation and reducing the explanatory variables. Test results showed that the benchmark curve obtained by the first-order difference GMM estimation was robust. The empirical results proved the core point of this research: the cubic curve better illustrates the nexus of complex CO_2 emissions and economic growth.

4.2. Decoupling Status of 29 Provinces in China

Due to the statistical caliber error caused by the division of jurisdictions in Chongqing and Sichuan, Chongqing and Sichuan Province were merged into the Sichuan region, and data on Tibet were not available. Therefore, this study only covered 29 provinces. Table 4 shows the decoupling elasticity coefficients of CO_2 in 29 provinces from 2007 to 2016.

Table 4. Decoupling elastic coefficient of CO_2 in 29 provinces of China for the period 2007–2016.

Province/Year	2007	2008	2009	2010	2011	2012	2013	2014	2015	2016
Beijing	WD	SD	WD	WD	SD	WD	SD	SD	WD	SD
Tianjin	WD	WD	END	WD	EC	WD	WD	SD	SD	SD
Hebei	WD	END	EC	END	WD	WD	END	SD	WD	WD
Shanxi	WD	END	EC	WD	WD	WD	END	SD	SD	WD
Neimenggu	WD	END	EC	WD	END	SD	SD	WD	WD	WD
Liaoning	WD	END	END	WD	SD	WD	END	WD	SD	RC
Jilin	WD	WD	WD	WD	END	SD	WD	WD	SD	SD
Heilongjiang	WD	END	WD	EC	EC	WD	SD	WD	WD	WD
Shanghai	WD	WD	EC	EC	SD	END	SD	SD	SD	SD
Jiangsu	WD	EC	EC	END	SD	WD	END	WD	WD	WD
Zhejiang	EC	WD	WD	WD	WD	SD	EC	WD	SD	SD
Anhui	WD	WD	EC	EC	WD	EC	END	WD	WD	WD
Fujian	WD	END	END	WD	END	WD	WD	WD	SD	SD
Jiangxi	EC	END	END	SD	WD	WD	END	WD	WD	WD
Shandong	WD	WD	WD	EC	WD	WD	SD	WD	WD	WD
Henan	WD	WD	EC	EC	WD	WD	WD	WD	SD	SD
Hubei	WD	WD	WD	END	EC	WD	SD	WD	SD	WD
Hunan	WD	WD	EC	EC	EC	SD	SD	WD	SD	EC
Guangdong	WD	WD	EC	EC	EC	WD	SD	SD	WD	WD
Guangxi	EC	END	END	END	EC	WD	WD	WD	SD	WD
Hainan	WD	EC	EC	END	END	SD	WD	WD	WD	SD
Sichuan	SD	END	EC	END	EC	WD	SD	WD	SD	SD
Guizhou	WD	SD	EC	WD	WD	EC	WD	WD	SD	WD
Yunnan	WD	WD	END	WD	WD	WD	SD	SD	SD	WD
Shaanxi	WD	EC	EC	END	WD	EC	WD	WD	SD	SD
Gansu	EC	END	SD	END	WD	END	WD	WD	SD	SD
Qinghai	WD	END	EC	SD	EC	END	EC	WD	WD	EC
Ningxia	WD	END	WD	END	END	SD	WD	WD	WD	SD
Xinjiang	EC	EC	END	EC	END	END	END	EC	WD	EC

Note: WD, weak decoupling; SD, strong decoupling; RD, recessive decoupling; END, expansive negative decoupling; SND, strong negative decoupling; WND, weak negative decoupling; EC expansive coupling; RC, recessive decoupling.

During the period 2007–2016, the provinces of China experienced rapid economic growth, with an average annual GDP growth rate of 11% in 29 provinces. Except for Liaoning Province, which decreased slightly from 2015 to 2016, the per capita GDP of each province showed an increasing trend from 2007 to 2016. Therefore, the calculation results contain five types of decoupling. As shown in Table 4, before 2014, absolute decoupling rarely occurred in all provinces, and no provinces experienced continuous absolute decoupling. The alternating states of expansive coupling, expansive negative decoupling, and weak decoupling were more common. The western region had the highest frequency of alternating, and the eastern region had the lowest. Regarding per capita GDP, the lower the initial value in 2007, the stronger the alternating frequency. For example, in Beijing and Shanghai, where the values were higher, the weak decoupling state was more common. However, the western provinces with low per capita GDP, such as Qinghai, Ningxia, and Xinjiang, had a high frequency of alternation and rare weak decoupling, and expansive negative decoupling was dominant. Except for a few other developed provinces, most provinces had a higher frequency of expansive negative decoupling, and the growth rate of CO_2 emissions may have been more than two times the economic growth rate. This shows that the rapid economic growth of most provinces represents a low-efficiency and extensive economic growth mode before 2014, bringing great harm to the environment.

Taking 2014 as the starting point for observation, the provinces no longer had expansive negative decoupling, weak decoupling states became dominant, and a stable strong decoupling state was still relatively rare. The weak and strong decoupling states often alternated, and the provincial economy was still growing rapidly. However, the increase in CO_2 emissions slowed down and returned to a reasonable state, and the economic growth style shifted from extensive to relatively intensive. Comparing the GDP of each province from 2014 to 2016 with the inflection point of the EKC model, it was found that the per capita GDP of other provinces and cities in China did not exceed the inflection point of per capita GDP except Shanghai in 2015–2016 and Tianjin in 2016, which suggests that CO_2

emissions would decrease as the economy continues to grow. This confirms the strong decoupling state of Shanghai for four consecutive years and Tianjin's stability for three consecutive years. It also shows that other provinces in China will continue to maintain the alternating state of strong and weak decoupling until the per capita GDP exceeds the estimated turning point of 85,000 yuan.

The findings of this study are basically in line with some relevant studies. For instance, Peng [37] used time series data for the period 1980–2008 to evaluate the decoupling effect of China's CO_2 emissions. The results indicated that expansive negative decoupling and weak decoupling of expansion are the most common conditions in Chinese provinces, while strong decoupling has never occurred. It is noteworthy that Peng's [37] findings reflect that for his sample period of 1980–2008, no provinces had a high enough income level beyond the inflection point. In a recent study, Bai et al. [38] used the panel data of Chinese provinces from 2006 to 2015 to calculate the carbon emissions decoupling index in the transportation sector. Consistent with the results of this research, their study also found that the decoupling state in the eastern region was significantly better than that in the central and western regions. According to the empirical study of this paper, China has five decoupling states. Comparatively, in Bai's [39] study, apart from the five decoupling states presented in this paper, there was also evidence for the strong negative decoupling state, which to some extent reflects the periodical characteristics of the transportation industry [39]. It is also noteworthy that the recessive decoupling state only appeared once in the measurement of this study (i.e., Liaoning in 2016), and it was also detected only once in the study of Bai et al. [38] (i.e., Gansu in 2015). This similarity suggests that the state of recessive decoupling is indeed relatively rare in China. Dong [40] used China's carbon dioxide emissions from 1965 to 2016 to verify the existence of emissions using structural break technology [40]. It also confirms that natural gas and renewable energy have an important impact on reducing CO_2 emissions. Chen [41] used provincial panel data from 1995 to 2012 to test the EKC hypothesis. The empirical results showed that there was an inverted U-shaped EKC curve for the eastern part of China, but not the central and western regions [41]. This conclusion verifies the inflection point value measured in this paper. Since the central and western regions are less developed, their per capita GDP is far from the turning point of carbon emissions decreasing with economic growth in 2012. Incomplete data leads to incomplete curves, and the EKC hypothesis cannot be verified.

5. Conclusions and Policy Recommendations

5.1. Conclusions

This study quantitatively investigated the decoupling effects between CO_2 emissions and economic growth in China. Using the GMM method, this study verified the existence of an essentially inverted U-shaped EKC relationship for CO_2 emissions. Furthermore, by employing the Tapio decoupling model, the decoupling status of individual provinces for the period 2007–2016 was evaluated. The main conclusions are as follows.

First, there was an essentially inverted U-shaped relationship between CO_2 emissions and economic growth. The estimation results were robust to different regression specifications and valid after potential endogeneity was well controlled for. In this regard, this study verifies an EKC relationship of CO_2 emissions and suggests that reasonable economic growth is critical for China to eventually accomplish its goal of sustainable development, as the peak of CO_2 emissions can be achieved only when the level of economic development is high enough.

Second, the inflection point of carbon dioxide emissions was relatively high (corresponding real GDP was about 85,000 yuan, measured at the 2000 constant price). Other things being equal, per capita CO2 emissions would not decline until per capita GDP reaches approximately 85,000 yuan. Because 85,000 yuan was the estimated level of GDP per capita corresponding to the turning point of CO_2 emissions, according to Xia and Zhong [27], combining the estimation of the results of Tapio and EKC models, it could be concluded that the level of economic growth must cross this threshold level corresponding to the turning point to achieve absolute decoupling. This calculated turning point was basically consistent when the cubic term of the logarithmic per capita GDP was introduced as a regressor. And the result of cubic term confirms the robustness of the results in this paper. It is noteworthy that income levels in the majority of provinces were still considerably lower than this turning point, suggesting that the peak of CO_2 emissions may not be easily achieved in the near future if the economic development styles in most provinces remain unchanged.

Third, the status of the Tapio decoupling was dependent on the inflection point of annual per capita GDP and can be broadly divided into two stages. In the first stage, before approaching the inflection point of 85,000 yuan (real value, with 2000 as the base year), the decoupling would be unstable, and most provinces would experience both strong and weak decoupling. In the second stage, after the per capita GDP exceeds the inflection point of 85,000 yuan, the provinces would reach a relative stable status with strong decoupling.

5.2. Policy Recommendations

Based on the above research conclusions, in order to reduce carbon emissions and reach China's emission reduction targets, achieve the 13th Five-Year Plan's low-carbon green development, and contribute to global carbon emissions control, this paper proposes the following policy recommendations:

First, according to the empirical results, the decoupling status of China's provinces in 2007–2016 was differentiated. Among them, weak decoupling occurred 135 times, strong decoupling occurred 56 times, expansive coupling appeared 47 times, expansion negative decoupling appeared 41 times, and recessive decoupling only appeared once. Therefore, to curb China's carbon emissions effectively, different provinces should adopt different CO_2 emission reduction policies on the basis of their decoupling status. Specifically, for the provinces with weak decoupling, such as the 13 provinces of Hebei, Shaanxi, Inner Mongolia, etc., in 2016, the development style is, in general, intensive expansion. The CO_2 emissions of these provinces should be further reduced, and more efforts should be made to move towards strong decoupling while maintaining relatively rapid development. For the provinces that have strong decoupling, such as the 12 provinces of Beijing, Tianjin, Jilin, etc., in 2016, the level of economic development is either relatively high or relatively backward. Among them, the main task for the provinces with higher levels of economic development is to further maintain the status quo and promote its own development model to the whole country. In contrast, the relatively poor provinces should strive to achieve fast economic growth while ensuring that the ecological carrying capacity is not exceeded. For the provinces with expansive negative decoupling, including the seven provinces of Hebei, Shaanxi, Liaoning, etc., in 2013, their carbon emissions growth is abnormally high, and the emission reduction capacity lags far behind the economic growth level, suggesting that the economic development mode is unsustainable. In this regard, these provinces should formulate more stringent emission reduction policies and strengthen carbon emission reduction capacity. For the provinces with expansive coupling, such as Hunan, Qinghai and Xinjiang in 2016, their emission reduction effects have already begun to appear, but carbon emissions are still growing at a faster rate. These provinces need to continue strictly implementing emission reduction policies and regulations and move toward a relative decoupling state. For the provinces with recessive coupling, such as Liaoning in 2016,their level of economic development needs to be promoted, and especially, the problem of negative economic growth must be solved. For these provinces, it is possible to relax the requirements for carbon emission control, appropriately adopt a fiscal policy of reducing taxes and tax reductions, and reduce the environmental governance costs of enterprises.

Second, from the regression results with the cubic term of logarithmic per capita GDP, the estimated coefficients of thermal power generation and technology input are large in magnitude, suggesting that they have relatively large impacts on CO_2 emissions. Therefore, China should adjust its energy consumption structure and enhance investments in science and technology in the field of carbon emissions reduction. When it comes to energy consumption, China should speed up R&D and the utilization of new energy sources, improve energy utilization efficiency, accelerate the adjustment of the energy structure, further reduce dependence on fossil fuels, and actively promote the utilization of renewable energy. Investment in technology is not limited to research on new energy technology, but also includes research on carbon emission regulation and capture technology, carbon pollution control technology, etc. Increasing investment in science and technology with regard to carbon emissions will not only weaken the impact of the greenhouse effect in China, but also contribute to the global governance of carbon emissions as a major country and establish a positive image in the international community.

Last but not least, the distinction between economic growth and economic development should be emphasized. Government regulators need to recognize that they are not the same. The connotation of economic development goes far beyond the scope of economic growth. Economic growth is only determined by the growth of economic indicators such as GDP. On the other hand, economic development not only requires more economic indicators, but also needs to take into account indicators such as the ecological environment. A connotation of economic development is to achieve sustainable economic development. Past experience shows that short-term, rapid, high-polluting economic growth will bring irreparable harm to the environment and make sustainable economic development impossible. Therefore, China must regard carbon emission reduction as an important strategy for sustainable economic development and an important starting point to build a low-carbon economy. China should work hard to accelerate the upgrading of economic development and undertake intensive economic growth.

Although this study quantitatively investigates the decoupling between carbon emissions and economic growth for individual provinces and China as a whole, there are still some limitations, which could also be possible future research directions. It is noteworthy that this study is purely empirical, but the theoretical mechanisms for the existence of the decoupling effects between carbon emissions and economic growth are also important. Therefore, follow-up studies could try to build theoretical models to thoroughly explain the empirical findings of this research. In addition, given the remarkable gaps in economic and social development across different cities within a province, the utilization of city-level data could reveal the decoupling effects more accurately for better and more reasonable policy-making.

Author Contributions: Y.H. designed the proposed control strategy; Z.H. conducted the experimental works, modeling, and simulation; H.W. helped with writing the manuscript.

Funding: This work was supported by the National Natural Science Foundation of China (71761137001, 71403015, 71521002), the Key Research Program of the Beijing Social Science Foundation (17JDYJA009), the Beijing Natural Science Foundation (9162013), the National Key Research and Development Program of China (2016YFA0602801, 2016YFA0602603), and the Special Fund for the Joint Development Program of Beijing Municipal Commission of Education.

Conflicts of Interest: The authors declare that there is no conflict of interest.

References

1. United Nations Climate Change Website. Available online: https://unfccc.int/news/finale-cop21 (accessed on 16 May 2019).
2. Ozkan, F.; Ozkan, O. An analysis of CO_2 emissions of Turkish industries and energy sector. *Reg. Sect. Econ. Stud.* **2012**, *12*, 65–85.
3. Global Carbon Project Website. 2017 Global Carbon Budget. Available online: https://www.globalcarbonproject.org/carbonbudget/archive.htm (accessed on 16 May 2019).
4. World Resources Report. Available online: https://www.wri.org/our-work/project/world-resources-report/wrr (accessed on 16 May 2019).
5. The Central People's Government of the People's Republic of China. Available online: http://www.gov.cn/zhengce/2017-05/07/content_5191604.htm (accessed on 16 May 2019).
6. Zhang, Q.; Liao, H.; Hao, Y. Does one path fit all? An empirical study on the relationship between energy consumption and economic development for individual Chinese provinces. *Energy* **2018**, *150*, 527–543.
7. Wang, Z.; Yang, L. Delinking indicators on regional industry development and carbon emissions: Beijing–Tianjin–Hebei economic band case. *Ecol. Indic.* **2015**, *48*, 41–48.

8. Tapio, P. Towards a theory of decoupling: Degrees of decoupling in the EU and the case of road traffic in Finland between 1970 and 2001. *Transp. Policy* **2005**, *12*, 137–151.
9. Yu, Y.; Chen, D.; Zhu, B.; Hu, S. Eco-efficiency trends in China, 1978–2010: Decoupling environmental pressure from economic growth. *Ecol. Indic.* **2013**, *24*, 177–184.
10. OECD. *Indicators to Measure Decoupling of Environmental Pressures for Economic Growth*; OECD: Paris, France, 2002.
11. Fader, M.; Gerten, D.; Krause, M.; Lucht, W.; Cramer, W. Spatial decoupling of agricultural production and consumption: Quantifying dependences of countries on food imports due to domestic land and water constraints. *Environ. Res. Lett.* **2013**, *8*, 014046.
12. Eggleston, S.; Buendia, L.; Miwa, K.; Ngara, T.; Tanabe, K. *2006 IPCC Guidelines for National Greenhouse Gas Inventories*; Institute for Global Environmental Strategies: Hayama, Japan, 2006.
13. Kuznets, S. Economic growth and income inequality. *Am. Econ. Rev.* **1955**, *45*, 1–28.
14. Grossman, G.M.; Krueger, A.B. *Environmental Impacts of a North American Free Trade Agreement (No. w3914)*; National Bureau of Economic Research: Cambridge, MA, USA, 1991.
15. Song, T.; Zheng, T.; Tong, L. An empirical test of the environmental Kuznets curve in China: A panel cointegration approach. *China Econ. Rev.* **2008**, *19*, 381–392.
16. Pal, D.; Mitra, S.K. The environmental Kuznets curve for carbon dioxide in India and China: Growth and pollution at crossroad. *J. Policy Model.* **2017**, *39*, 371–385.
17. Richmond, A.K.; Kaufmann, R.K. Is there a turning point in the relationship between income and energy use and/or carbon emissions? *Ecol. Econ.* **2006**, *56*, 176–189.
18. Stern, D.I. The environmental Kuznets curve after 25 years. *J. Bioecon.* **2017**, *19*, 7–28.
19. Sheng, Y.; Ou, M.; Liu, Q. Methods of Measuring Decoupling of Resource Environment: Speed Decoupling or Quantity Decoupling? *China Popul. Resour. Environ.* **2015**, *25*, 99–103. (In Chinese)
20. Auffhammer, M.; Carson, R.T. Forecasting the path of China's CO_2 emissions using province-level information. *J. Environ. Econ. Manag.* **2008**, *55*, 229–247.
21. Caviglia-Harrisa, J.L.; Chambersa, D.; Kahnb, J.R. A comprehensive analysis of the EKC and environmental degradation. *Ecol. Econ.* **2008**, *68*, 1149–1159.
22. Zhang, Y.J.; Da, Y.B. The decomposition of energy-related carbon emission and its decoupling with economic growth in China. *Renew. Sustain. Energy Rev.* **2015**, *41*, 1255–1266.
23. Zhao, X.; Zhang, X.; Shao, S. Decoupling CO_2 emissions and industrial growth in China over 1993–2013: The role of investment. *Energy Econ.* **2016**, *60*, 275–292.
24. Wang, Q.; Chen, X. Energy policies for managing China's carbon emission. *Renew. Sustain. Energy Rev.* **2015**, *50*, 470–479.
25. Zhang, M.; Wang, W. Decouple indicators on the CO_2 emission-economic growth linkage: The Jiangsu Province case. *Ecol. Indic.* **2013**, *32*, 239–244.
26. Tang, Z.; Shang, J.; Shi, C.; Liu, Z.; Bi, K. Decoupling indicators of CO_2 emissions from the tourism industry in China: 1990–2012. *Ecol. Indic.* **2014**, *46*, 390–397.
27. Xia, Y.; Zhong, M.C. Relationship between EKC hypothesis and the decoupling of environmental pollution from economic development: Based on China prefecture-level cities'decoupling partition. *China Popul. Resour. Environ.* **2016**, *26*, 8–16.
28. Hao, Y.; Wei, Y.M. When does the turning point in China's CO_2 emissions occur? Results based on the Green Solow model. *Environ. Dev. Econ.* **2015**, *20*, 723–745.
29. Liu, Y.; Kuang, Y.; Huang, N.; Wu, Z.; Wang, C. CO_2 emission from cement manufacturing and its driving forces in China. *Int. J. Environ. Pollut.* **2009**, *37*, 369–382.
30. Cheng, Z.; Li, L.; Liu, J. Industrial structure, technical progress and carbon intensity in China's provinces. *Renew. Sustain. Energy Rev.* **2018**, *81*, 2935–2946.
31. Wang, Q.; Jiang, X.T.; Li, R. Comparative decoupling analysis of energy-related carbon emission from electric output of electricity sector in Shandong Province, China. *Energy* **2017**, *127*, 78–88.
32. Kaika, D.; Zervas, E. The Environmental Kuznets Curve (EKC) theory—Part A: Concept, causes and the CO2 emissions case. *Energy Policy* **2013**, *62*, 1392–1402.
33. Jalil, A.; Feridun, M. The impact of growth, energy and financial development on the environment in China: A cointegration analysis. *Energy Econ.* **2011**, *33*, 284–291.

34. Wang, S.; Zhou, D.; Zhou, P.; Wang, Q. CO_2 emissions, energy consumption and economic growth in China: A panel data analysis. *Energy Policy* **2011**, *39*, 4870–4875.
35. Li, T.; Wang, Y.; Zhao, D. Environmental Kuznets curve in China: New evidence from dynamic panel analysis. *Energy Policy* **2016**, *91*, 138–147.
36. Wang, Z.; Bao, Y.; Wen, Z.; Tan, Q. Analysis of relationship between Beijing's environment and development based on Environmental Kuznets Curve. *Ecol. Indic.* **2016**, *67*, 474–483.
37. Peng, J.W.; Huang, X.J.; Zhong, T.Y.; Zhao, Y.T. Decoupling analysis of economic growth and energy carbon emissions in China. *Resour. Sci.* **2011**, *33*, 626–633.
38. Bai, C.; Chen, Y.; Yi, X.; Feng, C. Decoupling and decomposition analysis of transportation carbon emissions at the provincial level in China: Perspective from the 11th and 12th Five-Year Plan periods. *Environ. Sci. Pollut. Res.* **2019**, in press.
39. Timilsina, G.R.; Shrestha, A. Factors affecting transport sector CO_2 emissions growth in Latin American and Caribbean countries: An LMDI decomposition analysis. *Int. J. Energy Res.* **2009**, *33*, 396–414.
40. Dong, K.; Sun, R.; Dong, X. CO_2 emissions, natural gas and renewables, economic growth: Assessing the evidence from China. *Sci. Total Environ.* **2018**, *640*, 293–302.
41. Chen, Y.; Zhao, J.; Lai, Z.; Wang, Z.; Xia, H. Exploring the effects of economic growth, and renewable and non-renewable energy consumption on China's CO_2 emissions: Evidence from a regional panel analysis. *Renew. Energy* **2019**, in press.

energies

MDPI

Article

Does Energy-Growth and Environment Quality Matter for Agriculture Sector in Pakistan or not? An Application of Cointegration Approach

Abbas Ali Chandio [1,*], Yuansheng Jiang [1,*], Abdul Rauf [2], Amir Ali Mirani [3], Rashid Usman Shar [3], Fayyaz Ahmad [4] and Khurram Shehzad [2]

1 College of Economics, Sichuan Agricultural University, Chengdu 611130, China
2 School of Economics and Management, Southeast University, Nanjing 211189, Jiangsu, China; abdulrauf@seu.edu.cn (A.R.); 233189917@seu.edu.cn (K.S.)
3 College of Management, Sichuan Agricultural University, Chengdu 611130, Sichuan, China; aamirmirani07@gmail.com (A.A.M.); rashidusman12@outlook.com (R.U.S.)
4 School of Economics, Lanzhou University, Lanzhou 730000, Gansu, China; fayyaz@lzu.edu.cn
* Correspondence: alichandio@sicau.edu.cn (A.A.C.); yjiang@sicau.edu.cn (Y.J.)

Received: 14 March 2019; Accepted: 12 May 2019; Published: 16 May 2019

Abstract: The main objective of this paper is to examine the long-term effects of financial development, economic growth, energy consumption (electricity consumption in the agriculture sector), foreign direct investment (FDI), and population on the environmental quality in Pakistan during the period of 1980 to 2016. We use CO_2 emissions from the agriculture sector as a proxy indicator for environmental quality. We employ various unit root tests (e.g., ADF, PP, ERS, KPSS) and structural break unit root tests (Z&A, CMR) to check the stationarity and structural break in the data series. Cointegration tests, i.e., Johansen, Engle-Granger, and ARDL cointegration approaches are used to ensure their robustness. Results showed that significant long-term cointegration exists among the variables. Findings also indicated that an increase in financial development and foreign direct investment (FDI) improves environmental quality, whereas the increase in economic growth and electricity consumption in the agriculture sector degrades environmental quality in Pakistan. Based on the findings, we suggest policymakers should provide a conducive environment for foreign investment. Moreover, it is also suggested that a reliance on fossil fuels be reduced and a transition to renewable energy sources be encouraged to decrease the environmental pollution in the country.

Keywords: financial development; carbon emissions; energy consumption; environment quality cointegration; Pakistan

1. Introduction

The Food and Agriculture Organization (FAO) of the United Nations [1] examined the main factors of greenhouse gas (GHG) emissions with respect to agriculture, fishery and forestry sectors which had doubled their emissions in the past 50 years and could increase by as much as 30% in the future. Agriculture-related emissions from livestock and crops increased from 4.7 billion tons of carbon dioxide equivalent in 2001 to more than 5.3 billion tons in 2011, an increase of 14%. The increase is largely due to the increase in total agricultural output of developing countries [1]. The agriculture sector performs a vital role in the economy of Pakistan, functioning as the backbone of the country's economy. The farming sector not only provides food and raw materials but also creates employment opportunities for a large proportion of the population and provides food, fiber, (fuel from plants) and other products used to sustain and improve their living standards.

According to Pakistani statistics [2], agriculture accounted for 18.9% of the gross domestic product (GDP) and it is a source of livelihood for almost 42% of the rural population. The agriculture sector of Pakistan is made up of five subsectors including major crops, minor crops, livestock, fishing, and forestry, respectively. The major crops (e.g., wheat, rice, sugarcane, maize, and cotton) accounted for a 23.60% value addition in the agriculture sector and a 4.45% contribution to the gross domestic product (GDP). Likewise, the minor crops accounted for 10.80% of agriculture value addition and 2.04% of GDP. Similarly, livestock, fishing and forestry accounted for shares of 58.92%, 2.10% and 2.09% in the agriculture sector respectively, and 11.11%, 0.40% and 0.39% of GDP [2]. Accordingly, the enormous input from these subsectors to the agriculture segment may responsible for producing carbon dioxide (CO_2) in Pakistan. The negative effects of carbon dioxide (CO_2) emission from the agricultural sector, especially from fossil fuels, as well as the increase of greenhouse gases (GHGs) on the earth's surface, pose challenges for all countries of the world, regardless of economy size and the volume of population. Hence, all countries are responsible for the accumulation of such greenhouse gases (GHGs).

The earthquake in Haiti, floods in Pakistan and Australia, the tsunami in Japan and wildfires in Russia were among the most recent past major disasters that could be the consequence of environmental degradation. These conditions have caused damage to natural resources such as forests and wildlife, land and agricultural output, infrastructure and, above all, to human life. Economists and environmental experts believe that these catastrophic events are the main source of disruption to economic and financial development and have a significant impact on the environment [3].

Most developing countries started to work towards environmentally sustainable financial activities. However, economic growth activities often lead to an increase in the use of energy, which in turn contributes to the burning of fossil fuels and subsequently a rise in carbon dioxide (CO_2). This toxic substance increases the amount greenhouse gases (GHGs) and contributes to global warming. The hazards and consequences of climate change and global warming have led to the establishment of environmental friendly advocacy organizations. These organizations have made a significant contribution to the global green movement, promoting conditions in which human beings and the natural environment can come together to meet socio-economic and environmental needs [4]. Furthermore, financial development is seen as an alternative to achieving a quality environment, the challenge remains that carbon dioxide emissions are linked to the consumption of energy as a catalyst for the development and economic growth. In this case, reducing carbon dioxide emissions necessarily means slowing down the growth of the economy, while the country will not be keen to insist on economic growth. This requires innovative solutions through which the twin goals of better economic growth and a sustainable environment can be achieved. As stated in [3] this issue has been in existence since 1960, and since then there has been increased consciousness of the degradation of the environment and its more harmful influences on climate change and the environment among policymakers, ecological activists, and economists both at national and international levels. Several countries initially proposed regulatory policies and rules to address environmental pollution and degradation in pursuing of economic development.

The present study is different from previous studies in various aspects, and it has four contributions to the emerging economic literature, which is related to the studies of environmental quality: (1) we considered carbon emissions from the agriculture sector with reference to some further economic indicators in Pakistan, where its economy is enormously based on its agriculture output. (2) We used various unit root tests such as the Augmented Dickey–Fuller (ADF), the Phillips–Perron (PP), the Elliot, Rothenberg and Stock point optimal (ERS), the Kwiatkowski, Phillips, Schmidt and Shin (KPSS), Zovit Andrews and the clemente montanes reyes (CMR) tests are also utilized to consider the structural breaks. (3) For a long-term relationship, the ARDL approach is employed to check the short-term and long-term relationships between financial development, economic growth, energy consumption (electricity consumption in the agriculture sector), FDI, population and CO_2 emissions in Pakistan. (4) For the purpose of robustness, cointegration tests (Johansen and Engle-Granger tests) are applied for approving the long term cointegrating combinations among the variables.

The purpose of this paper is to analyze the long-term cointegrating association between financial development and CO_2 emissions in Pakistan over the period 1980–2016 by using the Johansen cointegration test, Engle-Granger cointegration and autoregressive distributed lag (ARDL) bounds testing cointegration approaches. Only a few studies in the past have investigated the impact of financial development on CO_2 emissions from the agriculture sector as an indicator of environmental quality. Because of the scarcity of the study, the study can fill this gap and contribute to the growing literature.

The remainder of this paper is organized in this manner: the literature review is stated in Section 2, and materials and econometric methods are portrayed in Section 3. Moreover, the empirical results and discussion are enclosed in Section 4, whereas Section 5 concludes the recent study and grants some policy implications along with future recommendations.

2. Literature Review

In Pakistan, several studies have been done in the past to see the impact of financial development, power and economic on CO_2 emissions. Some of the major studies in this regard done by [5–12]. An investigation has been conducted by in [6] that investigated the long-run cointegration association between monetary instability and ecological degradation in Pakistan for the period 1971–2009 using time-series analysis. The study found that financial instability increase environmental pollution in Pakistan. The study in [11] inspected the effect of financial development, growth, trade, and energy on CO_2 emissions in Pakistan between 1980 and 2015. It was reported that financial development, economic growth, consumption of power and skills are the increasing factors of CO_2 emissions. Furthermore, it was obtained that there is a long-run association between CO_2 emissions, financial development, energy consumption, capital, trade and economic growth in case of Pakistan. In the existing literature, some researchers found the insignificant impact of financial development on CO_2 emissions [13–15]. A research has been conducted by [3] examined the impact of growth, coal, financial development and trade on environmental quality in South Africa by using time-series data (1965–2008). Hence, results indicated that a rise in economic growth raises energy emissions, whereas financial development reduces it. Their findings also revealed that consumption of coal has a significant contribution to decline environment in the South African economy. By reducing the growth of energy pollutants, trade openness improves environmental quality for the case of South Africa.

Applying time-series analysis, [15] studied Turkey by using financial development, energy use, economic growth, trade openness, and CO_2 emissions data from the period 1960–2007. The results of the analysis revealed that economic growth and trade openness have significant effects causing environmental pollution but financial development has no significant impact on environmental quality. Using time-series analysis, [16] examined the impact of financial and economic development as well as energy on CO_2 emissions in China. They found the inverse effect of financial development on environmental pollution telling that the development of the financial sector has not taken place at the expense of environmental pollution in China. Additionally, an investigation has been conducted by [17] investigated the relations between economic growth, energy consumption, financial development, trade openness and CO_2 emissions over the period of 1975–2011 in case of Indonesia. They accomplished that energy use and economic growth increase CO_2 emissions, whereas trade openness and financial development compact it. As studied by the [18] examined the interplay between financial development, energy use and GDP on CO_2 emissions. Using time-series data for Turkey for the period 1976–1986, results of the analysis revealed that financial development develops environmental quality while energy use and economic growth reduce it. The study of [19] has investigated the interplay between energy consumption, economic growth, and CO_2 emissions by applying time-series data for eight Asian countries covering the period 1991–2013. The study proved that the growth of economic and consumption of energy have affected environmental degradation.

Inspecting the five western provinces of China, [20] established that the effect of tourism on the environment is negative for Gansu, Shanxi, Qinghai, and Ningxia. Overall, the negative impact of economic growth and energy consumption is more significant than tourism on CO_2 emission in the long run. According to [21,22] investigated an interrelationship between economic growth, level of energy consumption, financial development and oil prices in context of Italy for 1960 to 2014, where he found a long run cointegration among the variables under ARDL approach, and elaborated that estimators for oil prices and real economic growth have a noteworthy impact on level of energy usage. However, in short run results under the VAR technique, only real economic growth is an impacting factor for energy consumption. Furthermore, [23] broadened the literature with respect to Belt and Road Initiative countries for 1980–2016, where a panel of 47 nations acknowledged that financial development, energy consumption, capital formation, economic output and urbanization detrimentally fronting to the environmental abatement excluding trade openness which has a favorable link with CO_2 emissions. Similarly, [24] explored an EKC hypothesis considering to BRI 65 countries, results offered that mean group model authenticate it in all six regions. Likewise, the pooled mean group only confirmed the EKC hypothesis in developed European region but unacceptable for others.

Indeed, developing, emerging and advanced economies are converging to diminish the scale of CO_2 emissions without disturbing to the pace of sustainable progression. After reforms and open up the economy in China, the structure of its economic development has been transformed very swiftly, the operational segments of growth, i.e., agriculture, industry and service sectors tremendously sponsor to bolster the degree of economic progress in this age of competitiveness. The revealed estimates enlightened that agriculture, industry, services sectors, energy consumption, and trade detrimentally deflate to the natural environment of China [25]. Next, an exploration has been conducted [26] considering industrial growth, energy usage, services sector output and CO_2 emissions in China over the period of 1971–2016. The estimations divulged that industrial growth, services sector and level of energy utilization have an adverse effect on ecology, whereas the economic output is effectual for the environmental quality in the long run for China. However, in short-term industrial growth, the service sector and economic output harmfully effect on the environment. In addition, scrutiny has been warranted for Pakistan over the time range of 1984–2016, where a long-run interconnection was found between the variables. As per testified outcomes, gas and electricity consumption have a positive influence on the agriculture sector proportion of GDP in Pakistan [27].

Some important knowledge has been analyzed and a contribution to the existing body of literature made by distinguishing our current study and using CO_2 emissions from the agricultural sector as a substitute for environmental quality, the inclusion of population and money market financial indicators in simulating the association between financial development and environmental quality for the case of Pakistan.

3. Material and Econometric Methods

The theoretical basis of the present study comes from the expanded theory of production, which considers energy use to be an additional productive input in addition to workforce and capital. Once energy use is included in the production function, there is a case for it to be directly related to carbon dioxide emissions (CO_2). The expanded production doctrine also provides a framework for the use of development of the financial sector as a model of technological progress. This is based on greater financial development that can increase output and economic growth. Recent empirical works have employed expanded production theory to simulate the association amongst financial development, consumption of energy and carbon dioxide emissions (CO_2) [28–31]. However, the use of emissions from the agricultural sector makes the study very different from the available literature. In addition, modeling the log–log model specification compared to a simple linear-linear specification would reduce the sharpness of time series data and thus provide efficient results [32].

The empirical model specifications of this current study followed the emerging literature related to financial development and carbon dioxide emissions (CO_2), which provide empirical evidence to explore the links between growth, energy, financial development and carbon dioxide emissions (CO_2). In addition to the use of agricultural emissions, the study has added the population to further distinguish our empirical work from earlier studies [3,16,25]. These authors have included financial development in their empirical analysis. Following them, the functional form for carbon dioxide emissions (CO_2) in Pakistan can be specified as follows:

$$CO_{2t} = f(Y_t, EC_t, FD_t, FDI_t, POP_t) \tag{1}$$

The study used the log-linear specification in order to examine the interplay amongst dependent variable and independent variables. This study has formulated the log-linear model and it is specified as follows:

$$lnCO_{2t} = \lambda_0 + \lambda_1 lnY_t + \lambda_2 lnEC_t + \lambda_3 lnFD_t + \lambda_4 lnFDI_t + \lambda_5 lnPOP_t + \varepsilon_t \tag{2}$$

where $lnCO_2$ is the usual log of carbon dioxide (CO_2) from the agriculture sector, lnY stands for the natural log of economic growth, $lnEC$ represents the natural log of energy consumption (electricity consumption in agriculture sector), $lnFD$ symbolizes natural log of financial development, $lnFDI$ indicates natural log of foreign direct investment net inflows, $lnPOP$ represents natural log of population, $\lambda_1, \lambda_2, \lambda_3, \lambda_4, \lambda_5$ are coefficients to be estimated, λ_0 represents the constant term and ε_t denotes the stochastic error term, respectively. The present empirical work is based on the annual time series data to examine the effects of financial development and economic growth on agricultural CO_2 emissions in Pakistan. Data over the period 1980 to 2016 have been taken from the World Development Indicators (WDI, 2016), Food and Agriculture Organization (FAO, 2014) and Pakistan economic survey (GOP, 2016). Table 1 reports the description of the selected study variables.

Table 1. Study variables name, symbols, measurement and data sources.

Variable Name	Symbol	Variable Measurement	Data Source
CO_2 emissions	CO_2	CO_2 emissions from the agriculture sector (Gg)	(FAO, 2014)
Economic growth	Y	In constant 2010 US$	(WDI, 2016)
Electricity consumption	EC	Electricity consumption in agriculture sector (Gwh)	(GOP, 2016)
Financial development of the private sector	FD	Domestic credit to the private sector (% of GDP)	(WDI, 2016)
Foreign direct investment	FDI	Net inflows (% of GDP)	(WDI, 2016)
Population	POP	Total population (million)	(GOP, 2016)

Notes: GOP = Government of Pakistan; FAO = The Food and Agriculture Organization of the United Nations; WDI = World development Indicators.

Estimation Technique

Autoregressive Distributed Lag (ARDL)

The ARDL modelling approach proposed by [33] is used to check whether a long-run cointegration exists amongst the selected study variables or not. The autoregressive distributed lag (ARDL) modelling technique has some advantages over the traditional methods [34,35]. First, both the short-run and long-run parameters can be assessed at the same time. Second, this method can be employed even if the selected study variables are stationary at I(0), I(1) or a combination of both. Third, the ARDL modelling approach has been found much more efficient when dealing with a small sample size [29]. The ARDL-bound test cointegrations equations are given by:

$$\begin{aligned} \Delta lnCO2_t = {} & \delta_0 + \delta_1 \sum_{i=1}^{p} \Delta lnCO2_{t-1} + \delta_2 \sum_{i=1}^{p} \Delta lnY_{t-1} + \delta_3 \sum_{i=1}^{p} \Delta lnEC_{t-1} \\ & + \delta_4 \sum_{i=1}^{p} \Delta lnFD_{t-1} + \delta_5 \sum_{i=1}^{p} \Delta lnFDI_{t-1} + \delta_6 \sum_{i=1}^{p} \Delta lnPOP_{t-1} \\ & + \phi_1 lnCO2_{t-i} + \phi_2 lnY_{t-i} + \phi_3 lnEC_{t-i} + \phi_4 lnFD_{t-i} + \phi_5 lnFDI_{t-i} \\ & + \phi_6 lnPOP_{t-i} + \mu_t \end{aligned}$$

$$
\begin{aligned}
\Delta lnY_t \; &= \delta_0 + \delta_1 \sum_{i=1}^{p} \Delta lnY_{t-1} + \delta_2 \sum_{i=1}^{p} \Delta lnCO2_{t-1} + \delta_3 \sum_{i=1}^{p} \Delta lnEC_{t-1} + \delta_4 \sum_{i=1}^{p} \Delta lnFD_{t-1} \\
&+ \delta_5 \sum_{i=1}^{p} \Delta lnFDI_{t-1} + \delta_6 \sum_{i=1}^{p} \Delta lnPOP_{t-1} + \phi_1 lnY_{t-i} + \phi_2 lnCO2_{t-i} \\
&+ \phi_3 lnEC_{t-i} + \phi_4 lnFD_{t-i} + \phi_5 lnFDI_{t-i} + \phi_6 lnPOP_{t-i} + \mu_t \\
\Delta lnEC_t \; &= \delta_0 + \delta_1 \sum_{i=1}^{p} \Delta lnEC_{t-1} + \delta_2 \sum_{i=1}^{p} \Delta lnY_{t-1} + \delta_3 \sum_{i=1}^{p} \Delta lnCO2_{t-1}. \\
&+ \delta_4 \sum_{i=1}^{p} \Delta lnFD_{t-1} + \delta_5 \sum_{i=1}^{p} \Delta lnFDI_{t-1} + \delta_6 \sum_{i=1}^{p} \Delta lnPOP_{t-1}. \\
&+ \phi_1 lnEC_{t-i} + \phi_2 lnY_{t-i} + \phi_3 lnCO2_{t-i} + \phi_4 lnFD_{t-i} + \phi_5 lnFDI_{t-i} \\
&+ \phi_6 lnPOP_{t-i} + \mu_t \\
\Delta lnFD_t \; &= \delta_0 + \delta_1 \sum_{i=1}^{p} \Delta lnFD_{t-1} + \delta_2 \sum_{i=1}^{p} \Delta lnEC_{t-1} + \delta_3 \sum_{i=1}^{p} \Delta lnY_{t-1} \\
&+ \delta_4 \sum_{i=1}^{p} \Delta lnCO2_{t-1} + \delta_5 \sum_{i=1}^{p} \Delta lnFDI_{t-1} + \delta_6 \sum_{i=1}^{p} \Delta lnPOP_{t-1} \\
&+ \phi_1 lnFD_{t-i} + \phi_2 lnEC_{t-i} + \phi_3 lnY_{t-i} + \phi_4 lnCO2_{t-i} + \phi_5 lnFDI_{t-i}. \\
&+ \phi_6 lnPOP_{t-i} + \mu_t \\
\Delta lnFDI_t \; &= \delta_0 + \delta_1 \sum_{i=1}^{p} \Delta lnFDI_{t-1} + \delta_2 \sum_{i=1}^{p} \Delta lnFD_{t-1} + \delta_3 \sum_{i=1}^{p} \Delta lnEC_{t-1}. \\
&+ \delta_4 \sum_{i=1}^{p} \Delta lnY_{t-1} + \delta_5 \sum_{i=1}^{p} \Delta lnCO2_{t-1} + \delta_6 \sum_{i=1}^{p} \Delta lnPOP_{t-1}. \\
&+ \phi_1 lnFDI_{t-i} + \phi_2 lnFD_{t-i} + \phi_3 lnEC_{t-i} + \phi_4 lnY_{t-i} + \phi_5 lnCO2_{t-i} \\
&+ \phi_6 lnPOP_{t-i} + \mu_t \\
\Delta lnPOP_t \; &= \delta_0 + \delta_1 \sum_{i=1}^{p} \Delta lnPOP_{t-1} + \delta_2 \sum_{i=1}^{p} \Delta lnFDI_{t-1} + \delta_3 \sum_{i=1}^{p} \Delta lnFD_{t-1}. \\
&+ \delta_4 \sum_{i=1}^{p} \Delta lnEC_{t-1} + \delta_5 \sum_{i=1}^{p} \Delta lnY_{t-1} + \delta_6 \sum_{i=1}^{p} \Delta lnCO2_{t-1} \\
&+ \phi_1 lnPOP_{t-i} + \phi_2 lnFDI_{t-i} + \phi_3 lnFD_{t-i} + \phi_4 lnEC_{t-i} + \phi_5 lnY_{t-i} \\
&+ \phi_6 lnCO2_{t-i} + \mu_t
\end{aligned}
\tag{3}
$$

where δ_0 represents the constant term, μ_t stands for the error term, the dynamics for error correction in the short run are denoted by δ whereas the long-run links is presented in the next half of the equation symbolized by ϕ. The ARDL modeling approach employees F-statistics test to decide the presence of a long-run cointegration amongst the constructed study variables. The null hypothesis suggests the there is no a long-run cointegration against the alternative hypothesis of there exists a long-run cointegration among the variables. [33,36] proposed LCB (Lower Critical Bound) and the UCB (Upper Critical Bound) for large samples and small samples and large samples. A long-run cointegration among the variables exists if the computed F-statistics is greater than UCB value than the null hypothesis can be rejected and accepted the alternative hypothesis that a long-run cointegration exist. Furthermore, the null hypothesis cannot be rejected if the calculated F value is lower than LCB value and suggested that a long-run cointegration does not exist. However, if the calculated F value lies between the UCB and LCB, the result is inconclusive. In the present empirical study, we used the AIC (Akaike Information Criterion) for selection of the lag length. After the optimal lag length selections and model estimation, if there exists the long-run cointegration association so the short and long-run ARDL model equations are the following:

$$\begin{aligned}
\Delta lnCO2_t &= \delta_0 + \delta_1 \sum_{i=1}^{p} \Delta lnCO2_{t-1} + \delta_2 \sum_{i=1}^{p} \Delta lnY_{t-1} + \delta_3 \sum_{i=1}^{p} \Delta lnEC_{t-1} \\
&+ \delta_4 \sum_{i=1}^{p} \Delta lnFD_{t-1} + \delta_5 \sum_{i=1}^{p} \Delta lnFDI_{t-1} + \delta_6 \sum_{i=1}^{p} \Delta lnPOP_{t-1} \\
&+ \psi_1 ECT_{t-1} + \varepsilon_t \\
\Delta lnY_t &= \delta_0 + \delta_1 \sum_{i=1}^{p} \Delta lnY_{t-1} + \delta_2 \sum_{i=1}^{p} \Delta lnCO2_{t-1} + \delta_3 \sum_{i=1}^{p} \Delta lnEC_{t-1} \\
&+ \delta_4 \sum_{i=1}^{p} \Delta lnFD_{t-1} + \delta_5 \sum_{i=1}^{p} \Delta lnFDI_{t-1} + \delta_6 \sum_{i=1}^{p} \Delta lnPOP_{t-1} \\
&+ \psi_2 ECT_{t-1} + \varepsilon_t \\
\Delta lnEC_t &= \delta_0 + \delta_1 \sum_{i=1}^{p} \Delta lnEC_{t-1} + \delta_2 \sum_{i=1}^{p} \Delta lnY_{t-1} + \delta_3 \sum_{i=1}^{p} \Delta lnCO2_{t-1} + \delta_4 \sum_{i=1}^{p} \Delta lnFD_{t-1} \\
&+ \delta_5 \sum_{i=1}^{p} \Delta lnFDI_{t-1} + \delta_6 \sum_{i=1}^{p} \Delta lnPOP_{t-1} + \psi_3 ECT_{t-1} \\
&+ \varepsilon_t \\
\Delta lnFD_t &= \delta_0 + \delta_1 \sum_{i=1}^{p} \Delta lnFD_{t-1} + \delta_2 \sum_{i=1}^{p} \Delta lnEC_{t-1} + \delta_3 \sum_{i=1}^{p} \Delta lnY_{t-1} + \delta_4 \sum_{i=1}^{p} \Delta lnCO2_{t-1} \\
&+ \delta_5 \sum_{i=1}^{p} \Delta lnFDI_{t-1} + \delta_6 \sum_{i=1}^{p} \Delta lnPOP_{t-1} + \psi_4 ECT_{t-1} + \varepsilon_t \\
\Delta lnFDI_t &= \delta_0 + \delta_1 \sum_{i=1}^{p} \Delta lnFDI_{t-1} + \delta_2 \sum_{i=1}^{p} \Delta lnFD_{t-1} + \delta_3 \sum_{i=1}^{p} \Delta lnEC_{t-1} + \delta_4 \sum_{i=1}^{p} \Delta lnY_{t-1} \\
&+ \delta_5 \sum_{i=1}^{p} \Delta lnCO2_{t-1} + \delta_6 \sum_{i=1}^{p} \Delta lnPOP_{t-1} + \psi_5 ECT_{t-1} + \varepsilon_t \\
\Delta lnPOP_t &= \delta_0 + \delta_1 \sum_{i=1}^{p} \Delta lnPOP_{t-1} + \delta_2 \sum_{i=1}^{p} \Delta lnFDI_{t-1} + \delta_3 \sum_{i=1}^{p} \Delta lnFD_{t-1} \\
&+ \delta_4 \sum_{i=1}^{p} \Delta lnEC_{t-1} + \delta_5 \sum_{i=1}^{p} \Delta lnY_{t-1} + \delta_6 \sum_{i=1}^{p} \Delta lnCO2_{t-1} + \psi_6 ECT_{t-1}. \\
&+ \varepsilon_t.
\end{aligned} \tag{4}$$

where ECT_{t-1} represents the error correction term and it is denoted for the long-run equilibrium speed of adjustment. To check the good fitness of the empirical model, this study used the various diagnostic tests, including the serial correlation and heteroskedasticity test, while CUSUM (Cumulative Sum of Recursive Residuals) and CUSUMSQ (Cumulative Sum of Squares of Recursive Residuals) are also applied to check the stability of the model over the period.

4. Results and Discussions

4.1. Descriptive Statistics, Correlation Matrix, and Unit Root Test Analysis

Table 2 reports the basic statistical description of the study variables and results show that $lnCO_2$, lnY, lnEC, lnFDI, lnPOP are normally distributed but lnFD does not follow a normal distribution as suggested by Jarque-Bera test. Though, ARDL approach can solve the problem of non-normality. Likewise, the results of the correlation matrix are also shown in Table 2 and reveal that economic growth, electricity consumption in the agriculture sector, FDI and population have a strong positive and significant correlation with CO_2 emissions while financial development has negative and significant relation with CO_2 emissions, respectively.

Table 2. Descriptive summary and correlation matrix.

	$LNCO_2$	LNY	LNEC	LNFD	LNFDI	LNPOP
Mean	10.9673	25.4007	8.6769	3.2039	−0.2858	18.6722
Median	11.0014	25.4125	8.6736	3.2065	−0.3683	18.7000
Maximum	11.5157	26.1520	10.0508	4.2008	1.2997	19.0790
Minimum	10.4369	24.4944	7.6333	1.8048	−2.2762	18.1730
Std. Dev.	0.3192	0.4733	0.5872	0.4699	0.8055	0.2674
Skewness	−0.0674	−0.2236	0.3443	−0.2623	−0.1500	−0.2489
Kurtosis	1.7805	1.9846	3.3054	5.2553	2.8962	1.9112
Jarque-Bera	2.3206	1.8977	0.87517	8.2663	0.1554	2.2098
Probability	0.3133	0.3871	0.6455	0.0160	0.9252	0.3312
Observations	37	37	37	37	37	37
$LNCO_2$	1.0000					
	—					
LNY	0.9940 ***	1.0000				
	(0.0000)	—				
LNEC	0.9062 ***	0.9183 ***	1.0000			
	(0.0000)	(0.0000)	—			
LNFD	−0.3920 ***	−0.3376 **	−0.3541 **	1.0000		
	(0.0164)	(0.0410)	(0.0315)	—		
LNFDI	0.6928 ***	0.7132 ***	0.6334 ***	−0.0485	1.0000	
	(0.0000)	(0.0000)	(0.0000)	(0.7755)	—	
LNPOP	0.9955 ***	0.9981 ***	0.9082 ***	−0.3401 **	0.7022 ***	1.0000
	(0.0000)	(0.0000)	(0.0000)	(0.0394)	(0.0000)	—

Source: Authors' computation. **Note:** ***, ** Significant at 1% and 5% levels, respectively.

4.2. Empirical Results and Discussion

Before testing the cointegration association amongst the study variables, our first step is to examine their integration order. Although, if the variable is integrated in a dissimilar order, i.e., I(1) or I(0), the ARDL approach can be used. In doing so, the present empirical study uses several renowned unit root methods, for instance, ADF, PP, DF-GLS (ESR) and KPSS in order to firstly check the stationarity of data. Table 3 reports the outcomes of these renowned unit root approaches exhibits that all the study variables are stationary at the combination of I(0) and I(1). This validates the use of autoregressive distributed lag (ARDL) bound test approach suggested by [33,37].

Similarly, the results of the Z&A and CMR breakpoint unit root tests are summarized in Table 4. The results indicated that most of the variables had a unit root problem at level but became stationary at 1st difference as the test statistics are significant at the given level of significance. On the other hand, DLNY is stationary at level. Therefore, the estimations confirmed that our variables were stationary at the required levels, even in the existence of structural breaks, and the bounds testing method could be employed. The ARDL bounds test is employed to explore the presence of a long-run cointegration. In this study we have checked the cointegration of all variables and outcomes are described in Table 5. The ARDL cointegration test outcomes of first equation F_{CO2} (CO_2/Y, EC, FD, FDI, POP) disclose that there exists significant (at 5% level) a long-run cointegrating association between variables when CO_2 emissions was used as the dependent variable. Likewise, in both equations second and third F_Y (Y/CO_2, EC, FD, FDI, POP) and F_{EC} (EC/Y, CO_2, FD, FDI, POP) indicate that there no-cointegration exist amongst variables when economic growth and electricity consumption in agriculture sector were used as the dependent variables. Moreover, in the fourth equation of ARDL bounds test, we used financial development as a dependent variable F_{FD} (FD/EC, Y, CO_2, FDI, POP), results display that there exist a long-run cointegrating link between the variables. Similarly, the results of the fifth equation of ARDL bounds test F_{FDI} (FDI/FD, EC, Y, CO_2, POP) show that there is no long-run cointegration exist among variables when the foreign direct investment was used as the dependent variable.

Table 3. Results of unit root tests.

Intercept/Trend	Variables	ADF	PP	ERS	KPSS
At level					
Intercept	$LNCO_2$	0.067279	0.067279	1.086835	0.732173 **
	LNY	−1.306952	−2.373318	0.591036	0.731082 **
	LNEC	0.151339	−0.449389	0.444442	0.691326 **
	LNFD	−1.247226	−0.989193	−3.126253 ***	0.389965 *
	LNFDI	−2.165385	−2.054144	−1.807425	0.567661 **
	LNPOP	−1.735334	−3.504824 **	1.725230	0.729719 **
Intercept and trend	$LNCO_2$	−2.896777	−3.012182	−3.000342	0.132629 **
	LNY	−3.415602 *	5.743567 **	−1.718915	0.160034 **
	LNEC	−2.083088	−4.010968 ***	−2.103183	0.126498 **
	LNFD	−2.068319	−1.851540	−3.816892 ***	0.148328 *
	LNFDI	−2.649033	−2.773817	−2.746427	0.135530 *
	LNPOP	−5.104828 ***	−4.291706 ***	−3.335802 ***	0.193353 **
At first difference					
Intercept	$DLNCO_2$	−5.909376 ***	−5.909376 ***	−5.982860 ***	0.058512
	DLNY	−3.575677 ***	−3.544302 ***	−2.811139 ***	0.374628 *
	DLNEC	−11.77023 ***	−11.89641 ***	−2.487519 **	0.119807
	DLNFD	−4.612180 ***	−4.612180 ***	−8.962681 ***	0.367610 *
	DLNFDI	−5.824703 ***	−6.420425 ***	−5.737412 ***	0.176450
	DLNPOP	−2.052846	−1.275548	−0.432306	0.650474 **
Intercept and trend	$DLNCO_2$	−5.811907 ***	−5.811907 ***	−5.905037 ***	0.058439
	DLNY	−3.658144 **	−3.659494 **	−3.668914 ***	0.112724
	DLNEC	−11.69367 ***	−11.79462 ***	−2.569443	0.101405
	DLFD	−4.661032 ***	−4.661032 ***	−8.997397 ***	0.359870 ***
	DLNFDI	−5.741518 ***	−6.786958 ***	−5.799335 ***	0.150092 **
	DLNPOP	−0.300854	−0.674946	−1.556024	0.168522 **

Source: Authors' computation. **Notes:** ADF; PP; ERS and KPSS indicate the Augmented Dickey–Fuller test; the Phillips–Perron test; the Elliot, Rothenberg and Stock point optimal test and the Kwiatkowski, Phillips, Schmidt and Shin test, respectively. ***, ** and * Significant at 1%, 5% and 10% levels, respectively.

Table 4. Results of Zivot-Andrews and CMR structure break unit root tests.

	Zivot-Andrews Structure Break Unit Root Test				CMR Structure Break Unit Root Test			
Variables	Level		1st difference		Level		1st difference	
	T-statistics	Breaks	T-statistics	Breaks	T-statistics	Breaks	T-statistics	Breaks
$LNCO_2$	−1.07	1995	−9.81	1996	10.487	1997	0.691	1994
LNY	−5.74	2004	-	-	7.119	2005	−3.389	1990
LNEC	−1.81	2012	−12.28	2011	5.021	2012	0.741	2010
LNFD	−3.41	2011	−9.71	1993	−3.389	2013	−0.259	1991
LNFDI	−2.85	1992	−6.03	2009	5.962	1989	−0.398	2009
LNPOP	−3.09	2009	−6.28	2010	8.943	2000	−12.107	1993

Notes1: Z&A test produced critical values are as; −4.58, −4.93, and −5.34 at 1%, 5%, and 10% respectively. **Notes2:** CMR denotes for "Clemente Montanes Reyes" structure break unit root test, where it produced a critical value −3.560 at 5%.

Table 5. Results of cointegration bounds test.

Model for Estimation		F-Statistics	Decision
F_{CO2} (CO_2/Y,EC,FD,FDI,POP)	ARDL(1, 1, 0, 0, 1, 1)	4.949047 **	Cointegration exist
F_Y (Y/CO_2,EC,FD,FDI,POP)	ARDL(1, 1, 0, 0, 0, 0)	1.778916	No-cointegration exist
F_{EC} (EC/Y,CO_2,FD,FDI,POP)	ARDL(1, 0, 0, 0, 0, 0)	3.348705	No-cointegration exist
F_{FD} (FD/EC,Y,CO_2,FDI,POP)	ARDL(1, 0, 0, 1, 1, 0)	8.578359 ***	Cointegration exist
F_{FDI} (FDI/FD,EC,Y,CO_2,POP)	ARDL(1, 0, 0, 0, 0, 1)	2.887597	No-cointegration exist
F_{POP} (POP/FDI,FD,EC,Y,CO_2)	ARDL(1, 1, 0, 0, 1, 0)	19.73612 ***	Cointegration exist
Critical Value Bounds	I0 Bound	I1 Bound	
1%	3.15	5.23	
5%	3.12	4.25	
10%	3.93	3.79	

Source: Authors' computation. **Note:** ***, ** Significant at 1% and 5% levels, respectively.

The last equation for ARDL bounds test F_{POP} (POP/FDI, FD, EC, Y, CO_2) indicates a long-run cointegration exists among variables when the population is used as the dependent variable. To check the robustness of our long-run cointegrating results, we employed the Johansen cointegration test by using trace statistics and max–eigenvalue statistics. The estimated outcomes of (trace and max–eigenvalue) test are shown in Table 6. The trace and max–eigenvalue statistics values are greater than the critical value at 5% significance level; showing a long-run co-integration relationship among the variables. Additionally, the Engle-Granger (EG) cointegration test [38] is utilized to measure the further robustness of Johansen cointegration test outcomes. It is a dual-step errors-based test, so initially, dependent variable ($LNCO_2$) is regressed on explanatory variables (Y, EC, FD, FDI, POP) and computed the residuals from the equation. At that time, calculated residuals are further analyzed by the ADF unit root test in Table 7, where residuals are stationary at their level. It is an indication for at the first stage that variables are cointegrated. Moreover, the validation through the second step will be guaranteed the long run cointegration among the variables effectually. Next, the 1st difference of the residuals is regressed on its lagged based residuals in simple OLS approach in Table 8. The estimates of calculated residuals (New-1) in OLS regression results is statistically significant at 5%, which ensured that there is long-run cointegration among the set of variables. Hence, rejecting the null hypothesis instead of the alternative is evidence the dataset series are certainly cointegrated.

Table 6. Johansen cointegration test using Trace statistics and Max–Eigenvalue statistics.

Hypothesized		**Trace**	**0.05**	
No. of CE(s)	Eigenvalue	Statistic	Critical Value	Prob
None *	0.896802	184.6633	95.75366	0.0000
At most 1 *	0.743880	107.4457	69.81889	0.0000
At most 2 *	0.474210	61.13391	47.85613	0.0018
At most 3 *	0.449581	39.27687	29.79707	0.0030
At most 4 *	0.392142	18.97631	15.49471	0.0143
At most 5	0.058530	2.050617	3.841466	0.1521
Hypothesized No. of CE(s)	Eigenvalue	Max-Eigen Statistic	0.05 Critical Value	Prob
None *	0.896802	77.21766	40.07757	0.0000
At most 1 *	0.743880	46.31176	33.87687	0.0010
At most 2	0.474210	21.85705	27.58434	0.2278
At most 3	0.449581	20.30056	21.13162	0.0650
At most 4 *	0.392142	16.92569	14.26460	0.0185
At most 5	0.058530	2.050617	3.841466	0.1521

Source: Authors' computation. **Note:** * 5% level, statistical significance.

Table 7. First step in Engle-Granger cointegration test to calculating the residuals unit root.

ADF Test Statistic at a Level for (Calculated Residuals)	
t-Statistic	Prob.*
−4.032361	0.0035 ***
Test critical values:	
1% level	−3.626784
5% level	−2.945842
10% level	−2.611531

Source: Authors' computation; ***, * Significant at 1%, 5% and 10% levels, respectively.

Table 8. Second step in Engle Granger cointegration test for significance evaluation.

Variable	Coefficient	Std. Error	t-Statistic	Prob.
D(LNEC)	0.002775	0.0108	0.256973	0.799
D(LNFD)	−0.02262	0.007154	−3.16153	0.0037 ***
D(LNFDI)	−0.00632	0.006445	−0.98062	0.3349
D(LNPOP)	−0.27731	0.696161	−0.39834	0.6933
D(LNY)	0.402064	0.187295	2.146686	0.0403 **
NEW(-1)	−0.40261	0.152648	−2.63748	0.0133 **
C	0.017448	0.01602	1.08916	0.2851

Source: Authors' computation. **Note:** ***, ** Significant at 1% and 5% levels, respectively.

Table 9 displays the estimated outcomes of both long and short-run of ARDL approach. The CO_2 emissions from the agricultural sector is used as a dependent variable while economic growth, electricity consumption in the agriculture sector, financial development, FDI and population have been used as independent variables in this empirical study. The estimated outcomes show that economic growth is significant and inversely linked to CO_2 emissions from the agricultural sector at 1% significance level in the long run. The estimated coefficient of economic growth shows that a 1% increase in economic growth reduces CO_2 emissions from agriculture by 0.45%; this means that economic growth improves the environmental quality in Pakistan. The empirical outcomes of this study support the theoretical arguments in the literature that the adoption of cleaner energy sources boosts up the economic growth that improves environmental quality. Our estimated findings are in line with the results of previous studies. Reference [39] found that the economic growth and electricity consumption degrade environmental quality in belt and road initiative (BRI) countries. Reference [40] revealed that economic growth is inversely associated with CO_2 emissions, indicating that economic growth improves environmental quality in Nigeria. But, our findings are contrary to the results of [21] who reported that economic growth has a positive and significant effect on CO_2 emissions. The long-run coefficient of the agricultural electricity consumption is positive but it is non-significant. The result of the positive effect of electricity consumption in the agriculture sector on CO_2 emissions is in line with, and supports the results of earlier research [41,42]. The result shows that presently electricity is a critical factor for the level of CO_2 emissions, which is highly alarming in Pakistan. High-level use of energy causes high environmental degradation [43]. The carbon-free sources of energy such as nuclear and wind, related innovative technology is also favorable to improve the quality of the environment [44]. Likewise, the coefficient of financial development is negative and highly significant at a 1% significance level in the long run. Financial development coefficient outcomes show that a 1% rise in financial development has the capacity to reduce the CO_2 emissions from the agricultural sector and improve the environmental quality almost 0.02%. The findings of financial development are in line with the results of earlier researchers. [24] revealed that financial development significantly enhances the environmental degradation in the One Belt and One Road region. [18,43] reported that financial development improves environmental quality in Turkey. Similarly, FDI coefficient results indicate a positive significant and dominant effect on CO_2 in the long run in Pakistan. FDI results indicated that FDI contributes to environmental degradation. Additionally, the population coefficient is positive and significantly associated with CO_2 emissions in the long run, showing that a 1% increase in population could increase environmental pollution by 1.42%. The population growth will increase the land openness for residential construction, agriculture, and other related economic activities.

The finding of this paper is intuitive with the previous study of [45]. Table 9 reports the estimated outcomes of the short run ARDL technique. Outcomes of the short-run cointegration show that economic growth has a positive but statistically non-significant effect on CO_2 emissions, indicating that economic growth does not have any statistical influence to cause environmental degradation in Pakistan. Whereas, financial development has a strong negative association (−0.023) with CO_2 in the short-run analysis. Results of financial development indicate that a 1% increase in financial development reduces the CO_2 emissions from the agricultural sector and improves environmental quality. FDI has a positive and non-significant effect on CO_2 emissions in Pakistan in the short run. The findings of this study are consistent with the outcomes of [38] Saud et al. (2018) which stated that an increase in financial development and FDI improve the quality of the environment. Additionally, the results display a strong positive association (9.022) among the population and CO_2 emissions in Pakistan in the short-run. Results of the population indicate that a 1% increase in population will increase CO_2 emissions by 9.02% in the short-run.

Table 9. Estimated long-run and short-run coefficients of ARDL model.

Variable	Coefficient	Std. Error	t-Statistic	Prob.
Long-run estimation				
LNY	−0.451404 ***	0.163762	−2.756465	0.0105
LNEC	0.008475	0.013542	0.625863	0.5369
LNFD	−0.027816 ***	0.007907	−3.517833	0.0016
LNFDI	0.044922 ***	0.009849	4.561101	0.0001
LNPOP	1.425103 ***	0.386299	3.689118	0.0010
Constant	−4.829268	4.283987	−1.127284	0.2699
Trend	0.015515 ***	0.004183	3.709142	0.0010
Short-run Dynamics				
D(LNY)	0.157755	0.182425	0.864770	0.3951
D(LNEC)	0.007047	0.011039	0.638382	0.5288
D(LNFD)	−0.023128 ***	0.007517	−3.076716	0.0049
D(LNFDI)	0.009470	0.006154	1.538790	0.1359
D(LNPOP)	9.022650 ***	3.144515	2.869329	0.0081
DTrend	0.012900 ***	0.004982	2.589351	0.0155
ECM (−1)	−0.831464 ***	0.155434	−5.349315	0.0000
R-squared	0.998298			
Adjusted R-squared	0.997730			
F-statistic	59.170			
Prob(F-statistic)	0.000000			
Durbin-Watson stat	1.873053			

Source: Authors' computation. **Note:** *** Significant at 1% level.

To test the stability of the ARDL model this study used various diagnostic tests, for example, Breusch-Godfrey for serial correlation, White for heteroscedasticity, CUSUM and CUSUMQS for the stability of the parameters, outcomes are described in Table 10. The diagnostic test results display that the ARDL model has successfully passed all diagnostic tests. Moreover, the results of CUSUM and CUSUMQS presented in Figures 1 and 2, indicating that the values of the parameters are stable over the period.

Table 10. Diagnostic tests for the stability of the ARDL model.

Breusch-Godfrey Serial Correlation LM Test: Serial Autocorrelation			
F-statistic	0.166770	Probability	0.8474
Obs*R-squared	0.507160	Probability	0.7760
White Heteroskedasticity Test:			
F-statistic	0.047588	Probability	0.8286
Obs*R-squared	0.050317	Probability	0.8225
Ramsey RESET Test: Model Misspecification			
F-statistic	1.270951	Probability	0.2703

Source: Authors' computation. **Note:** * Significant at 5% level.

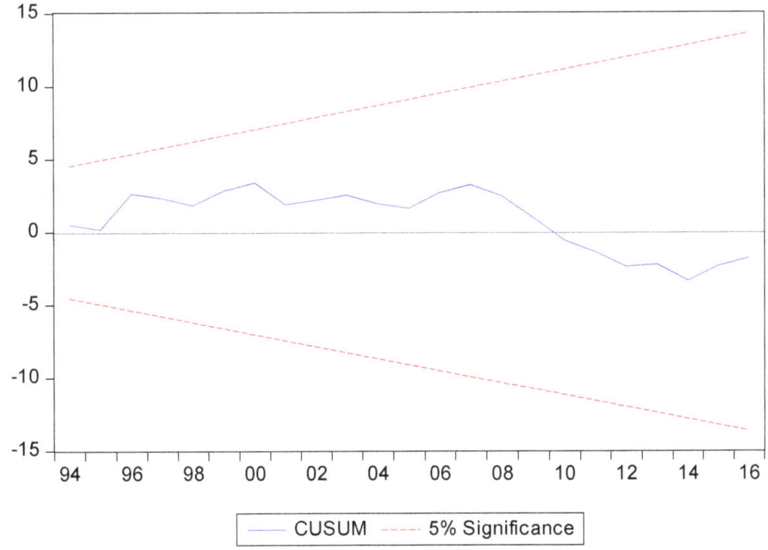

Figure 1. The plot of the cumulative sum of recursive residuals.

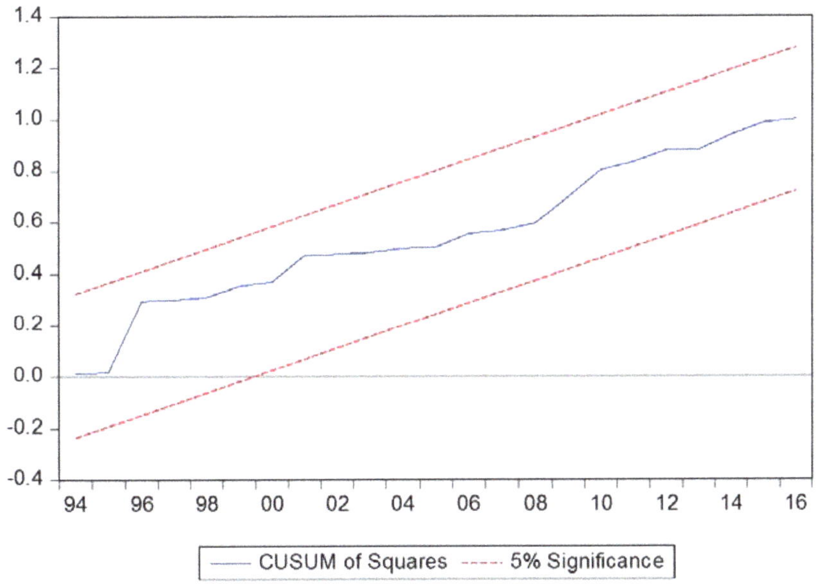

Figure 2. The plot of the cumulative sum of squares of recursive residuals.

In order to test the direction of causality between the variables, the study conducted the pair-wise Granger causality test. The Granger causality approach has three categories such as bidirectional causality, unidirectional causality, and no causality. Table 11 reports the pair-wise Granger causality outcomes. The results of the pair-wise Granger causality test show that the null hypothesis that economic growth does not Granger cause CO_2 emissions is rejected at 10% significance level, implying that economic growth does Granger cause CO_2 emissions. However, the null hypothesis that CO_2 emissions do not Granger cause economic growth is not rejected, meaning that CO_2 emissions do not Granger cause economic growth. There is evidence of unidirectional causality running from LnY → $LnCO_2$ at the 10% significance level. The results of the Granger causality test failed to reject the null hypothesis that energy consumption (electricity consumption in the agriculture sector) does not Granger cause CO_2 emissions. However, CO_2 emissions Granger cause energy consumption (electricity consumption in the agriculture sector) at a 1% level of significance. There is evidence of unidirectional causality running from CO_2 → LnEC. The Granger causality test results display that the null hypothesis that financial development does not Granger cause CO_2 emissions is no rejected, implying financial development does not Granger-cause CO_2 emissions. However, the null hypothesis of CO_2 emissions does not Granger-cause financial development is rejected at a 5% level of significance, implying CO_2 emissions does Granger-cause financial development. Thus, a unidirectional causality has been identified from CO_2 → LnFD at the 5% significance level. Moreover, the null hypotheses that the population does not Granger-cause CO_2 emissions is rejected at 5% significance level. There is evidence of bidirectional causality between LnPOP ↔ $LnCO_2$.

Table 11. Granger causality between CO_2 and its determinants.

Null Hypothesis	F-statistic	Probability
LnY does not Granger Cause $LnCO_2$	3.34546	0.0764 *
$LnCO_2$ does not Granger Cause LnY	0.31787	0.5767
LnEC does not Granger Cause $LnCO_2$	1.70653	0.2005
$LnCO_2$ does not Granger Cause LnEC	10.3927	0.0028 ***
LnFD does not Granger Cause $LnCO_2$	2.79801	0.1038
$LnCO_2$ does not Granger Cause LnFD	4.14764	0.0498 **
LnFDI does not Granger Cause $LnCO_2$	0.32867	0.5703
$LnCO_2$ does not Granger Cause LnFDI	2.22259	0.1455
LnPOP does not Granger Cause $LnCO_2$	4.00315	0.0537 **
$LnCO_2$ does not Granger Cause LnPOP	5.69914	0.0229 **

Source: Authors' computation. **Note:** *, **, *** indicate rejection of null hypothesis at 10%, 5% and 1% levels of significance, respectively.

5. Conclusions, Recommendations and Future Implications

This paper examined the effects of financial development, economic growth, electricity consumption in the agriculture sector, FDI and population on the environmental quality in Pakistan for the period 1980 to 2016. We used CO_2 emissions from the agriculture sector as a proxy indicator for environmental quality. Several unit root tests (ADF, PP, ERS, KPSS) and structural break unit root tests (Z&A, CMR) are applied to test the stationarity and structural break in the dataset series. Cointegration approaches, i.e., Johansen cointegration, Engle-Granger, and ARDL cointegration approaches ensure their robustness.

The ARDL bounds method establish the long-run cointegration association between financial development, economic growth, electricity consumption in the agriculture sector, FDI, population and CO_2 emissions. The ARDL bounds method, Engle-Granger, and Johansen cointegration tests outcomes confirmed the presence of a long-term cointegrating connection among the variables. The long-run coefficients of economic growth and financial development have negative effects on CO_2 emissions. These findings indicate that a 1% increase in economic growth and financial development will reduce CO_2 emissions growth and improve the environmental quality in Pakistan by 0.45% and 0.02% respectively. Whereas, the results of the long-run coefficients of electricity consumption in the agriculture sector, FDI and population have positive impacts on CO_2 emissions. This indicates that a 1% increase in energy consumption (electricity consumption in the agriculture sector) and FDI net inflows will degrade environmental quality by 0.008% and 1.42% while a 1% increase in population could increase environmental pollution by 1.42% in the long-run in Pakistan. Furthermore, in order to check the direction of causality amongst the study variables, the study applied the pairwise Granger causality test. The Granger causality test results showed a unidirectional causality between economic growth and CO_2 emissions. However, there was a bi-directional causality between population and CO_2 emissions.

Based on the findings, our study suggested that the Government and policymakers should further increase financial development and economic growth, since such development may further improve the quality of environment in the country. Additionally, the use of energy and CO_2 emissions are directly associated with each other, therefore, our study also suggested that the efficient energy consumption from fossil sources and a conversion to renewable energy sources, so as to reduce environmental pollution in the country.

As perceived from the outcomes, the CO_2 mitigation guidelines grounded on energy usage and gross demotic product (income) unaccompanied may not determine to be productive as financial expansion is an essential fragment of the greenhouse gas (GHG) mitigation strategy. Consequently, financial growth is extracted to get better environmental quality with regard to the agriculture sector in Pakistan. Thus, the policy implications may retrieve from the recent study as, to utilize the financial segment across the banking system, and to reassure energy-efficient and green portfolio investments. Subsequently, monetary regulatory policy can be outlined to pose minor interest charges and other markdowns for environmentally friendly manufacturing practices by business corporations/organizations. However, in the recent time period, the Pakistani financial division and its various sectors have had a low volume portion and would have to experience an extremely stretched mode before attaining its optimal point.

In this respect, the state government can support the financial markets by launching a solid strategic agenda that generates enduring worth for (GHG) emissions cuts and constant provisions for the expansion of novel technological tools that may guide a low carbon-concentrated country. Additionally, well-organized capital and financial markets can be an alternative appreciated policy choice that might be accepted. Hence, this is due to which companies can shrink their liquidity perils and can activate the needed funds via portfolio divergence, that would be enormously advantageous in developing a wide-ranging technology foundation in the long run.

Lastly, this recent study spreads the room for future investigations, where the investigators can practice our methodological procedure to catch the greater awareness of economic development, energy usage and environmental quality interrelationships with regard to the agriculture sector in nations other than Pakistan. Supplementary, current ARDL approach may exchange with nonlinear ARDL (NARDL) or can be upgraded by building an index of financial development in place of exercising a sole element as a deputation for financial advancement. The on-hand study has employed the cumulative CO_2 emissions dataset for Pakistan; however, in future exploring the linkages between income, financial expansion and CO_2 emissions amount at a disaggregate scale (industry wise) may offer some improved understandings. Consequently, it may assist policy architects to articulate environment-friendly monetary and fiscal policies.

Author Contributions: All authors contributed to writing, editing, analysis, idea generation, and data collection. Conceptualization: A.A.C. and A.R.; Formal analysis, A.A.C. and A.A.M.; Funding acquisition, Y.J.; Investigation, A.A.C., R.U.S., F.A and A.R.; Methodology, A.A.C. and A.R.; Software, A.A.C. and A.R.; Supervision, Y.J.; Writing–original draft, A.A.C., A.A.M., and R.U.S.; Writing–review and editing, A.A.C., A.R., F.A, K.S and R.U.S.

Funding: This research was funded by Double Supporting Plan of Sichuan Agricultural University: Research Group Funding for Rural Finance and Development.

Conflicts of Interest: The authors declare that there is no conflict of interest.

References

1. FAO. Food and Agriculture Organization of the United Nations. 2014. Available online: http://www.fao.org/home/en/ (accessed on 2 March 2019).
2. GOP. *Economic Survey 2017–18*; Finance Division, Economic Advisors: Wing, Islamabad, 2018.
3. Shahbaz, M.; Tiwari, A.K.; Nasir, M. The effects of financial development, economic growth, coal consumption and trade openness on CO2 emissions in South Africa. *Energy Policy* **2013**, *61*, 1452–1459. [CrossRef]

4. Apak, S.; Atay, E. Renewable Energy Financial Management in the EU's Enlargement Strategy and Environmental Crises. *Procedia-Soc. Behav. Sci.* **2013**, *75*, 255–263. [CrossRef]
5. Nasir, M.; Rehman, F.U. Environmental Kuznets Curve for carbon emissions in Pakistan: An empirical investigation. *Energy Policy* **2011**, *39*, 1857–1864. [CrossRef]
6. Shahbaz, M. Does financial instability increase environmental degradation? Fresh evidence from Pakistan. *Econ. Model.* **2013**, *33*, 537–544. [CrossRef]
7. Munir, S.; Khan, A. Impact of Fossil Fuel Energy Consumption on CO_2 Emissions: Evidence from Pakistan (1980–2010). *Pak. Dev. Rev.* **2014**, *53*, 327–346. [CrossRef]
8. Ahmed, K.; Shahbaz, M.; Qasim, A.; Long, W. The linkages between deforestation, energy and growth for environmental degradation in Pakistan. *Ecol. Indic.* **2015**, *49*, 95–103. [CrossRef]
9. Javid, M.; Sharif, F. Environmental Kuznets curve and financial development in Pakistan. *Renew. Sustain. Energy Rev.* **2016**, *54*, 406–414. [CrossRef]
10. Shahbaz, M.; Shahzad, S.J.H.; Ahmad, N.; Alam, S. Financial development and environmental quality: The way forward. *Energy Policy* **2016**, *98*, 353–364. [CrossRef]
11. Siddique, H.M.A. Impact of Financial Development and Energy Consumption on CO_2 Emissions: Evidence from Pakistan. *Bull. Bus. Econ.* **2017**, *6*, 68–73.
12. Ullah, A.; Khan, D.; Khan, I.; Zheng, S. Does agricultural ecosystem cause environmental pollution in Pakistan? Promise and menace. *Environ. Sci. Pollut. Res.* **2018**, *25*, 13938–13955. [CrossRef]
13. Dogan, E.; Seker, F. The influence of real output, renewable and non-renewable energy, trade and financial development on carbon emissions in the top renewable energy countries. *Renew. Sustain. Energy Rev.* **2016**, *60*, 1074–1085. [CrossRef]
14. Omri, A.; Daly, S.; Rault, C.; Chaibi, A. Financial Development, Environmental Quality, Trade and Economic Growth: What Causes What in MENA Countries. *Energy Econ.* **2015**, *48*, 242–252. [CrossRef]
15. Ozturk, I.; Acaravci, A. The long-run and causal analysis of energy, growth, openness and financial development on carbon emissions in Turkey. *Energy Econ.* **2013**, *36*, 262–267. [CrossRef]
16. Jalil, A.; Feridun, M. The impact of growth, energy and financial development on the environment in China: A cointegration analysis. *Energy Econ.* **2011**, *33*, 284–291. [CrossRef]
17. Shahbaz, M.; Hye, Q.M.A.; Tiwari, A.K.; Leitão, N.C. Economic growth, energy consumption, financial development, international trade and CO_2 emissions in Indonesia. *Renew. Sustain. Energy Rev.* **2013**, *25*, 109–121. [CrossRef]
18. Dar, J.A.; Asif, M. Does financial development improve environmental quality in Turkey? An application of endogenous structural breaks based cointegration approach. Manag. *Environ. Qual. Int. J.* **2018**, *29*, 368–384. [CrossRef]
19. Jamel, L.; Derbali, A. Do energy consumption and economic growth lead to environmental degradation? Evidence from Asian economies. *Cogent Econ. Financ.* **2016**, *4*, 1170653. [CrossRef]
20. Ahmad, F.; Draz, M.U.; Su, L.; Ozturk, I.; Rauf, A. Tourism and Environmental Pollution: Evidence from the One Belt One Road Provinces of Western China. *Sustainability* **2018**, *10*, 3520. [CrossRef]
21. Magazzino, C. GDP, energy consumption and financial development in Italy. *Int. J. Energy Sect. Manag.* **2018**, *12*, 28–43. [CrossRef]
22. Magazzino, C. Energy consumption, real GDP, and financial development nexus in Italy: An application of an auto-regressive distributed lag bound testing approach. In Proceedings of the WIT Transactions on Ecology and the Environment—Energy Quest 2016, Ancona, Italy, 6–8 September 2016; Brebbia, C.A., Polonara, F., Magaril, E.R., Passerini, G., Eds.; WIT Press: Southampton, UK, 2016; pp. 21–32.
23. Rauf, A.; Liu, X.; Amin, W.; Ozturk, I.; Rehman, O.U.; Sarwar, S. Energy and Ecological Sustainability: Challenges and Panoramas in Belt and Road Initiative Countries. *Sustainability* **2018**, *10*, 2743. [CrossRef]
24. Rauf, A.; Liu, X.; Amin, W.; Ozturk, I.; Rehman, O.; Hafeez, M. Testing EKC hypothesis with energy and sustainable development challenges: A fresh evidence from Belt and Road Initiative economies. *Environ. Sci. Pollut. Res.* **2018**, *25*, 32066–32080. [CrossRef]
25. Rauf, A.; Zhang, J.; Li, J.; Amin, W. Structural changes, energy consumption and Carbon emissions in China: Empirical evidence from ARDL bound testing model. *Struct. Chang. Econ. Dyn.* **2018**, *47*, 194–206. [CrossRef]

26. Rauf, A.; Liu, X.; Amin, W.; Rehman, O.U.; Sarfraz, M. Nexus between Industrial Growth, Energy Consumption and Environmental Deterioration: OBOR Challenges and Prospects to China. In Proceedings of the 2018 5th International Conference on Industrial Economics System and Industrial Security Engineering (IEIS), Toronto, ON, Canada, 3–6 August 2018; pp. 1–6.
27. Chandio, A.A.; Jiang, Y.; Rehman, A. Energy consumption and agricultural economic growth in Pakistan: Is there a nexus? *Int. J. Energy Sect. Manag.* **2018**. [CrossRef]
28. Shahbaz, M.; Islam, F.; Butt, M.S. Finance–growth–energy nexus and the role of agriculture and modern sectors: Evidence from ARDL bounds test approach to cointegration in Pakistan. *Glob. Bus. Rev.* **2016**, *17*, 1037–1059. [CrossRef]
29. Irfan, M.; Shaw, K. Modeling the effects of energy consumption and urbanization on environmental pollution in South Asian countries: a nonparametric panel approach. *Qual. Quant.* **2017**, *51*, 65–78. [CrossRef]
30. Guan, X.; Zhou, M.; Zhang, M. Using the ARDL-ECM approach to explore the nexus among urbanization, energy consumption, and economic growth in Jiangsu Province, China. *Emerg. Mark. Financ. Trade* **2015**, *51*, 391–399. [CrossRef]
31. Hafeez, M.; Chunhui, Y.; Strohmaier, D.; Ahmed, M.; Jie, L. Does finance affect environmental degradation: Evidence from One Belt and One Road Initiative region? *Environ. Sci. Pollut. Res.* **2018**, *25*, 9579–9592. [CrossRef] [PubMed]
32. Iheanacho, E. The impact of financial development on economic growth in Nigeria: An ARDL analysis. *Economies* **2016**, *4*, 26. [CrossRef]
33. Pesaran, M.H.; Shin, Y.; Smith, R.J. Bounds testing approaches to the analysis of level relationships. *J. Appl. Econ.* **2001**, *16*, 289–326. [CrossRef]
34. Engle, R.F.; Granger, C.W.J. Co-integration and Error Correction: Representation, Estimation, and Testing. *Econometrica* **1987**, *55*, 251–276. [CrossRef]
35. Johansen, S. Estimation and Hypothesis Testing of Cointegration Vectors in Gaussian Vector Autoregressive Models. *Econometrica* **1991**, *59*, 1551–1580. [CrossRef]
36. Narayan, P.K.K. The saving and investment nexus for China: Evidence from cointegration tests. *Appl. Econ.* **2005**, *37*, 1979–1990. [CrossRef]
37. Pesaran, M.H.; Shin, Y. An autoregressive distributed lag modelling approach to cointegration analysis. *Econ. Soc. Monogr.* **1998**, *31*, 371–413.
38. Saud, S.; Chen, S.; Haseeb, A. Impact of financial development and economic growth on environmental quality: An empirical analysis from Belt and Road Initiative (BRI) countries. *Environ. Sci. Pollut. Res.* **2019**, *26*, 2253–2269. [CrossRef] [PubMed]
39. Maji, I.K.; Habibullah, M.S.; Saari, M.Y. Emissions from agricultural sector and financial development in Nigeria: An empirical study. *Int. J. Econ. Manag.* **2016**, *10*, 173–187.
40. Kasman, A.; Duman, Y.S. CO_2 emissions, economic growth, energy consumption, trade and urbanization in new EU member and candidate countries: A panel data analysis. *Econ. Model.* **2015**, *44*, 97–103. [CrossRef]
41. Zhang, L.; Gao, J. Exploring the effects of international tourism on China's economic growth, energy consumption and environmental pollution: Evidence from a regional panel analysis. *Renew. Sustain. Energy Rev.* **2016**, *53*, 225–234. [CrossRef]
42. Choi, Y.; Zhang, N.; Zhou, P. Efficiency and abatement costs of energy-related CO_2 emissions in China: A slacks-based efficiency measure. *Appl. Energy* **2012**, *98*, 198–208. [CrossRef]
43. Chen, Y.; Zhang, S.; Xu, S.; Li, G.Y. Fundamental trade-offs on green wireless networks. *IEEE Commun. Mag.* **2011**, *49*, 30–37. [CrossRef]
44. Islam, F.; Shahbaz, M.; Ahmed, A.U.; Alam, M.M. Financial development and energy consumption nexus in Malaysia: A multivariate time series analysis. *Econ. Model.* **2013**, *30*, 435–441. [CrossRef]
45. Munir, K.; Ameer, A. Effect of economic growth, trade openness, urbanization, and technology on environment of Asian emerging economies. *Manag. Environ. Qual. Int. J.* **2018**, *29*, 1123–1134. [CrossRef]

energies

MDPI

Article

Measuring Impact of Uncertainty in a Stylized Macroeconomic Climate Model within a Dynamic Game Perspective †

Valentijn Stienen * and Jacob Engwerda

Department of Econometrics and Operations Research, Tilburg University, 5037 AB Tilburg, The Netherlands; j.c.engwerda@uvt.nl

* Correspondence: v.f.stienen@uvt.nl

† This paper is a revised version of the discussion paper, published in Tilburg University CentER DP Series, No. 2018-007. This discussion paper is partially presented at the first IFAC workshop on Integrated Assessment Modelling for Environmental Systems (IAMES2018), Brescia, Italy, 10–11 May 2018 and is published in the corresponding conference proceedings *IFAC-PapersOnLine 51*, 138–143.

Received: 13 November 2019; Accepted:13 January 2020; Published: 18 January 2020

Abstract: In this paper, we present a stylized dynamic interdependent multi-country energy transition model. The goal of this paper is to provide a starting point for examining the impact of uncertainty in such models. To do this, we define a simple model based on the standard Solow macroeconomic growth model. We consider this model in a two-country setting using a non-cooperative dynamic game perspective. Total carbon dioxide (CO_2) emission is added in this growth model as a factor that has a negative impact on economic growth, whereas production can be realized using either green or fossil energy. Additionally, a factor is incorporated that captures the difficulties of using green energy, such as accessibility per country. We calibrate this model for a two-player setting, in which one player represents all countries affiliated with the Organization for Economic Cooperation and Development (OECD) and the other player represents countries not affiliated with the OECD. It is shown that, in general, the model is capable to describe energy transitions towards quite different equilibrium constellations. It turns out that this is mainly caused by the choice of policy parameters chosen in the objective function. We also analyze the optimal response strategies of both countries if the model in equilibrium would be hit by a CO_2 shock. Also, here we observe a quite natural response. As the model is quite stylized, a serious study is performed to the impact several model uncertainties have on the results. It turns out that, within the OECD/non-OECD framework, most of the considered uncertainties do not impact results much. However, the way we calibrate policy parameters does carry much uncertainty and, as such, influences equilibrium outcomes a lot.

Keywords: differential games; environmental engineering; uncertain dynamic systems; linearization; economic systems; open-loop control systems

JEL Classification: Q43; Q54; Q56; Q58; C61; C72; C73

1. Introduction

Climate change is a key topic on the agenda of most of the world's leading presidents. Reports of the European Environment Agency (EEA) [1] and the Intergovernmental Panel on Climate Change (IPCC) [2] show that the average global temperature is rising. For example, Figure 1a shows the global land and ocean temperature anomalies, with 1940 as a base year. From this figure we can see that the average change in temperature per decade is approximately +0.07 °C. Besides, according to data from the National Centers for Environmental Information (NCEI) [3], the total CO_2 emission has been increasing exponentially over time, see Figure 1b.

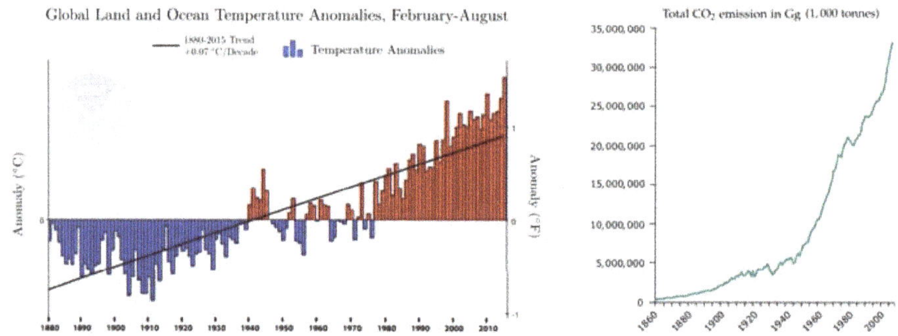

(**a**) Global temperature anomalies (EEA)

(**b**) Global CO_2 emission (NCEI)

Figure 1. Climate facts.

According to the IPCC reports, with 90% probability, a doubling (compared to its value in the year 2000) of CO_2 concentration will lead to an increase of the average world temperature by 1.5 °C. This increase will affect all countries. The broadly accepted consensus is therefore that actions are needed to reduce the level of CO_2 emission all over the world. For instance, by using more green energy instead of fossil energy. However, nowadays fossil fuel reserves are abundant. This means that it is not easy to convince countries to restrict their use of fossil energy and begin to expand their green energy use. Currently, using green energy is typically more expensive than using fossil energy. In particular, countries which experience a period of economic growth could be rather skeptic about changing their climate policy to a more green policy. They have to invest in green energy resources, which costs money and could deteriorate their economic growth. There are some policies that try to mitigate this problem. For instance, introducing a carbon tax, subsidizing the use of green energy and forming coalitions of countries to get cooperation gains. Each of these policies has its advantages and disadvantages. For instance, a possible disadvantage of a carbon tax is that it will only work well, if it is implemented over the whole world. Next to this comes the difficulty to price this tax for legally emitting CO_2. Another policy is to introduce tradable permits that give companies the right to emit a certain amount of CO_2 per year. Again, difficult questions arise about, for instance, the distribution of these permits over the world. With rapid advances in computing power over the last decade, large-scale models have become essential to decision-making in public policy. However there are also risks in using these models. A central issue in the economics of climate change is understanding and dealing with the vast array of uncertainties. These range from those regarding economic and population growth, emission intensities and new technologies, the carbon cycle, and climate response, to the costs and benefits of different policy objectives. Most of the time policy makers must make decisions based upon the outcome of a model that assumes a lot of (possibly) uncertain parameters. Typically, some sensitivity analyses on particular parameters are performed to give the policymaker an indication of the uncertainty involved. However, this may not give a good representation of the uncertainties involved in the model. What we often want is to give a measure of uncertainty and to provide information about a possible probability distribution of the outcome(s) of the model. This is hardly possible to realize. A more down-to-earth approach is performing an elaborate uncertainty analysis consisting of (see, e.g., [4]): (i) stochastic parameters, where parameters are assumed to be random variables following specific probability distributions; (ii) stochastic relations, where relations are assumed to contain a stochastic element; (iii) deterministic, worst-case scenario, where a new variable is added to the system which can be viewed as *nature* that is always counteracting the objective(s) of players; (iv) scenario analyses, where scenarios consisting of combinations of different assumptions about possible states of the world are considered. Scenario analyses involve performing model runs for different combinations of assumptions and comparing the results; (v) extending the model: this means that some parts of the model are reconsidered and extended where necessary.

Already several studies exist that try to incorporate uncertainty into energy system models, e.g., Pizer [5] presents a framework for determining optimal climate change policies under uncertainty. The authors use econometric estimates for some parameters, which are then used to solve the model. They compare the results with those derived from an analysis with best-guess parameter values. Their aim is to show that incorporating uncertainty within a climate model can significantly change the optimal policy recommendations. In particular, they suggest that analyses which ignore uncertainty can lead to inefficient policy recommendations. Gillingham et al. [6] investigate model and parametric uncertainties for population, total factor productivity, and climate sensitivity. Estimates for probability density functions of key output variables are derived, including CO_2 concentrations, temperature, damages, and the social cost of carbon (SCC). The authors investigate uncertainty in outcomes for climate change using multiple integrated assessment models (IAMs). An IAM is used to assess policy options for climate change by combining the scientific and economic aspects of climate change. Details can be found in [7]. This multi-model intercomparison approach is also considered in [8,9]. Furthermore, Fragtos et al. [10] develop a stochastic model of the world energy system that is designed to produce joint empirical distributions of future outcomes. A representation of all important variables is derived using causal chains, with time series analysis for providing patterns of variation over time. Tol [11] investigates the question whether uncertainty about climate change is too large for running an expected cost benefit analysis. The approach is to test whether the uncertainties about climate change are infinite. This is done by calculating the expectation and variance of the marginal costs of CO_2 emissions. In short, the author concludes here that climate change is an area that tests decision analytic tools to the extreme. In this paper we differ from the above-mentioned papers by several aspects. All models above are trying to quantify uncertainty within an IAM that does not incorporate interrelations between players. The models are developed to optimize a policy for a country, without incorporating the interrelations between countries. However, for instance, the use of fossil energy by one country (and therefore the total CO_2 emission of that country) is an externality to other countries. As there is no supranational agency that controls these emissions, we consider a dynamic game framework where countries either cooperate, or do not cooperate, in their decisions on CO_2 emissions. One of the main reasons for choosing a dynamic framework is the important property of CO_2 that once it is in the atmosphere, part of it stays there for a long period of time (estimates range from 30 to 95 years for 50% of the CO_2). In this way, we can incorporate both the impact of long- and short-term strategies. As such, this paper belongs to the literature that uses the framework of differential games to formulate and analyze intertemporal many decision-maker problems in the economics and management of pollution (see, e.g., [12,13] for surveys on this literature and [14] for the economic impact of many issues related to and resulting from global warming).

One of the first papers that treat global warming as a multi-agent problem is [15]. In that paper, the authors develop a discrete-time, dynamic, multi-agent, general-equilibrium model (RICE) incorporating climate and economy. They compare a cooperative and a non-cooperative approach in which all countries choose climate policies to maximize global (respectively own) consumption. The energy transition model we will develop here uses the same basic economic framework. On a more detailed level, both models differ as we focus here on a different problem. The most striking differences between both models are that we distinguish between the use of green energy and fossils, use a continuous-time framework and a more explicit relationship between the impact of CO_2 emissions on production, and we do not model consumption and temperature effects of CO_2 emissions explicitly. Furthermore, our model is closed using a different welfare function. As the focus of this paper is to explore which factors have a major impact on the transition of fossils towards using green energy, our welfare function considers that a certain fraction of output must be realized using energy, the use of green energy might be more costly than using fossils, and that CO_2 emissions are disliked.

One of the first models that address climate negotiations as a game is developed in [16]. The model, called World Induced Technical Change Hybrid (WITCH), captures the economic interrelations between world regions. It is designed to analyze the optimal economic and environment policies in each world region as the outcome of a dynamic game. In this WITCH model, investment decisions of countries are also interrelated. As emphasized before, the goal of our research is to provide a starting point in examining uncertainty in climate models from a dynamic game perspective such as the WITCH model.

Our benchmark model is closely related to a similar model as used in [17] to analyze the impact of pollution over time on the fossil fuel/green energy ratio in a dynamic world characterized by four players that have different interests. Results obtained with that model seem to be quite plausible, and, therefore, the question in that paper was already posed how robust the presented results are with respect to different sorts of model uncertainties. This paper tries to provide some additional information on this issue. For that purpose, we reconsider a somewhat simplified version of that model in a two-player context. One player represents all countries affiliated with the Organization for Economic Cooperation and Development (OECD) and the other player represents countries not affiliated with the OECD, called non-OECD countries. Using a number of the uncertainty approaches mentioned above under (i)–(v), we investigate which factors (parameters, relations, scenarios, etc.) impact equilibria and strategies most. That is, we want to get a broad overview of the uncertainty involved, by applying and evaluating multiple uncertainty approaches as described above. Results of this study can be used to conclude which parts of similar models need special attention when calibrating. The outline of the rest of the paper is as follows. In Section 2, we create our simple dynamic linear two country growth model along the lines of [17] based on the standard Solow growth model introduced in [18]. We integrate the impact of CO_2 emission on economic growth in this model to get a world energy model. Using an extensive model calibration, we arrive at our benchmark model. In Section 3, we perform some experiments with this benchmark model. This to illustrate the basic operation of the model and explain the outcome of the model by investigating the use of the different forms of energy for both players under different scenarios. Next, in Section 4, we perform an extensive uncertainty analysis of this model. The approaches (i), (ii), and (iv) for measuring uncertainty in a model, discussed above, are used to analyze this impact. Section 5 concludes. The appendix contains elaborations on several issues.

2. The Model

In this section, we formulate our benchmark endogenous growth model. The model is based upon the standard Solow exogenous growth model introduced in [18]. The model is obtained along the lines of [17]. Therefore, we do not provide all details here again. Below, we start by introducing the control, state, and output variables of the dynamic model. Then, we discuss the basic model equations that describe the dynamic system, and the welfare function that each player wants to maximize. Then, we adjust the model so that the production function satisfies constant returns to scale. We end up with a nonlinear model, which means that we cannot solve it directly. Instead, we assume that both countries operate within the neighborhood of the equilibrium of this nonlinear model. If a shock occurs to one of the variables, i.e., the model is out of this equilibrium, it is assumed that both players want to return to the equilibrium as soon as possible. Finally, we approximate the dynamics around the equilibrium of the (nonlinear) model by a linear model. This model is then used for our benchmark results about optimal strategies.

In this paper, we consider a two-player setting. With Y_i denoting the output, F_i the production/use of fossil energy, G_i the production/use of green energy, K_i the amount of capital, L_i the total population, T_i the state of technology, E_i the total CO_2 emission, and A_i measuring the total factor productivity, all in country i ($i \in \{1,2\}$); the basic model equations are as follows,

$$
\begin{aligned}
Y_i(t) &= A_i(K_i(t))^{\alpha_i}(L_i(t))^{\beta_i}(E_i(t))^{\gamma_i}(T_i(t))^{\kappa_i}, \quad \alpha_i, \beta_i, \kappa_i \geq 0, && (1)\\
\dot{K}_i(t) &= s_iY_i(t) + s_{ij}Y_j(t) - \delta_iK_i(t) + \tau_iT_i(t), && (2)\\
\dot{T}_i(t) &= g_iT_i(t) + g_{ij}T_j(t) + \epsilon_iK_i(t), && (3)\\
\dot{E}_i(t) &= \zeta_iF_i(t) + \zeta_{ij}F_j(t) - \xi_iE_i(t), && (4)\\
\dot{L}_i(t) &= \eta_iL_i(t), \text{ with } j = 2 \text{ if } i = 1, \text{ and } j = 1 \text{ if } i = 2. && (5)
\end{aligned}
$$

That is, in Equation (1) we assume that production is provided by a Cobb–Douglas function which, in particular, depends on total CO_2 emission levels and the state of technology. Notice that, as CO_2 emissions may have a negative influence on the production, γ_i could be a negative number. The change in capital (2) is endogenous and depends on domestic and foreign production output, depreciation of the current capital, and domestic technology. CO_2 emissions are included here as a separate growth factor to model its effect on economic growth as predicted by the IPCC reports. Technological progress depends on both domestic and foreign technology and the amount of domestic capital (3). The change in CO_2 emission is endogenous too and increases due to domestic and foreign use of fossil fuels and depreciation of the current stock of CO_2 emission (4). We assume that the increase in CO_2 emission due to the domestic use of fossil fuels is proportional to the amount of used fossil fuels. Finally, labor supply is assumed to grow at a constant rate η_i (5).

Furthermore, with $U(t) := \mu_iY_i(t) - (F_i(t) + G_i(t))$ and $E(t) := E_1(t) + E_2(t)$, we assume both countries like to minimize the following objective function,

$$O_i = \int_0^\infty e^{-\theta_i t}\left(U^2(t) + \pi_iE^2(t) + \rho_iG_i^2(t)\right) dt. \tag{6}$$

Here, μ_i is the proportion of output in country i that can only be produced with the use of energy. This means that $\mu_iY_i(t)$ is the required energy at time t. Therefore, $F_i(t) + G_i(t)$, ideally, needs to be equal to $\mu_iY_i(t)$. In mathematical terms, we want $U(t)$ to be as close to zero as possible. Therefore, we minimize $U^2(t)$. In this objective function, the weight of meeting these energy requirements is set equal to 1 to emphasize the need for realizing this objective. Factor ρ_i represents the disadvantages of using green energy for country i. It captures, for instance, the possibly higher price of using green energy in a country. Furthermore, each country has its own availability of resources. It might be difficult to use green energy, because there are no resources in the neighborhood. Note that we multiply this parameter with $E^2(t)$ instead of $E(t)$. This is done in order to make larger deviations from the equilibrium increasingly less preferred than small deviations from the equilibrium. Notice that this interpretation makes it superfluous to introduce a separate penalty for using fossil energy in the objective function like in [17]. Factor π_i expresses that the higher the CO_2 emission, the more it is disliked. For instance, it may be used to express that emitting lots of CO_2 entails costs implied by environmental changes. Note that, again, we square the variable $G(t)$ for similar reasons as for $E(t)$. The values of both ρ_i and π_i imply a priority among the terms in the objective. For the calibration of these two parameters, we refer to Appendix A. For convenience, we rewrite the objective as a maximization problem. Minimizing (6) is the same as maximizing next total discounted welfare:

$$W_i = \int_0^\infty e^{-\theta_i t}\left(-U^2(t) - \pi_iE^2(t) - \rho_iG_i^2(t)\right) dt. \tag{7}$$

Under the assumption that the Cobb–Douglas production functions satisfy constant returns to scale (i.e., $\alpha Y(K, L, E, T) = Y(\alpha K, \alpha L, \alpha E, \alpha T)$, or, in this specific case, the production function parameters satisfy $\alpha_i + \beta_i + \kappa_i + \gamma_i = 1$), above equations can be rescaled in terms of effective labor. Therefore, to achieve constant returns to scale, we define our new set of variables as follows,

$y_i := \log(\frac{Y_i}{L_i})$, $k_i := \log(\frac{K_i}{L_i})$, $t_i := \log(\frac{T_i}{L_i})$, $e_i := \log(\frac{E_i}{L_i})$, $f_i := \log(\frac{F_i}{L_i})$ and $g_i := \log(\frac{G_i}{L_i})$. Then, Equations (1)–(5) can be rewritten as,

$$\begin{aligned} y_i(t) &= \log(A_i) + \kappa_i t_i(t) + \alpha_i k_i(t) + \gamma_i e_i(t) \\ \dot{k}_i(t) &= -(\eta_i + \delta_i) + e^{-k_i(t)}\left(s_i e^{y_i(t)} + s_{ij} e^{y_j(t)+t(\eta_j-\eta_i)} + \tau_i e^{t_i(t)}\right) \\ \dot{t}_i(t) &= -\eta_i + g_i + e^{-t_i(t)}\left(g_{ij} e^{t_j(t)+t(\eta_j-\eta_i)} + \epsilon_i e^{k_i(t)}\right) \\ \dot{e}_i(t) &= -(\xi_i + \eta_i) + e^{-e_i(t)}\left(\zeta_i e^{f_i(t)} + \zeta_{ij} e^{f_j(t)+t(\eta_j-\eta_i)}\right). \end{aligned} \tag{8}$$

We also rewrite the objective (7) in terms of the new variables. First, we rewrite objective (7) in terms of labor:

$$W_i = \int_0^\infty L_i^2(t)e^{-\theta_i t}\left(-\left(\mu_i\frac{Y_i(t)}{L_i(t)} - \left(\frac{F_i(t)}{L_i(t)} + \frac{G_i(t)}{L_i(t)}\right)\right)^2 - \pi_i\left(\frac{E_i(t)}{L_i(t)} + \frac{E_j(t)}{L_j(t)}\frac{L_j(t)}{L_i(t)}\right)^2 - \rho_i\left(\frac{G_i(t)}{L_i(t)}\right)^2\right)dt.$$

$$= L_i^2(0)\int_0^\infty e^{(2\eta_i-\theta_i)t}\left(-\left(\mu_i\frac{Y_i(t)}{L_i(t)} - \left(\frac{F_i(t)}{L_i(t)} + \frac{G_i(t)}{L_i(t)}\right)\right)^2 - \pi_i\left(\frac{E_i(t)}{L_i(t)} + \frac{E_j(t)}{L_j(t)}\frac{L_j(0)}{L_i(0)}\right)^2 - \rho_i\left(\frac{G_i(t)}{L_i(t)}\right)^2\right)dt.$$

Note that for the second equality we use that η_i equals η_j. This assumption is explained in Appendix B. Now, we apply the monotone log transformation to this new objective. This means that we can write the objective in terms of the new variables as follows, i.e., maximizing (7) is the same as maximizing

$$w_i = \int_0^\infty e^{(2\eta_i-\theta_i)t}\left(-u^2(t) - \pi_i\left(e_i(t) + \sum_{j\neq i} p_i e_j(t)\right)^2 - \rho_i g_i^2(t)\right)dt, \tag{9}$$

where $u(t) = \mu_i y_i(t) - (f_i(t) + g_i(t))$. Furthermore, $p = [\Psi_o \ \Psi_{no}]$, where Ψ_i is the total number of people in country j divided by the total number of people in country i. Next, we calibrate our parameters in the above model (8) and (9). We choose to concentrate on the OECD countries and the non-OECD countries as our two parties involved. Note that we want to define a simple case of two (interrelated) parties for which information is widely available (to be able to calibrate the parameters). It is highly likely that within one of these groups there is no common interest. It might be necessary to include more players that do have common interests. This is beyond the scope of this research. This research can be seen a starting point in examining uncertainty in climate models from a dynamic game perspective. Therefore, we choose two parties for which information is widely available. There are two databases where most of the parameters are calibrated from. http://data.oecd.org from the OECD and http://data.worldbank.org from the World Bank. For the OECD countries, finding appropriate data is not a problem. For the non-OECD members this is, in particular for small countries, not always the case. As these small non-OECD countries are very small in all aspects concerning the variables involved (compared to more developed (higher-income) non-OECD countries), we exclude them from our analysis. Therefore, for calibration purposes, we only use information from the higher-income non-OECD countries. A detailed account for the calibrations of key parameters, initial variables and policy parameters can be found in Appendix A. Tables 1–3, below, report the results for the OECD (first row in each table) and non-OECD (second row in each table) countries. We use the acronym O (n-O) for the OECD (non-OECD) countries.

Table 1. Non-spillover parameters.

	A	α	β	γ	κ	η	δ	τ	ϵ	ξ
O	2042.5	0.23	0.76	−0.021	0.027	0.0073	0.062	0.018	0.114	0.023
n-O	499.1	0.35	0.69	−0.050	0.011	0.0073	0.075	0.031	0.031	0.023

Table 2. Initial variables calibration.

	y	k	t	e	f	g
O	10.55	12.30	5.45	2.29	8.13	6.67
n-O	9.85	10.58	5.18	2.46	8.38	5.89

Table 3. Policy parameter calibration.

	θ	μ	π	ρ
O	0.04	1.40	0.45/250	0.55/250
n-O	0.06	1.45	0.41/250	0.59/250

Clearly if, e.g., the OECD countries like to determine their optimal use of fossil and green energy over time by maximizing their welfare (7) subject to the dynamic constraints (1)–(5), these energy levels depend on the corresponding levels chosen by the non-OECD countries. The same observation applies of course for the non-OECD countries optimal energy levels. Therefore, additional assumptions are needed before we can conclude which energy levels are chosen by both sets of countries over time. A common assumption made within this context is that both sets of countries use such strategies that neither of them has an incentive to deviate from their strategy. That is, they use (open-loop) Nash strategies. Assuming that both OECD and non-OECD countries use such strategies to maximize their welfare, we derive in Appendix B the resulting strategies and, moreover, calculate the resulting steady state values of the variables. As one can see from the equations tabulated at the end of this Appendix B, even under these simplifying assumptions, the calculation of these steady state values is not a trivial task. It requires the solution of a set of 18 highly nonlinear equations. With some abuse of notation we will call these steady state values, that are obtained assuming players use Nash strategies, the equilibrium of the model. We use the notation s^e to indicate the steady state value of a variable s. Equilibrium values, using our benchmark parameters, are tabulated (again row-wise for both countries) in Table 4.

Table 4. Equilibrium variables.

	y^e	k^e	t^e	e^e	f^e	g^e
O	14.73	28.84	30.40	16.35	12.28	8.33
n-O	16.57	30.92	31.95	16.32	12.26	11.74

We want to briefly discuss two features of this equilibrium. Note that both countries use more fossil energy than green energy in the equilibrium. There are two model features that play a role in this phenomenon. First, we do not include changing parameters over time. For instance, it might be the case that it becomes easier to access green energy over time. However, this effect is beyond the scope of this research. Second, the initial calibration of the ratio between π and ρ (for details, see Appendix A) may play a role. If we adjust this ratio, for instance, by multiplying π^{initial} with a factor (keeping ρ the same), we end up with different results. As an example, we plot in Figure 2 the equilibrium share of green energy for different factors. The initial calibration for π is denoted by π^{initial}.

Note that a higher π means that both countries dislike emitting CO_2 more. This effect is also observed in Figure 2, where we observe that the total share of green energy increases when π increases. Furthermore, Figure 2 clearly illustrates the sensitivity of the model equilibrium outcomes with respect to the choice of these preference parameters, π versus ρ in the utility function. The non-smooth behavior is probably due to numerical issues, in the sense that the calculation of the full equilibria did not occur yet for certain values of α. Note that this figure only shows the possible variation in equilibrium outcomes due to changes in the ratio between π and ρ. In Section 4, we limit the investigation to the more realistic options of this ratio. In Section 4, we also show that uncertainty in these policy parameter choices is the major cause for variability in equilibrium outcomes of the model. On the other hand, we will see that its impact on implied optimal out-of-equilibrium strategies is not that large and more in line with the impact other sources of uncertainty have.

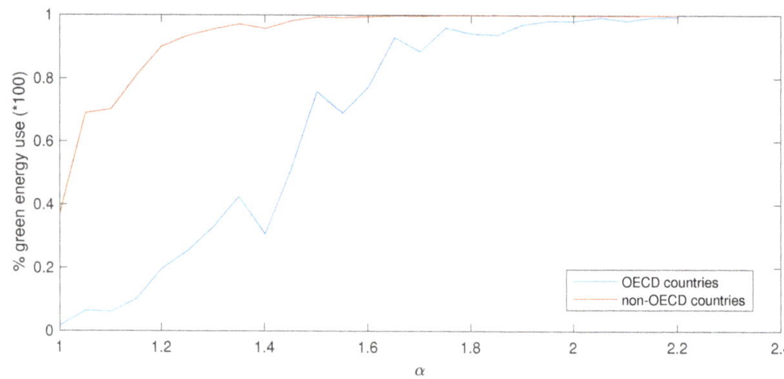

Figure 2. Percentage green energy with $\pi = \alpha \cdot \pi^{\text{initial}}$ with $\alpha \in [1, 2.5]$.

Second, we observe that the non-OECD countries use much more green energy in the equilibrium than the non-OECD countries. One of the causes of this phenomenon is the difference in the number of working people for both countries. For the number of working people in OECD (non-OECD) we use the number 837,816,057 (227,833,932), as discussed in Appendix A. This results in $p = [0.27, 3.68]$ (see objective (9)). In other words, the weight in the objective of the non-OECD countries on the CO_2 emission per capita of OECD countries is 13 times as high as the weight in the objective of the OECD countries on the CO_2 emission per capita of non-OECD countries. Therefore, the CO_2 emission of the OECD countries already negatively affects the objective of the non-OECD countries. To minimize the impact of the total CO_2 emission on their objective, the non-OECD countries may decide to increase their share of green energy. Therefore, in our model, the difference in number of working people per country is one of the causes of the non-OECD countries using more green energy than the OECD countries. Another cause for the discrepancy between the green energy use of both countries is the fact that total CO_2 emission of both countries is equally disliked for both countries. In mathematical terms, the total CO_2 emission in the objective of country i is $e_i + p_i e_j$, where j represents the country not equal to i. However, one may argue that a country should not care that much about the total CO_2 emission of another country. One reason might be that an other country cannot influence this CO_2 emission directly. This can be quantified by replacing the $e_i + p_i e_j$ with $e_i + M p_i e_j$, where $M \in [0, 1]$ represents the proportion of foreign CO_2 emission that is disliked by an other country. In Figure 3, we show for all $M \in [0, 1]$ the total share of green energy for both countries. Note that $M = 1$ is the original case.

We observe that using $M \leq 50\%$, results in approximately the same green energy use for both countries. A higher M results in equilibria in which the non-OECD countries use a greater percentage of green energy than the OECD countries. Note that for the rest of this paper we keep using $M = 1$.

To see how the developed model performs if, e.g., shocks occur in the emission level of carbon dioxide, we assume that both countries operate within the neighborhood of the steady-state values mentioned above. We can approximate the dynamics around the equilibrium of the nonlinear model by the next linear model (see Appendix C):

$$\begin{aligned} y_{li}(t) &= \kappa_i t_{li}(t) + \alpha_i k_{li}(t) + \gamma_i e_{li}(t), \\ \dot{k}_{li}(t) &= \tilde{s}_i(y_{li}(t) - k_{li}(t)) + \tilde{s}_{ij}(y_{lj}(t) - k_{li}(t)) + \tilde{\tau}_i(t_{li}(t) - k_{li}(t)), \\ \dot{t}_{li}(t) &= \tilde{g}_{ij}(t_{lj}(t) - t_{li}(t)) + \tilde{\epsilon}_i(k_{li}(t) - t_{li}(t)), \\ \dot{e}_{li}(t) &= \tilde{\zeta}_i(f_{li}(t) - e_{li}(t)) + \tilde{\zeta}_{ij}(f_{lj}(t) - e_{li}(t)). \end{aligned} \tag{10}$$

The corresponding parameters are provided, row-wise again, for both countries in Tables 5–7.

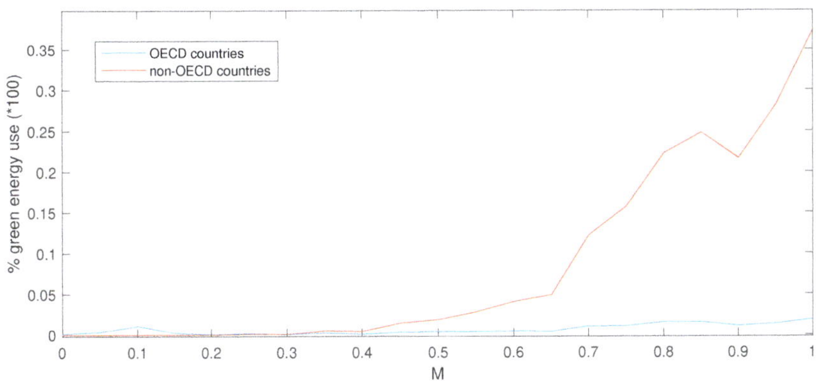

Figure 3. Percentage green energy with $e_i + Me_j$ with $M \in [0,1]$.

Table 5. Parameter calibration for non-spillover parameters, linearized model.

	α	γ	κ	$\tilde{\tau}$	$\tilde{\epsilon}$	ζ
O	0.23	−0.021	0.027	0.0854	0.0240	0.023
n-O	0.35	−0.050	0.011	0.0869	0.0111	0.023

Table 6. Parameter calibration for spillover parameter: $\tilde{s}$ and $\tilde{g}$, linearized model.

$\tilde{s}$ $(*10^{-6})$	O	n-O	$\tilde{g}$	O	n-O
O	0.1612	0.0367		0.0170	0.0236
n-O	0.0054	0.1499		0.0375	−0.0050

Table 7. Parameter calibration for spillover parameter: $\tilde{\zeta}$, linearized model.

$\tilde{\zeta}$	O	n-O
O	0.0009	0.0297
n-O	0.0009	0.0295

In particular, notice that output gap dynamics in non-OECD countries are more than twice as vulnerable for CO_2 emissions as OECD countries. Therefore, a priori one may expect that the impact of a CO_2 emission shock will have much more consequences in terms of policies in non-OECD countries than in OECD countries. This will be clearly illustrated in the simulation study performed in the next section too.

The corresponding objective function for both players can then be approximated by carrying out a second-order Taylor expansion of the welfare functions w_i (9). This results in a quadratic cost criterion (see Appendix D for details):

$$\tilde{J}_i := \frac{1}{2}\int_0^{\infty} [x^T(t)\ u^T(t)] H_i^{''} \begin{bmatrix} x(t) \\ u(t) \end{bmatrix} dt, \quad i = 1,2, \tag{11}$$

where $x^T = [k_{l1}\ k_{l2}\ t_{l1}\ t_{l2}\ e_{l1}\ e_{l2}]$ is the state variable of our model (10); $u^T = [f_{l1}\ g_{l1}\ f_{l2}\ g_{l2}]$ the corresponding control variable and matrix $H_i^{''}$ is as reported in Appendix D.

Thus, in conclusion, the almost optimal response of both OECD and non-OECD countries when the model in equilibrium is disturbed can be determined by solving above linear quadratic differential game (10) and (11). Again, to determine this response, assumptions have to be made on whether both sets of countries will cooperate or not to fight the disturbance. We consider both options and discuss them in some more detail in the next section.

3. Benchmark Model Simulations

In this section we illustrate, by considering a couple of scenarios, how models (8) and (9) will approximately respond if it is out of equilibrium. We visualize the responses by strategy curves. These curves visualize how the model responds to a shock by showing the *percentage* increase (or decrease) of all variables. Therefore, the vertical axes represent percentages. To that end, we perform two different kind of shocks to the equilibrium: symmetric shocks and asymmetric shocks. Symmetric shocks are shocks that hit both countries at the same time. Asymmetric shocks are shocks that occur to just one of both countries. Furthermore, we distinguish between two forms of cooperation. We have a cooperative situation and a non-cooperative situation. In the cooperative situation, we discuss a regime where both countries form a coalition. In the non-cooperative situation, we discuss the regime where both countries play actions in the Nash sense. Within the context of this paper, we only analyze emission shocks.

To perform the simulations, we use the algorithm developed in [19] to solve N-player affine linear-quadratic open-loop differential games. Clearly, the use of open-loop strategies is made to simplify the analysis. A discussion of pros and cons using this setting can be found in, e.g., [17]. In particular, we recall some observations from literature suggesting that the difference between open-loop and feedback policies in practice might not be that large (see, e.g., [20,21]).

3.1. *Asymmetric Emission Shock*

We start with an asymmetric positive CO_2 emission shock, which hits the non-OECD countries. Such a shock impacts the model outcomes in two ways. First, it has a direct effect on the welfare functions of both countries via an increase of total CO_2 emission levels. Notice that this negative impact can only be mitigated by reducing the total amount of fossil energy that is used (cf. (4)). The second effect is that it directly reduces output in the non-OECD countries. Therefore, instantaneously, less energy is required in order to meet the production requirements in non-OECD countries. However, to return to its equilibrium value, in the long end, more energy is required again. Together with the discounting effect, which makes that future cost are less important than current cost, this makes it intricate to predict the reaction of both countries in terms of their use of energy in general.

Figure 4 shows the total response over time of both countries in a non-cooperative setting in terms of energy consumption, f and g, for our parameter setting. The first point that stands out is the scaling of the reactions in both countries. As the shock only hits the non-OECD countries, we expect less pronounced reactions in the OECD countries. This is clearly demonstrated in both plots. On a more detailed level, we see that both countries immediately start to use more green energy and less fossil energy. This can be explained as follows. For non-OECD countries, initial production is affected by the CO_2 shock. As already indicated above, the only way to mitigate this impact is by reducing the amount of fossils used, and, to meet the energy requirements for production, increase the amount of green energy accordingly. For OECD countries, the only direct impact of the shock comes via the increased total emission level. To compensate this increase they start to reduce their domestic fossil energy use, and therefore increase their green energy use.

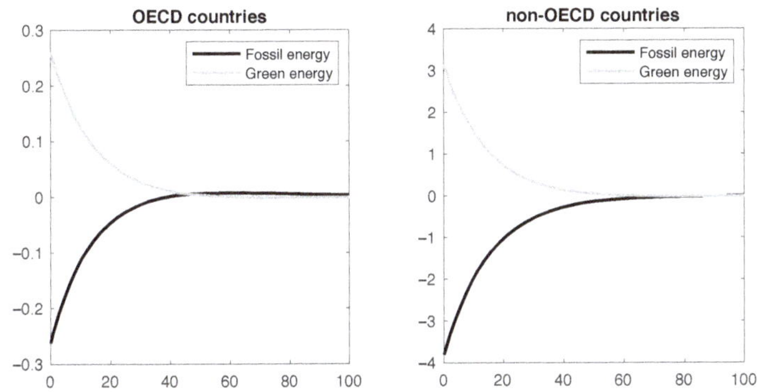

Figure 4. Control variables.

After some time, we observe even that OECD countries use amounts of fossils slightly above its equilibrium value. This behavior is explained by the choice of the discount factor. If we decrease the discount factor of the OECD countries to, for instance, 0.01, then the use of fossil energy by these countries remains below its equilibrium value anytime. As in that case, the future realization of CO_2 levels are also very relevant and, therefore, the instantaneous advantages of using fossils compared to green energy evaporate. Therefore, in short, the slight percentage increase in fossil energy use shown in Figure 4 is caused by a relative strong preference for short-term objective gains.

Figure 5 shows the corresponding evolution of capital, technology, and stock of emissions for both countries. Note that the line corresponding to capital coincides with the line for technology for both the OECD and non-OECD countries. We see that for both the non-OECD and OECD countries capital and technology are almost not affected by the emission shock. Moreover, as explained above, the lagged behavior of emission levels in the OECD countries is due to the increased use of green energy by the non-OECD countries, which gradually normalizes over time again.

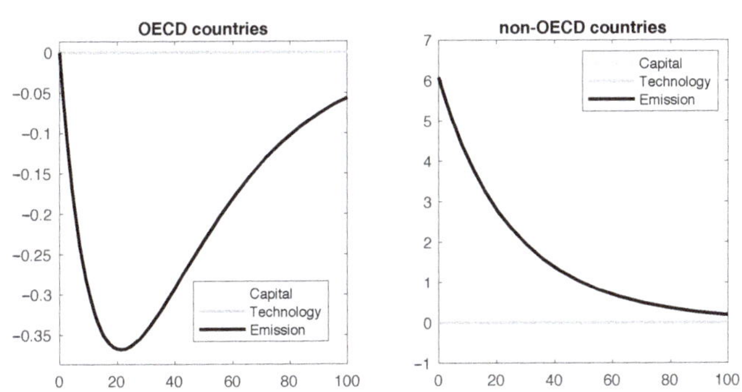

Figure 5. State variables.

Finally, Figure 6 displays the corresponding output drop occurring over time for both OECD and non-OECD countries.

Next we consider the case how both countries respond if they decide to fight the shock collectively. This is modeled by assuming that control instruments by both countries are determined such that the weighted sum of both welfare functions is collectively minimized. We assume weights to be equal, i.e., $\frac{1}{2}$. The main difference in the simulation results (strategies) is that the OECD countries increase their green energy use compared to the non-cooperative setting. This is due to the fact that the amount of CO_2 emission produced by OECD countries greatly affects the objective of non-OECD countries.

One reason for this is that the OECD countries have four times as much (working) population as the non-OECD countries (see Appendix A). Now that the OECD countries also care about this objective, they can directly reduce this effect by increasing their own green energy use even more (and therefore reducing the total CO_2 emission). Finally, Table 8 reports the total losses in the cooperative setting. Here we use the acronym NC (C) for the non-cooperative (cooperative) setting.

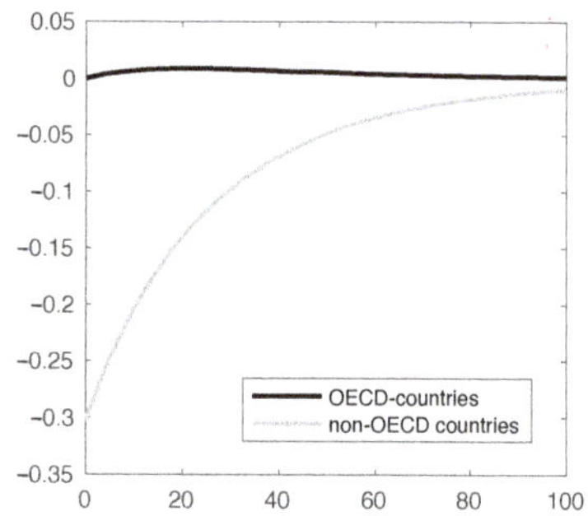

Figure 6. Output variables.

Table 8. Losses under asymmetric shock.

	NC	C	Loss Reduction (%)
O	0.0010	0.0024	−140.0
n-O	0.0139	0.0102	26.6

First, we see that the total loss in the cooperative setting is lower than the total loss in the non-cooperative setting. Second, we observe that cooperation for non-OECD countries would be profitable, where OECD countries would not profit from it. The higher costs of non-OECD countries in the non-cooperative setting are now shared costs between both countries. This means that a cooperative setting is likely to be only realistic when OECD countries are in any other way compensated for this cost-sharing.

3.2. Symmetric Emission Shock

Next, we consider a symmetric emission shock, meaning the OECD countries are now also hit by an emission shock. We suppose that the shock that hits the OECD countries is relatively as large as the shock that hits the non-OECD countries. To accomplish this, we base the shocks upon the calculated equilibrium emission value. Therefore, to have a relative shock for OECD countries of 1 an absolute shock of $(1/16.32) \cdot 16.35 = 1.0018$ is used. First, we consider a non-cooperative setting again.

Figure 7 illustrates the response of both countries in terms of energy consumption. Again, first notice the difference in scale of both responses. Having a closer look at both graphs, we see that both countries react in a similar way now. They both increase their fossil energy use in favor of their green energy use. The reaction by the OECD countries is in line with their reaction we observed in the asymmetric case. As their production is now directly hit by the CO_2 shock, they start to use more green energy in order to close the production gap.

In Figure 8, we again show the results for capital, technology and stock of emissions. Also these graphs are similar to the asymmetric shock case. The only difference is that the OECD countries experience now from the outset on an increase in the stock of emissions. Note that this is due to them being also hit by an emission shock.

Concerning output, we see from Figure 9 that the emission shock hits the non-OECD countries most; as output in the OECD countries is less vulnerable for CO_2 emissions (cf. Table 1, γ_i, and (1)), therefore less changes in energy consumption are required to close the production gap.

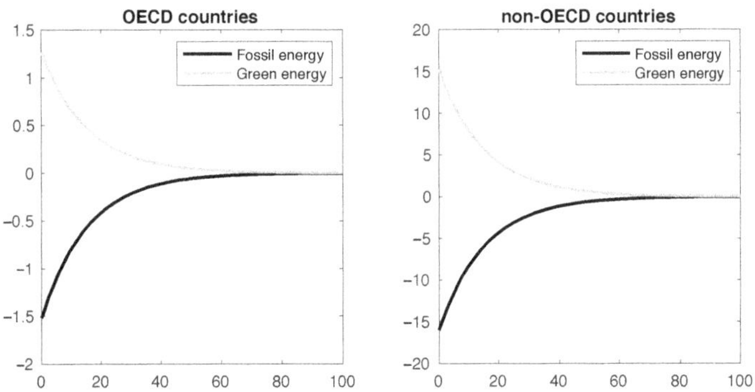

Figure 7. Control variables.

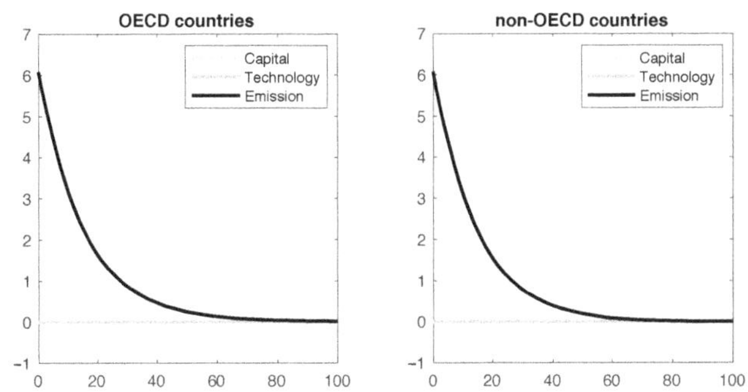

Figure 8. State variables.

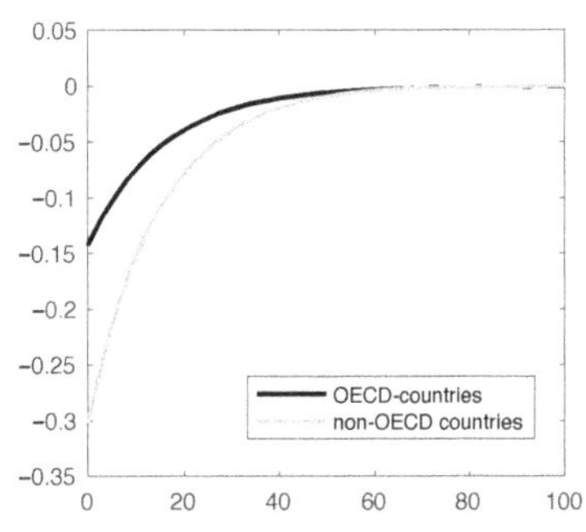

Figure 9. Output variables.

In a cooperative setting, we observe the same changes in response strategies as in the asymmetric shock case. The OECD countries increase their green energy use to accomplish a total CO_2 emission reduction. Table 9 reports losses for both countries and the corresponding cost reduction under both cooperation regimes.

Table 9. Losses under symmetric shock.

	NC	C	Loss Reduction (%)
O	0.0228	0.0593	−160.1
n-O	0.3331	0.2386	39.6

Again, we observe that the total loss in the cooperative setting is lower than in the non-cooperative setting. Second, we observe that the losses are higher for both countries compared to the losses in the asymmetric shock scenario. This makes sense, as both countries must deal with an additional shock now. We also see, similar as in the asymmetric scenario, that the non-OECD countries are the only one who profit from cooperation. Finally, we observe that the relative changes between the non-cooperative and cooperative setting are slightly larger than in the the asymmetric shock occurred. Note that this is likely also caused by the higher total loss when both countries are hit by an emission shock.

4. Uncertainty Analysis

Clearly in arriving at our linear adjustment model (10) and (11) several approximations are made, and the question is how sensitive results obtained in this linear model are to inaccuracies in the original model specification (8) and (9). In Sections 4.1–4.5 below, we consider some potential inaccuracies and analyze how they impact the results presented in the previous section. To that end we distinguish two kinds of impact. The impact on the equilibrium values and the impact on the optimal out of equilibrium strategies. In Section 4.1, we address the consequences of the assumption that our production functions satisfy constant returns to scale. Section 4.2 considers the effects of having stochastic parameters. In this section we investigate parameters occurring in the dynamics of the model. Section 4.3, on the other hand, looks at the parameters occurring in the objective of the model. Then, in Section 4.4, we assume that the realization of capital contains a stochastic term. Finally, in Section 4.5, we consider a scenario where both the initial use of green energy and the parameter ρ, which represents the disadvantage of using green energy, are correlated.

4.1. The Production Function

In this section, we reconsider the assumption that the production functions (1) satisfy constant returns to scale. After calibration it turned out that the sum of the involved parameters, $\alpha_i + \beta_i + \gamma_i + \kappa_i$, was equal to 0.905 for OECD countries and 1.03 for non-OECD countries. The corresponding tabulated numbers in Table 1 were obtained by normalizing these parameters for both countries. In this section, we consider how equilibrium values of (8) and (9) change if we fix all but one of these parameters to their calibrated value, and estimate the remaining parameter as the difference between one and the sum of the calibrated parameters, e.g., if we calibrate $\alpha_i = \bar{\alpha}_i$, $\beta_i = \bar{\beta}_i$, $\gamma_i = \bar{\gamma}_i$, we fix κ_i at $1 - \bar{\alpha}_i - \bar{\beta}_i - \bar{\gamma}_i$. We calculate for all four possible combinations corresponding equilibrium values of (8) and (9). Table 10 reports the average of all equilibrium variables for all these four possibilities.

Table 10. Weighted equilibrium variables.

	y	k	t	e	f	g
O	15.99	29.83	31.39	17.06	13.12	9.24
n-O	16.60	31.91	32.94	17.03	12.79	11.26

Computing the average absolute difference for both countries from the original equilibrium results in respectively a 6.2% (OECD) and 1.8% (non-OECD) difference. The largest percentual deviation is for the fossil energy use of the OECD countries. This difference is almost 11%.

4.2. Stochastic Parameters

Next, we consider the case that two of the key parameters in the model—ξ and θ—are only approximately known. Note that ξ represents the natural depreciation rate of CO_2 emissions, and θ represents the discount factor for future losses.

First, we look at ξ. We initially assumed, based on [22], a CO_2 lifetime of 30 years for 50% of the CO_2 emission today. The IPCC, on the other hand, estimates a CO_2 lifetime of 50 years for 50% of the CO_2 emission today. This results in a 20 year difference between the two studies. We estimate a distribution of the lifetime of CO_2 emission as shown in Figure 10.

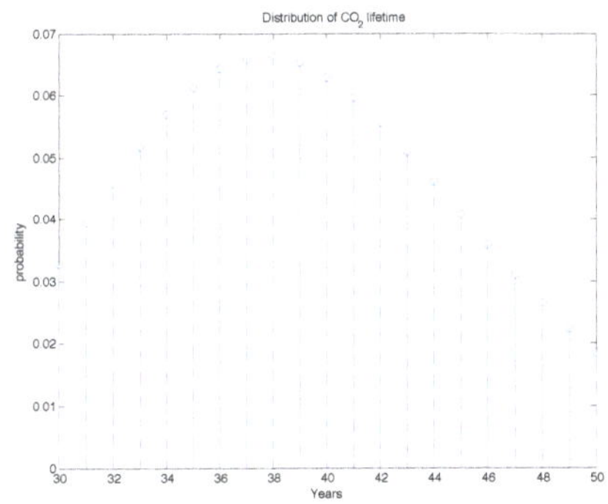

Figure 10. Estimated probability distribution ξ.

After performing many simulations, apparently this assumption does not have a large impact on the resulting equilibrium values and on the value of the objective function. Complementary to this approach we also compute the equilibrium values for the complete, specified range of CO_2 lifetimes. It turns out that the CO_2 lifetime is not affecting the equilibrium values much. Figure 11, shows the corresponding plot of the equilibrium values for OECD and non-OECD countries.

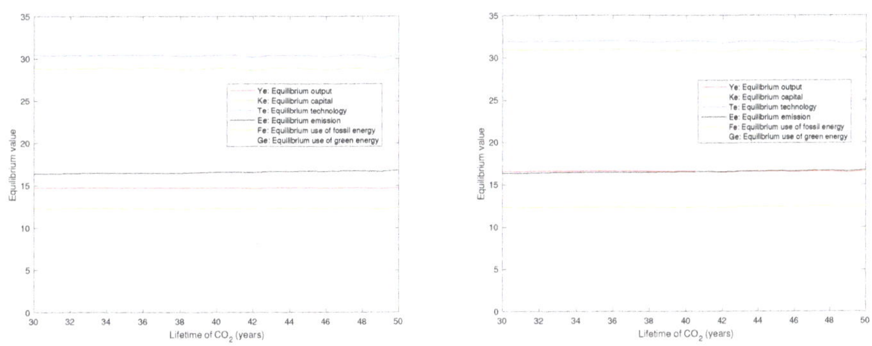

(**a**) OECD equilibria, CO_2 lifetimes (**b**) non-OECD equilibria, CO_2 lifetimes

Figure 11. Equilibrium values, CO_2 lifetimes.

As we can see in Figure 11, the equilibrium values are rather constant for the specified range of CO_2 lifetimes. By comparing equilibrium outcomes just for extreme choices of this parameter, we get the percentage changes tabulated in Table 11.

We see that the average percentage difference is less than 2% for both countries.

Finally, we also determine the effect on the optimal strategies that a change in this parameter has. As an example, Figure 12 shows the effect on the optimal control variables if the lifetime of CO_2 is assumed to be 50 years (New) instead of 30 years (Old) in the non-cooperative setting when an asymmetric shock only hits the non-OECD countries. We see that both trajectories do not change much.

Second, we investigate the impact of uncertainty with respect to the discount rate, θ. Note that we set the discount rate equal to 4% for the OECD countries and 6% for non-OECD countries. According to data from the Impact Data Source [23], there is a variability with a spread of 0.5 in these numbers.

Therefore, we consider the case that the discount rate for OECD (non-OECD) countries is normally distributed with a mean of 4% (6%), and a variance of 0.2%. Similar to the previous case we also calculate the equilibrium values on a grid where θ_1 ranges between 3.5 and 4.5 and θ_2 ranges between 5.5 and 6.5. Figure 13 shows the corresponding equilibrium values for the green energy consumption. Results of the other variables are visualized in Figures A2 and A3 in Appendix E.

Table 11. Percentage difference from original equilibrium.

y	k	t	e	f	g	**Average**
0.61%	1.13%	1.07%	2.72%	0.54%	0.85%	1.15%
0.79%	1.06%	1.02%	2.46%	1.79%	2.50%	1.60%

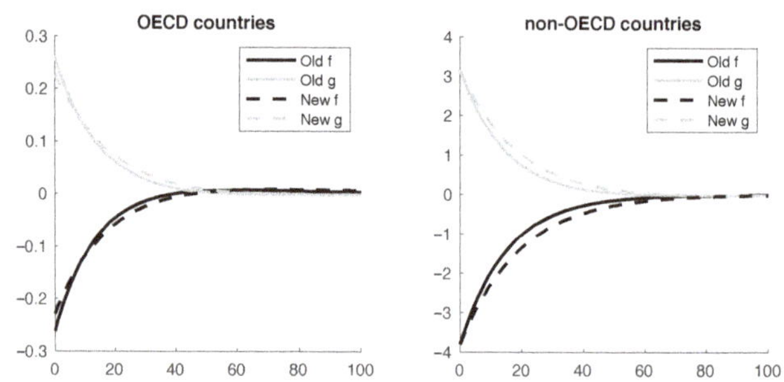

Figure 12. Control variables.

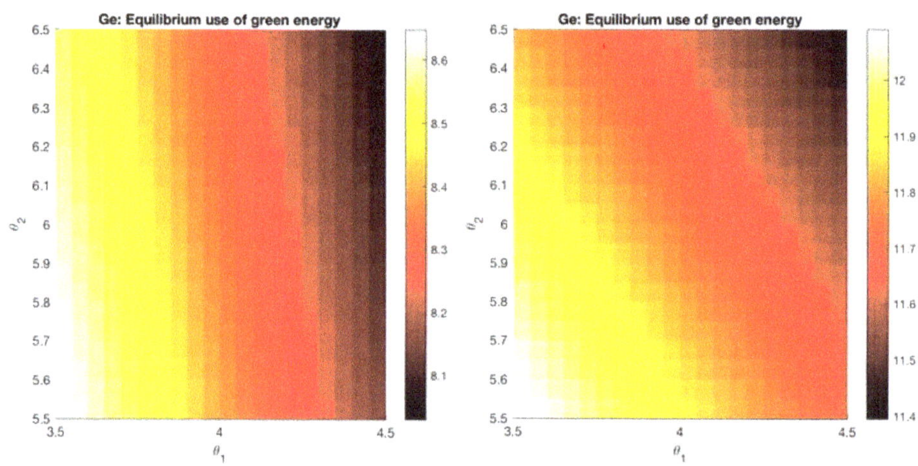

(**a**) OECD countries (**b**) non-OECD countries

Figure 13. Equilibrium values, green energy use.

In Figure 13, we observe that both countries have a higher equilibrium value of green energy use when both discount rates get smaller. Smaller discount rates mean that the short-term goals are becoming less important compared to the long-term goals (objective values). Second, we observe that the green energy use of the non-OECD countries depends more on the discount rate of the OECD countries than vice versa. One of the main reasons for this phenomenon is the parameter γ, which implicitly determines how much the output is affected by CO_2 emission. We calibrated that the non-OECD countries experience a larger (negative) impact on output when the total CO_2 emission increases in that country (see Appendix A). When the discount factor used in the OECD countries is very high, the OECD countries use the smallest amount of green energy (and therefore more fossil energy). This means that the total CO_2 emission in the atmosphere increases. If the non-OECD countries would also increase their fossil energy use, the total CO_2 emission in the atmosphere increases even more. This would have a large impact on their output. Therefore, the amount of green energy use for the non-OECD countries depends more on the discount rate of the OECD countries than vice versa (to reduce the total CO_2 emission in the atmosphere).

Similar simulations as for the ξ variable show that changes in the discount factor do not influence the optimal strategies of both players much. Finally, Table 12 below shows the maximal absolute percentage difference between all computed equilibria.

Table 12. Percentage difference from original equilibrium.

y	k	t	e	f	g	Average
1.01%	2.15%	2.04%	3.40%	5.29%	7.79%	3.61%
1.18%	2.01%	1.94%	3.43%	4.01%	4.31%	2.81%

We observe that the largest percentage difference occurs in the variables for green and fossil energy. As discussed, an alternative equilibrium may be reached which results in different values for these variables. From this table, we see that the average percentage difference in both countries is below 4%.

4.3. Stochastic Policy Parameters

A parameter which turns out to affect results more than those considered in the previous subsection is the parameter π_i in the objective function. This parameter measures the social cost of carbon emissions (SCC). That is the (monetized) damages associated with excess carbon emissions in a given year (relative to its equilibrium value). We use a distribution for this parameter π_i, based upon the results described in [24]. In this paper, 28 studies are listed with all their own estimates of the social cost of carbon. We exclude some of these studies because they present very extreme results and, according to this paper, the used methods to arrive at these results are questionable. Figure 14 shows how often each different value for the social cost of carbon is used in the set of studies. Note that we divide the values in sectors with a $15 range. The SCC is represented per metric ton of carbon dioxide emission (/tC).

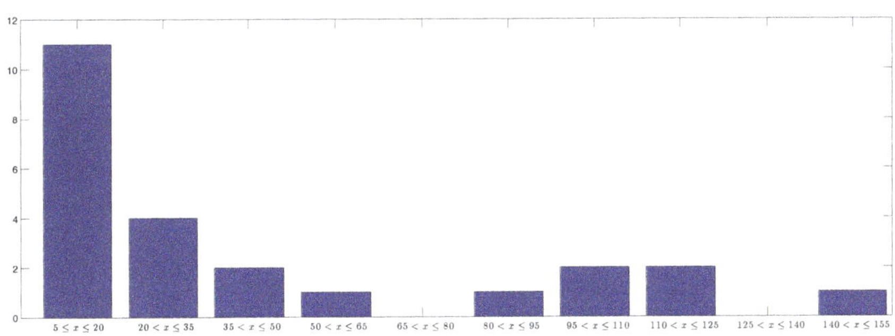

Figure 14. Distribution of the social cost of carbon (x), according to the studies discussed ($).

Our initial estimates for π, 0.45/250 for OECD countries, and 0.41/250 for non-OECD countries, are chosen in line with the most occurring range of SCC values, i.e., 5–20 $/tC. Figure 14 shows that all remaining studies estimate this cost to be higher. This means that it could be that we are underestimating this cost by our choice for π_i. We sample 100 values from the distribution implied by the histogram in Figure 14. For each trellis to the right we increase π_i by a certain amount. This amount is based upon [25]. In this report, the economic costs of premature deaths from Ambient Particulate Matter Pollution (APMP) and Household Air Pollution (HAP) are tabulated as a percentage of GDP for a list of countries. Classifying the list in OECD and non-OECD countries results in the statistics shown in Table 13.

Table 13. Economic cost of the specified pollution as a% of GDP.

	Minimum	Average	Maximum
O	0.3%	4.9%	19%
n-O	3.3%	17.0%	35.2%

We assume that our initially estimated π_i belongs to the average case in this table. For the OECD countries, we see that the economic cost of the specified pollution can be 4 times higher than average. This means that we allow π_{oecd} to become 4 times bigger. This extreme case will correspond to the last trellis of Figure 14. The same procedure is applied for the non-OECD countries. In this case $\pi_{\text{non-oecd}}$ can become approximately 2 times larger than initially calibrated.

After some extensive calculations, we obtain the histograms of equilibrium values plotted in Figure A4. From these histograms, we see that changing this parameter may have a large impact on the equilibrium values of e, f and g. To investigate the impact of π_i in more detail, Figures A5 and A6 show the equilibrium values for both countries if these parameters are chosen from a grid, where π_{oecd} ranges from $0.45/250$ to $4 \cdot 0.45/250$, and $\pi_{\text{non-oecd}}$ from $0.41/250$ to $2 \cdot 0.41/250$. The graphs also show some outliers for the variables y, k and t. Some further simulations show that these are due to slow convergence of parameters. In Table 14, we state the maximal percentage differences from the original equilibrium for all possible combinations of π.

Table 14. Percentage difference from original equilibrium.

y	k	t	e	f	g	Average
3.92%	6.50%	6.16%	41.94%	55.15%	110.73%	37.40%
2.98%	5.84%	5.64%	41.71%	52.53%	106.58%	35.88%

We see that most differences occur in the variables e, f, and g. Figures A7 and A8 visualize this impact from a different perspective. Here, we show equilibrium values for the e, f, and g variables for both countries, where we focus on the possible impact of the value of π in the other country. The first thing that draws the attention is that the figures corresponding to the non-OECD countries typically have a larger black band than the figures corresponding to the OECD countries. This means that for a given value of $\pi_{\text{non-oecd}}$, the variables of the non-OECD countries depend more heavily on the choice of π_{oecd} than vice versa. Recall that in the analysis about the discount factor, we also concluded that the amount of green energy used by the non-OECD countries depends more on the discount factor used by OECD countries than vice versa.

We note that the EU has set itself a long-term goal of reducing greenhouse gas emissions by 80–95% when compared to 1990 levels by 2050. If we assume that we must reach the average of this long-term goal (87.5% reduction) within 60 years, then we must reduce greenhouse gas emissions by 3.4% each year. This means that we reach this goal when our equilibrium emission value is smaller than $0.966^{36} \cdot 22.20 = 4.7064$ (using the data from 2014). Computing the equilibrium values using different values for π_{oecd}, we conclude that we never reach this e if $\pi_{\text{oecd}} \in (0.45/250, 4 \cdot 0.45/250)$ (see also Figure A7). This means, to keep track of this goal, other parameters play a significant role in reducing the greenhouse gas emissions. For instance, it should become easier to access green energy, green energy should be subsidized, fossil energy should become more expensive (e.g., by introducing a carbon tax).

To visualize the impact on strategies and state trajectories, Figures 15–17 show for the asymmetric emission shock these trajectories in case $\pi = [4 \cdot 0.45/250 \;\; 2 \cdot 0.41/250]$. For comparison reasons we also include the corresponding benchmark plots. We see that all variables return slightly earlier to their equilibrium values when using the higher values for π. As a result, we also see that OECD countries increase their green energy use more than in the original setting (in percentage). Furthermore, we see that the non-OECD countries react in the opposite way, by increasing their green energy use less than in the original setting. This is possible due to the increase in green energy use by the OECD countries, which reduces the total CO_2 emission in the atmosphere. Observe that the strategies differ by at most 1% (the green/fossil energy use of the non-OECD countries) from the original strategies. Finally, due to the larger reduction in CO_2 emission by the OECD countries, the impact of the shock on the output of the non-OECD countries is slightly smaller than in the benchmark case (see Figure 17).

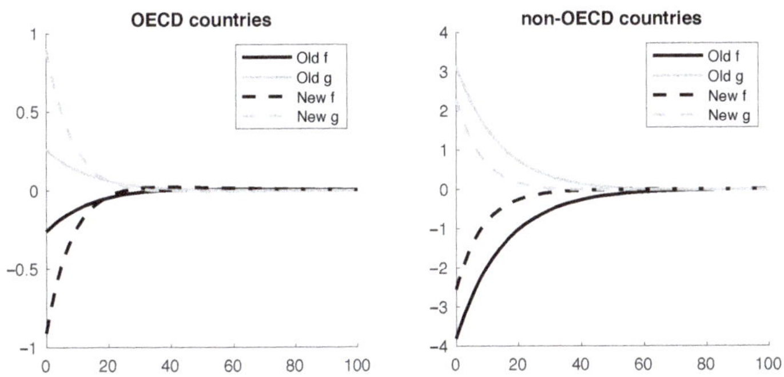

Figure 15. Control variables, simulation with π.

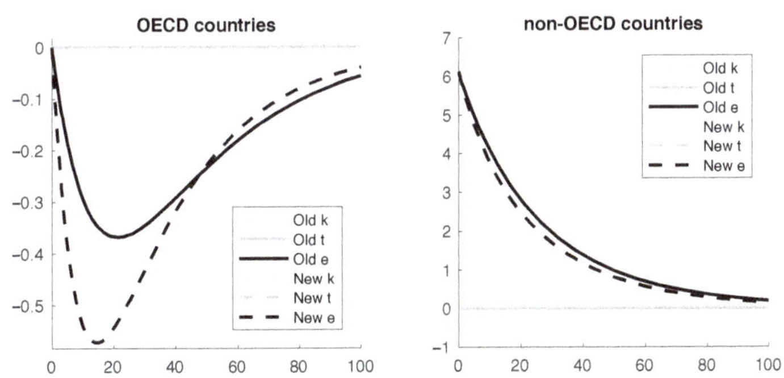

Figure 16. State variables, simulation with π.

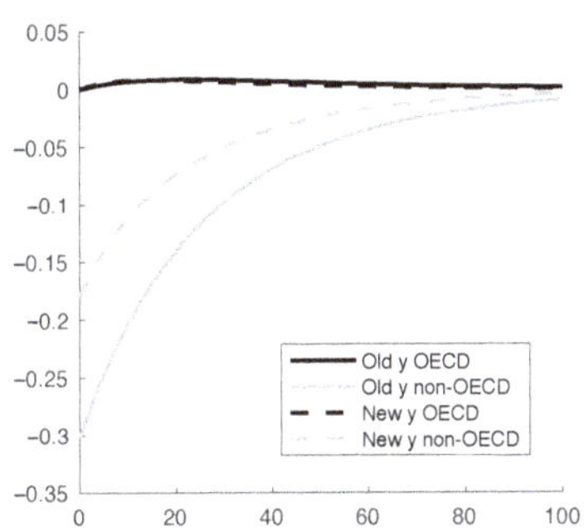

Figure 17. Output variables, simulation with π.

4.4. Stochastic Relations

One of the equations that might be oversimplified is the relation of the accumulation of capital. Therefore, it seems reasonable to include some uncertainty in the proposed equation. As we do not know much about the involved uncertainties, we assume that these are normally distributed, with mean zero and variance 0.1. Furthermore, we truncate this distribution at values that will cause the initial variable calibration for the largest k to deviate by more than 5%. This means that the distribution is truncated at -0.05 and 0.05. We then use this "restricted" normal distribution for sampling. Therefore, capital accumulation, $\dot{k}_i$, is assumed to be generated by the next equation,

$$\dot{k}_i = -(\eta_i + \delta_i) + e^{-k_i(t)}\left(s_i e^{y_i(t)} + s_{ij} e^{y_j(t)+t(\eta_j-\eta_i)} + \tau_i e^{t_i(t)}\right) + \Lambda,$$

where for every simulation constant Λ is drawn once from the *restricted* normal $N(0, 0.2)$ distribution. Roughly speaking, we observe that under this assumption equilibrium values, values for the objective function and strategies are affected similarly as in the considered stochastic parameter context. See, Figures A9 and A10 in Appendix E. We observe that, as expected, our original equilibrium values are close to the most occurring equilibrium values. Analogously to the stochastic parameter case, we also consider the realization of the equilibrium values if we vary Λ between ± 0.05. Figure 18 shows the results.

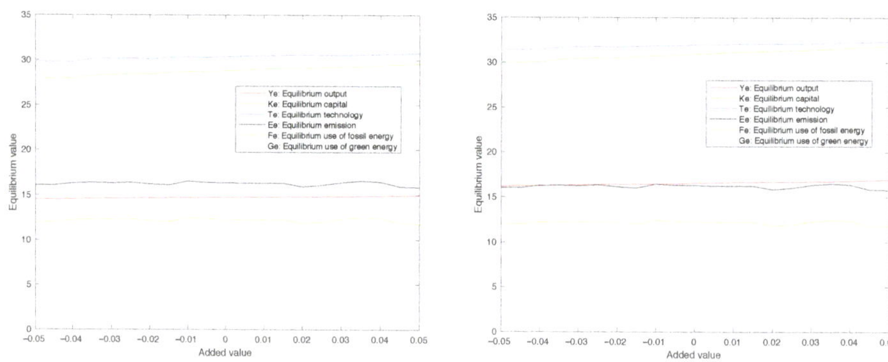

(**a**) OECD equilibria, added values (**b**) non-OECD equilibria, added values

Figure 18. Equilibrium values, added values.

We see that if we increase Λ, the equilibrium capital value of both countries increases. This is easonable, as capital accumulation is increased at any point in time with a constant. Therefore, equilibrium output increases too (see production function). We also see that both the equilibrium technology values increase as Λ increases. This is due to the fact that, for both countries, technology accumulation depends positively upon the capital value. Again, by considering a worst-case scenario from a noise realization perspective, we like to quantify the involved uncertainty. Therefore, we look at the maximal, absolute percentage difference of the equilibrium values of the simulated relation and the original equilibrium. The results are tabulated in Table 15.

We conclude that the average percentage difference in both countries is below 4%.

Table 15. Percentage differences from original equilibrium.

y	k	t	e	f	g	Average
1.49%	3.17%	1.75%	3.06%	4.13%	9.35%	3.83%
2.21%	3.11%	1.65%	3.18%	3.99%	8.69%	3.80%

4.5. Scenario Analysis

In this section, we want to investigate the impact of considering a larger value for the initial use of green energy in both countries. Initially, we estimated that the total amount of fossil energy used is 81.1% (OECD) and 92.3% (non-OECD) from the total energy used. In this section, we evaluate the outcomes of changing these percentages. First, we decrease both percentages by 5%. Note that, using more green energy, will typically be an outcome of good availability of resources and a reduced price. This is automatically taken into account by the new values for π and ρ, which are recalculated based upon the new initial variable calibrations. We calculate the new equilibrium variables under this scenario and find that this adjustment has no large impact on the values (or on the optimal strategies).

Also for this scenario analysis we calculate, for all possible combinations of ratios between 0 and 5% for both countries, corresponding equilibrium outcomes. This means that we look at the equilibrium results where the initial f and g are changed. We determine all equilibrium values when initial values of fossil energy use varies between 76.1 and 81.1% for OECD countries and between 87.3 and 92.3% for non-OECD countries. Next, we compute for all these values the corresponding equilibrium values. The results are visualized as dots in Figures A11 and A12 of Appendix E. To see the general structure in the equilibria more clearly, we fit a plane through the equilibrium values for each variable. This reduces the noise from the fact that the numerical computations for finding the equilibria may not have been converged yet. Note that the vertical axis has a small range, which means that a small amount of noise could already be seen in the plot. Again, the maximal percentage deviations from the original equilibrium are tabulated in Table 16.

Table 16. Percentage difference from origal equilibrium.

y	k	t	e	f	g	**Average**
1.55%	2.66%	2.40%	12.31%	4.26%	2.57%	4.29%
1.40%	2.17%	1.86%	13.92%	11.03%	6.86%	6.21%

From this table we see that the maximal percentage difference of both countries is on average 5%. Note that the maximal deviation relates to the total emission variables. This confirms the observation that the policy parameters have large impact on the equilibrium values, as shown in Section 4.3.

5. Concluding Remarks

In this paper, we consider a simplistic model that analyzes the ratio between fossil energy use and green energy use within a context of OECD and non-OECD countries. This model can be viewed as a simplified two-player version of the model considered in [17]. One of the open issues in that paper is to see how robust the obtained results are with respect to several uncertainties/modeling inaccuracies. For that purpose, we develop a simplified version here and determine the main factors that impact the model outcomes most. Starting from some basic economic relationships, we derive our nonlinear, two-country, growth model. We determine for this model its equilibrium, under the assumption that both players want to maximize their welfare in a non-cooperative setting. To see how both players will react to distortions, we derive the corresponding linear dynamics around the equilibrium. Some shock simulations with this benchmark model turn out to provide results that are not too unrealistic. We also consider the question if a coalition of OECD countries and non-OECD countries could be profitable for both countries. It turns out that this is not the case. The non-OECD countries will, in general, not profit from this, where the OECD countries will. Moreover, we observe that strategies performed under a cooperative regime are similar except for the fact that they lead to a faster convergence towards equilibrium values than those performed under a non-cooperative regime.

As already mentioned above, given the large number of uncertainties involved in modeling this kind of problems, our main objective is to perform an extensive uncertainty analysis. We start with analyzing the impact of normalizing the parameters in the production function to satisfy constant returns to scale. We observe that the equilibrium values may turn out to deviate on average 6% from the original equilibrium values. Furthermore, we find that small changes to the parameters used in the dynamics of the model do not affect the outcome of the model much. Adding, for instance, stochastics to such a particular parameter results in the worst-case, on average, in 4% deviation from the original equilibrium values. If we add stochastics to a complete state equation, we also may end up in an equilibrium in which the variables deviate on average 4% from the original equilibrium values. This means that both changing the set-up of one of the state equations in our model with a small amount and changing the *parameters* within such a state equation with a small amount, have a similar impact on the outcome of the model.

So far, the uncertainty involved seems to have no direct effect on the optimal strategies of both players in returning to the equilibrium after an emission shock. However, we also investigate the uncertainty involved in the parameters that occur in the objective function of both players. In particular, we investigate the effect on the outcome of the model by changing the preference rate for emitting CO_2. This parameter seems to have a slightly larger effect on the optimal strategies than the parameters we just discussed. Moreover, we show that it has a large impact on the equilibrium ratio between the use of fossil and green energy. The impact of it on equilibrium values for the remaining variables is in the order of the above discussed cases. The higher values of π result in strategies in which the variables return earlier to their equilibrium values.

In Table 17, a short overview is given where the approximate uncertainty is tabulated for each analysis. This uncertainty is divided in uncertainty in the equilibrium values and uncertainty in the optimal strategies of both players. The percentage in the left column of the equilibrium values is based upon the maximal percentage difference with the original equilibrium. The column with the strategies is based upon the maximal percentage change in using fossil or green energy.

Table 17. Overview of the uncertainties.

	Equilibrium	**Strategies**
Shares of income	$\approx 6\%$	$\approx 0\%$
Parameter (dynamics)	$\approx 4\%$	$\approx 0\%$
Parameter (objective)	$\approx 37\%$	$\approx 1\%$ [1]
Relation	$\approx 4\%$	$\approx 0\%$
Scenario	$\approx 5\%$	$\approx 0\%$

[1] As discussed earlier, the structure is approximately the same.

We conclude that the calibration of the parameters that occur in the objective of the players needs special attention. These parameters *carry* the most uncertainty for the outcome of the model. Both in the equilibrium and in the optimal strategies. Note that the strategies may only differ by 1% compared to the 37% of the equilibrium values. Second, we see that the structure of the optimal strategies after an emission shock occurred, does not variate much based upon the performed uncertainty analyses. Changing the parameters of the objective neither affects the path of the variables much. It only changes the size of the reaction of both players. The direction seems to be very stable against the uncertainties involved.

Potential lines for further research include extending the uncertainty analysis with a worst-case scenario expectation by players. This gives an extra dimension to the question what impact (not only model, but also player's) uncertainty has on equilibria and strategies. Research performed with similar models used for different applications usually show that one might expect that players engage into more short-term active strategies, the larger the worst-case expected level of uncertainty. Furthermore, we now have performed several uncertainty analyses separate from each other. This can be extended to analyses, where different uncertainty analyses are combined.

There are also several further research opportunities regarding the model we use. We develop a simple economic growth model, as we are focusing on the uncertainty involved in such models. This model can be extended to more realistic models. As an example, we represent the interdependencies between countries by a fixed factor. However, the interdependencies between countries may also be related to trade effects, which depend on the development of market prices rather than on a fixed part of the gross domestic product. This means that the parameters related to the interdependencies are time-dependent, therefore the model might be extended with time-dependent parameters.

Furthermore, we studied a two-player setting containing OECD and non-OECD countries. A general case in which more players are interrelated can be examined. If the number of players increases, the number of parameters increases as well. Therefore, more uncertainty may be present in the system, which means that recommendations about the model calibration phase might be even more important.

Author Contributions: The authors contributed equally to this work. All authors have read and agreed to the published version of the manuscript.

Funding: This research received no external funding.

Conflicts of Interest: The authors declare no conflict of interest.

Appendix A. Calibrations of Parameters and Initial Values

Appendix A.1. Non-Spillover Parameter Calibration

A: Total factor productivity, A, is in fact the last parameter we calibrate. First, all other initial values of the variables in our model are calibrated. Finally, A is taken such that the production function applies for the current state of the world (using the initial variable calibration).

α: According to data from the World Bank [26], the GDP of the OECD countries in 2014 was 49,289,717 million dollars. In the same year, the total investment in capital was 10,111,756 million dollars. If we divide those numbers, we get the fraction of GDP that is invested in capital, which gives us a good estimate for α_{oecd}. Similarly, for non-OECD countries total GDP in 2014 was 33,721,083 million dollars and total investment in capital was 12,093,681 million dollars. Again, the quotient gives us an estimate for $\alpha_{\text{non-oecd}}$.

β: The labor share in income, β_i, is estimated in a similar way using data from the World Bank [26]. For the OECD countries, the gross national income per capita in 2013 was 38,213 US dollars. In the same year, the disposable income per capita was 26,500 US dollars. If we divide those numbers, we get the fraction of labor income to total income, which gives us a good estimate for β_{oecd}. Similarly, for the non-OECD countries the gross national income per capita in 2013 was 21,082 US dollars and disposable income per capita was 15,000 US dollars. The quotient gives us the estimate for $\beta_{\text{non-oecd}}$.

γ: This calibration is based upon [27]. In this paper, the author estimates the economic damage due to climate change for several regions of the world. To get the results for our two countries, we use appropriate weights and calculate the weighted emission share of income.

κ: According to data from the World Bank [26], the expenditures on research and development as a percentage of GDP are 2.4% for OECD countries. This is used as an estimate for κ_{oecd}. For the non-OECD countries this percentage was 1.1%.

η: In this paper we restrict our analysis to the so-called, high-income non-OECD members, as low-income countries have a (relatively) small impact on global CO_2 emission. We use data from the data bank of the OECD [28]. Figure A1 shows for both countries the population growth. From this we observe that the assumption that both growth rates coincide is reasonable. We choose η_i equal to the data of 2014, i.e., $\eta_i = 0.73\%$.

δ: There is a lot of variety in the service life of different forms of capital. It is difficult to capture this depreciation of capital in one number. We use the percentage 6.24%, as obtained in [29]. This number is based upon a weighted average for OECD countries. Based upon data from the

World Bank [26], we assume the depreciation rate of natural capital for non-OECD countries to be 20% higher than the rate of the OECD countries.

τ: According to Claassen's logarithmic law of usefulness [30], $Y_i = \log_{10}(T_i)$, as α_i is the capital share in income, we have $K_i = \alpha_i \log_{10}(T_i)$. Therefore, the slope of this function at T_i is an estimate of τ_i. From the literature we recall that $T = 5$ is approximately an average starting point of technology. Because we assume that τ_i is a linear parameter, we approximate $\alpha_i \log_{10}(T_i)$ with a second order Taylor expansion around $T = 5$ for both countries. This yields the tabulated estimates for τ_i.

ϵ: According to data from the World Bank [26], the expenditures on research and development as a percentage of GDP are 2.4% (1.1%) for OECD countries (non-OECD countries). So, $T_{\text{oecd}} = 0.024 \cdot Y_{\text{oecd}} = 0.024 \cdot \frac{K_{\text{oecd}}}{\alpha_{\text{oecd}}} = \epsilon_{\text{oecd}} K_{\text{oecd}}$. And similarly for ϵ_{oecd}.

ξ: Until now, there is no consensus about the exact lifetime of CO_2 in our atmosphere. For instance, the IPCC estimates this lifetime at approximately a hundred years, where other studies start at thirty years. Based on [22], we use a CO_2 lifetime of 30 years for 50% of the CO_2 emission today. The rest of the CO_2 emission remains more than several hundreds of years in the atmosphere. Using these assumptions, we know that approximately 2.3% leaves the atmosphere without human intervention.

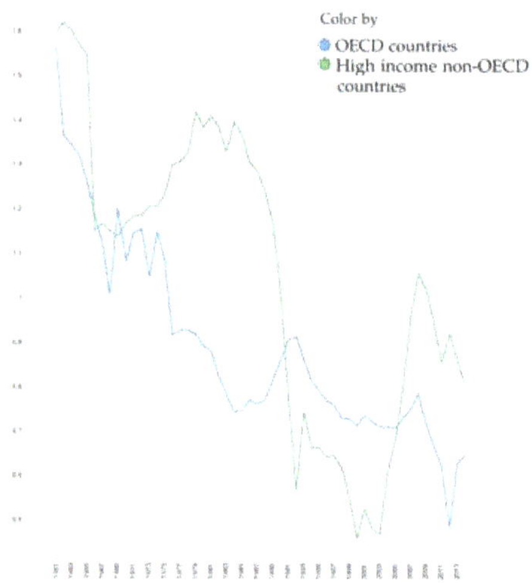

Figure A1. Population growth rate for OECD and (high-income) non-OECD members.

Appendix A.2. Spillover Parameter Calibration

s: Direct saving rates of capital in the country itself, s_{ii}, are estimated based on data of the World Bank [26] on gross national savings as a percentage of GDP. In 2013 these numbers are 20.4% (29.4%) for OECD countries (high-income non-OECD countries). For the cross-terms we use the corresponding data for foreign direct investment (net inflows) as a percentage of GDP for both countries. The results are shown in Table A1.

g: We estimate domestic technological progress due to the domestic state of technology by the growth of the number of researchers in R&D, as determined by the World Bank [26]. The increase in domestic technological progress due to foreign technology use is estimated by dividing the amount of high-technology exports by the domestic country's GDP (see Table A1).

ζ: The increase in CO_2 emission due to the domestic use of fossil fuels is assumed to be proportional to the amount of used fossil fuels. We set the proportion of CO_2 emission due to fossil energy use for non-OECD countries to 1.00 and for OECD countries to 0.85. We base these numbers on the engagement and implementation of CO_2 emission reducing production techniques. Furthermore, we base this on the number of international climate partnerships (e.g., UN Framework Convention on Climate Change, International Convention on the Prevention of Pollution from Ships, Global Methane Initiative, Global Data Center Energy Efficiency Task Force, Carbon Sequestration Leadership Forum, etc.). The cross-terms are obtained by the fact that CO_2 emission affects all countries at the same time. Thus, the increase of CO_2 emission in the OECD countries from the used fossil fuels in non-OECD countries is just the same number as for the non-OECD countries themselves, so $\zeta_{\text{oecd,non-oecd}} = 1.00$. The same reasoning holds for $\zeta_{\text{non-oecd,oecd}}$. The results are shown in Table A2.

Table A1. Parameter calibration for spillover parameter s and g.

s	O	n-O	g	O	n-O
O	0.204	0.009		0.017	0.005
n-O	0.055	0.294		0.177	−0.005

Table A2. Parameter calibration for spillover parameter: ζ.

ζ	OECD	Non-OECD
OECD	0.85	1.00
non-OECD	0.85	1.00

Appendix A.3. Initial Variable Calibration

All initial values for the variables we use in our model are expressed per (working) capita. For the number of working people in OECD (non-OECD), we use the number 837,816,057 (227,833,932).

y: Initial output is based upon data about the GDP per capita for 2014, obtained via the World Bank [26], see Table A3. Therefore, our initial value for the variable y is the logarithm of this number for both countries, as shown in Table 2.

k: Initial capital per capita for both countries is based on the capital intensities of both countries. We use the results derived in [31]. However, these results are calibrated for the year 2000. Therefore, we have to multiply these numbers with the average price increase in the period 2000–2014. In this time period, the prices increased with about 37.5%. The result is shown in Table A3. Again, by taking the logarithm of these numbers we obtain our initial values for the variable k, as shown in Table 2.

t: The initial values of t are calibrated using the total number of researchers measured in Full Time Equivalent (FTE) in 2013. For the OECD countries there were about 4,403,168 FTE researchers and for non-OECD, there were about 2,111,638 FTE researchers. We multiply these numbers with the gross average wage in the corresponding country. For OECD countries this wage is equal to $44,290 and for the non-OECD countries the gross average wage is equal to $19,077 (see the World Bank [26]). The numbers (per capita) are provided again in Table A3. Again, by taking the logarithm of these numbers we find our initial values for t, shown in Table 2.

e: The initial CO_2 emission per capita is calibrated using information from the World Bank [26]. In the last column of Table A3, the total CO_2 emission in metric tons per capita in 2014 is stated. This number is based on data from 2011 and on the CO_2 emission accumulation from the data bank of the OECD [28]. According to this database the CO_2 emission did not change significantly between 2011 and 2014. Therefore, the results of the 2011 database are used as an initial estimate for E/L. By taking the logarithm we obtain our initial values for e, stated in Table 2.

f, g: To calibrate the initial values for f, we need an estimate of the total energy used in each country, and we need data about what fraction corresponds to green energy. We use data from 2013 of the World Bank [26]. The corresponding estimates are given in Table A4. Again, taking the logarithm gives us the estimates for the initial values of f and g.

Table A3. Variable calibration.

	GDP/L	K/L	T/L	E/L
OECD	$38,349	219,563	233	9.9
non-OECD	$19,040	39,418	177	11.7

Table A4. Energy consumption data.

	Energy/L	% Fossil	F/L	G/L
OECD	4174	81.1%	3385	789
non-OECD	4712	92.3%	4349	363

Appendix A.4. Policy Parameter Calibration

θ: Calibrations of the discount factors are based upon suggestions taken from [23]. As an example, the discount factors obtained imply that OECD countries are more interested in future developments than the non-OECD countries.

μ: The proportion of output that can only be produced with the use of energy, μ, is calibrated with the assumption that in our base year (2014), this proportion holds for both countries. In other words, $\mu_i = \frac{f_i+g_i}{y_i}$, where the variables have their initial values.

ρ, π: In the objective, we have to calibrate three different weights. The first weight belongs to the energy requirements (u^2). We set this weight equal to 1. This means that we will set the weights corresponding to the total emission and the total green energy, respectively, π and ρ, relative to this 1. First, note that the energy requirements *must* hold. This means that the π and ρ must be sufficiently small so that they do not get priority above the energy requirements.

As discussed in the introduction of this paper, ρ represents the disadvantages of using green energy. For instance, the price of using green energy in non-OECD countries is higher than the price of using fossil energy. Each country has its own availability of resources, it may be difficult to use green energy, because there might be no resources in the neighborhood. This is confirmed by the International Institute for Environment and Development (IIED) [32], who concluded that a lot of non-OECD countries have little access to the green energy market. Furthermore, adapting green energy into their system still seems very difficult to achieve. According to data from the World Bank [26], about 92% of the energy used in non-OECD countries is fossil energy, where this percentage for OECD members lies just above 80%. Therefore, we will calibrate π and ρ using the following numbers.

$$\pi_i = \frac{1}{250} \cdot \left(\frac{g_i}{f_i + g_i}\right) \quad \text{and} \quad \rho_i = \frac{1}{250} \cdot \left(\frac{f_i}{f_i + g_i}\right) \tag{A1}$$

Note that the part between brackets represents the fraction of total energy that is currently satisfied via green energy. In short, a country that currently uses a lot of green energy is assumed to strongly dislike emitting CO_2 compared to a country that is using less green energy. Similarly, a country that currently uses a lot of fossil energy is assumed to have difficulties with accessing green energy compared to countries using less fossil energy. Dividing the fraction by 250 addresses the fact that the energy requirements are met and have a higher priority than the other two factors in the objective. This number is determined using two criteria. The first criterion is that the energy requirements should be met. In other words, $(\mu_i y_i - (f_i + g_i))^2$ should be converged to (approximately) zero. Note that it may not be exactly zero due to the fact that we use a certain number of iterations (in this case 20,000) to find the equilibrium. It is possible that the numerical computations did not converge to the equilibrium yet for all i. The second criterion is that using this factor, the equilibrium values itself should be (approximately) converged to one equilibrium. These two criteria are both satisfied using a factor of 250.

Appendix B. Equilibrium Calculation

In this appendix, we derive the necessary conditions that must be satisfied by state and control variables assuming the countries play open-loop Nash strategies. Let

$$J_i(x_0, u) := \int_0^\infty h_i(t, x(t), u(t))dt \tag{A2}$$

where the state variable $x(t)$ satisfies the differential equation:

$$\dot{x}(t) = f(t, x(t), u(t)), \quad x(0) = x_0 \tag{A3}$$

In our case the functions are

- $x(t) := \begin{bmatrix} \mathbf{k}(t) & \mathbf{t}(t) & \mathbf{e}(t) \end{bmatrix},$
- $u(t) := \begin{bmatrix} f_1(t) & g_1(t) & f_2(t) & g_2(t) \end{bmatrix},$
- $h_i(t, x(t), u(t)) := e^{(2\eta_i - \theta_i)t}\left((\mu_i y_i(t) - (f_i(t) + g_i(t)))^2 + \pi_i\left(e_i + p_i e_j(t)\right)^2 + \rho_i g_i^2(t)\right),$ with $i = 1, j = 2$, and $i = 2, j = 1$, respectively.
- $f(t, x(t), u(t)) := \begin{bmatrix} \dot{\mathbf{k}}(t) & \dot{\mathbf{t}}(t) & \dot{\mathbf{e}}(t) \end{bmatrix},$

where the bold printed letters mean that it is a row vector consisting of the functions for all countries i (so, $\mathbf{k} = [k_1\ k_2]$ etc.). We make next assumptions.

Assumption A1. *$f(t, x, u)$ and $h_i(t, x, u)$ are continuous functions on $\mathbb{R}^{1+n+m}$. Moreover, for both f and h_i all partial derivatives w.r.t. x and u exist and are continuous.*

Assumption A2. *The log of fossil fuel use per labor supply and the log of green energy use per labor supply will not grow forever, so the set of admissible control policies considered in this thesis is given by the set of locally square integrable functions:*

$$\mathcal{U} := \{(f_i(\cdot), g_i(\cdot)) \in L_{2,loc} | \lim_{t\to\infty} f_i(t) = f_i^e,\ \lim_{t\to\infty} g_i(t) = g_i^e, \\ \lim_{t\to\infty} k_i(t) = k_i^e,\ \lim_{t\to\infty} t_i(t) = t_i^e,\ \lim_{t\to\infty} e_i(t) = e_i^e\}$$

Now, let $(f_i, g_i) \in \mathcal{U},\ i = 1, ..., N$ be a set of open-loop Nash strategies. Consider next corresponding Hamiltonians for this game:

$$\begin{aligned} H_i := {} & e^{(2\eta_i - \theta_i)t}\left((\mu_i y_i(t) - (f_i(t) + g_i(t)))^2 + \pi_i\left(e_i(t) + p_i e_j(t)\right)^2 + \rho_i g_i^2(t)\right) \\ & + \lambda_{i,1}(t)\left(-(\eta_i + \delta_i) + e^{-k_i(t)}\left(s_i e^{y_i(t)} + \sum_{j\neq i} s_{ij} e^{y_j(t) + t(\eta_j - \eta_i)} + \tau_i e^{t_i(t)}\right)\right) \\ & + \lambda_{i,2}(t)\left(-\eta_i + g_i + e^{-t_i(t)}\left(\sum_{j\neq i} g_{ij} e^{t_j(t) + t(\eta_j - \eta_i)} + \epsilon_i e^{k_i(t)}\right)\right) \\ & + \lambda_{i,3}(t)\left(-(\xi_i + \eta_i) + e^{-e_i(t)}\left(\zeta_i e^{f_i(t)} + \sum_{j\neq i} \zeta_{ij} e^{f_j(t) + t(\eta_j - \eta_i)}\right)\right) \end{aligned}$$

Then, using Pontryagin's maximum principle (see, e.g., in [33]), if there exists an optimal control function for this problem, then there exist costate functions $\lambda_i(\cdot)$ satisfying the following set of equations:

$$y_i(t) = \log(A_i) + \kappa_i t_i(t) + \alpha_i k_i(t) + \gamma_i e_i(t) \tag{A4}$$

$$\dot{k}_i(t) = -(\eta_i + \delta_i) + e^{-k_i(t)}\left(s_i e^{y_i(t)} + \textstyle\sum_{j\neq i} s_{ij} e^{y_j(t)+t(\eta_j-\eta_i)} + \tau_i e^{t_i(t)}\right) \tag{A5}$$

$$\dot{t}_i(t) = -\eta_i + g_i + e^{-t_i(t)}\left(\textstyle\sum_{j\neq i} g_{ij} e^{t_j(t)+t(\eta_j-\eta_i)} + \epsilon_i e^{k_i(t)}\right) \tag{A6}$$

$$\dot{e}_i(t) = -(\xi_i + \eta_i) + e^{-e_i(t)}\left(\zeta_i e^{f_i(t)} + \textstyle\sum_{j\neq i} \zeta_{ij} e^{f_j(t)+t(\eta_j-\eta_i)}\right) \tag{A7}$$

$$\begin{aligned} -\dot{\lambda}_{i,1} = \frac{\partial H_i}{\partial k_i} &= e^{(2\eta_i-\theta_i)t}\Big(2\mu_i\alpha_i(\mu_i y_i(t) - (f_i(t)+g_i(t))\Big) \\ &+ \lambda_{i,1}(t)e^{-k_i(t)}\left(s_i(\alpha_i-1)e^{y_i(t)} - \textstyle\sum_{j\neq i} s_{ij}e^{y_j(t)+t(\eta_j-\eta_i)} - \tau_i e^{t_i(t)}\right) \\ &+ \lambda_{i,2}(t)\epsilon_i e^{-t_i(t)+k_i(t)} \end{aligned} \tag{A8}$$

$$\begin{aligned} -\dot{\lambda}_{i,2} = \frac{\partial H_i}{\partial t_i} &= e^{(2\eta_i-\theta_i)t}\Big(2\mu_i\kappa_i(\mu_i y_i(t) - (f_i(t)+g_i(t))\Big) \\ &+ \lambda_{i,1}(t)\left(s_i\kappa_i e^{-k_i(t)+y_i(t)} + \tau_i e^{-k_i(t)+t_i(t)}\right) \\ &- \lambda_{i,2}(t)e^{-t_i(t)}\left(\textstyle\sum_{j\neq i} g_{ij}e^{t_j(t)+t(\eta_j-\eta_i)} + \epsilon_i e^{k_i(t)}\right) \end{aligned} \tag{A9}$$

$$\begin{aligned} -\dot{\lambda}_{i,3} = \frac{\partial H_i}{\partial e_i} &= e^{(2\eta_i-\theta_i)t}\Big(2\mu_i\gamma_i(\mu_i y_i(t) - (f_i(t)+g_i(t)) + 2\pi_i\left(e_i + p_i e_j(t)\right)\Big) \\ &+ \lambda_{i,1}(t)s_i\gamma_i e^{-k_i(t)+y_i(t)} \\ &- \lambda_{i,3}(t)e^{-e_i(t)}\left(\zeta_i e^{f_i(t)} + \textstyle\sum_{j\neq i}\zeta_{ij}e^{f_j(t)+t(\eta_j-\eta_i)}\right) \end{aligned} \tag{A10}$$

$$\begin{aligned} 0 = \frac{\partial H_i}{\partial f_i} &= -2e^{(2\eta_i-\theta_i)t}(\mu_i y_i(t) - (f_i(t)+g_i(t))) \\ &+ \lambda_{i,3}(t)\zeta_i e^{-e_i(t)+f_i(t)} \end{aligned} \tag{A11}$$

$$0 = \frac{\partial H_i}{\partial g_i} = -2e^{(2\eta_i-\theta_i)t}(\mu_i y_i(t) - (f_i(t)+g_i(t)) - \rho_i g_i(t)) \tag{A12}$$

Note that the first four Equations (A4)–(A7) are the original model equations. Conditions (A11) and (A12) are the first-order conditions. They state that both players minimize their Hamiltonian function by the choice of the control variable at each point in time t along the optimal trajectory, where the actions of the other player are considered to be fixed. In particular, (A12) states that at any point in time, the gap between the energy required for the output, and the total produced energy should equal the instantaneous disadvantage of used green energy.

Conditions (A8)–(A10) define the development of the costate variables. They state that the rate of change of these variables equals the negative derivative of the corresponding Hamiltonian with respect to the associated state variable. Note that these costate variables allow (under some regularity conditions [33]) for a so-called shadow-price interpretation. That is, assuming again that the actions of the other player are fixed, one can consider at any point in time t the minimal cost to go if the system is in the optimal state $x^*(t)$, i.e.,

$$V_i(t, x^*(t)) = \min_u \int_t^\infty h_i(s, x(s), u(s))ds, \quad \text{where} \quad \dot{x}(s) = f(s, x(s), u(s)), \quad x(t) = x^*(t).$$

Then, $\lambda^*_{i,j}(t) = \frac{\partial V_i(t,x^*(t))}{\partial x_j}$. In economic literature, this condition is interpreted as that if p_j would be the price of x_j; then, in the optimal situation, this price p_j at time t should equal $\lambda^*_{i,j}(t)$ [33] (p. 159).

As, by assumption, $y_i(\cdot), k_i(\cdot), t_i(\cdot), e_i(\cdot), f_i(\cdot)$, and $g_i(\cdot)$ in the above equations converge if t goes to infinity, it follows that $e^{(\theta_i-2\eta_i)t} \cdot \lambda_i(t)$ converges to some stationary value. Moreover, we can calculate these stationary values using above equations if we assume one more thing:

Assumption A3. *We assume that the growth rate of the population for all countries is the same, i.e.,* $\eta_i = \eta_j, \forall i,j$.

- This assumption is necessary because we want to enforce that $e^{y_j(t)+t(\eta_j-\eta_i)}$ does not become infinity when t grows to infinity. There are two options to accomplish this. We either assume that the growth rate of the population for all countries is the same, which comes down to eliminating the $t(\eta_j - \eta_i)$-term. Or, we assume that the logarithms of all variables converge and then rewrite the model in terms of these variables. We choose the first option, because, in our case, we only differentiate between two countries: OECD and non-OECD countries. In the model calibration, we explain why we can indeed use this assumption.

Using this it follows, with $\lambda^e_i := \lim_{t\to\infty} e^{(\theta_i-2\eta_i)t}\lambda_i(t)$, that the stationary values $y^e_i, k^e_i, t^e_i, e^e_i, f^e_i$ and g^e_i solve the next set of algebraic equations.

$$
\begin{aligned}
y^e_i &= \log(A_i) + \kappa_i t^e_i + \alpha_i k^e_i + \gamma_i e^e_i \\
0 &= -(\eta_i + \delta_i) + e^{-k^e_i}\left(s_i e^{y^e_i} + \textstyle\sum_{j\neq i} s_{ij} e^{y^e_j} + \tau_i e^{t^e_i}\right) \\
0 &= -\eta_i + g_i + e^{-t^e_i}\left(\textstyle\sum_{j\neq i} g_{ij} e^{t^e_j} + \epsilon_i e^{k^e_i}\right) \\
0 &= -(\xi_i + \eta_i) + e^{-e^e_i}\left(\zeta_i e^{f^e_i} + \textstyle\sum_{j\neq i} \zeta_{ij} e^{f^e_j}\right) \\
0 &= (2\eta_i - \theta_i)\lambda^e_{i,1} + 2\mu_i\alpha_i(\mu_i y^e_i - (f^e_i + g^e_i)) \\
&\quad + \lambda^e_{i,1} e^{-k^e_i}\left(s_i(\alpha_i - 1)e^{y^e_i} - \textstyle\sum_{j\neq i} s_{ij} e^{y^e_j} - \tau_i e^{t^e_i}\right) \\
&\quad + \lambda^e_{i,2}\epsilon_i e^{-t^e_i + k^e_i} \\
0 &= (2\eta_i - \theta_i)\lambda^e_{i,2} + 2\mu_i\kappa_i(\mu_i y^e_i - (f^e_i + g^e_i)) \\
&\quad + \lambda^e_{i,1}\left(s_i\kappa_i e^{-k^e_i + y^e_i} + \tau_i e^{-k^e_i + t^e_i}\right) \\
&\quad - \lambda^e_{i,2} e^{-t^e_i}\left(\textstyle\sum_{j\neq i} g_{ij} e^{t^e_j} + \epsilon_i e^{k^e_i}\right) \\
0 &= (2\eta_i - \theta_i)\lambda^e_{i,3} + 2\mu_i\gamma_i(\mu_i y^e_i - (f^e_i + g^e_i)) + 2\pi_i\left(e^e_i + p_i e^e_j\right) \\
&\quad + \lambda^e_{i,1} s_i\gamma_i e^{-k^e_i + y^e_i} \\
&\quad - \lambda^e_{i,3} e^{-e^e_i}\left(\zeta_i e^{f^e_i} + \textstyle\sum_{j\neq i} \zeta_{ij} e^{f^e_j}\right) \\
0 &= \mu_i y^e_i - (f^e_i + g^e_i) - \tfrac{1}{2}\lambda^e_{i,3}\zeta_i e^{-e^e_i + f^e_i} \\
0 &= \mu_i y^e_i - (f^e_i + g^e_i) - \rho_i g^e_i
\end{aligned}
$$

As a remark, note that the last two equations imply that the equilibrium price of CO_2 emission equals $\frac{2}{\zeta_i e^{-e^e_i + f^e_i}}\rho_i g^e_i$. As expected, this price depends on the disadvantages obtained using green energy.

Appendix C. Linearization of the Model

Consider next deviations of variables from their equilibrium value: $y_{li}(t) := y_i(t) - y_i^e$, $k_{li}(t) := k_i(t) - k_i^e$, $t_{li}(t) := t_i(t) - t_i^e$ and $e_{li}(t) := e_i(t) - e_i^e$. Using these variables we rewrite the model Equations (8) as follows.

- $y_{li}(t) = y_i(t) - y_i^e = \log(A_i) + \kappa_i t_i(t) + \alpha_i k_i(t) + \gamma_i e_i(t) - (\log(A_i) + \kappa_i t_i^e + \alpha_i k_i^e + \gamma_i e_i^e)$

  $= \kappa_i t_{li}(t) + \alpha_i k_{li}(t) + \gamma_i e_{li}(t)$
- $\dot{k}_{li}(t) = \frac{\partial}{\partial t}(k_i(t) - k_i^e) = \dot{k}_i(t) = \frac{L_i(t)}{K_i(t)} \cdot \frac{\dot{K}_i(t)}{L_i(t)} - \eta_i$

  $= \frac{L_i(t)}{K_i(t)} \frac{1}{L_i(t)} \left(s_i Y_i(t) + \sum_{j \neq i} s_{ij} Y_j(t) - \delta_i K_i(t) + \tau_i T_i(t) \right) - \eta_i$

  $= -(\delta_i + \eta_i) + s_i \frac{L_i Y_i}{K_i L_i} + \sum_{j \neq i} s_{ij} e^{t(\eta_j - \eta_i)} \frac{L_i Y_j}{K_i L_j} + \tau_i \frac{L_i T_i}{K_i \cdot L_i}$

Next, we use the MacLaurin series expansion of $\log(x)$ around x^e: $x \approx x^e + x^e(\log(x) - \log(x^e))$, to approximate above expression,

$$\approx -(\delta_i + \eta_i) + s_i \left(\frac{L_i^e Y_i^e}{K_i^e L_i^e} + \frac{L_i^e Y_i^e}{K_i^e L_i^e} (\log(\frac{L_i Y_i}{K_i L_i}) - \log(\frac{L_i^e Y_i^e}{K_i^e L_i^e}) \right) +$$

$$\sum_{j \neq i} s_{ij} e^{t(\eta_j - \eta_i)} \left(\frac{L_i^e Y_j^e}{K_i^e L_j^e} + \frac{L_i^e Y_j^e}{K_i^e L_j^e} (\log(\frac{L_i Y_j}{K_i L_j}) - \log(\frac{L_i^e Y_j^e}{K_i^e L_j^e}) \right) +$$

$$\tau_i \left(\frac{L_i^e T_i^e}{K_i^e L_i^e} + \frac{L_i^e T_i^e}{K_i^e L_i^e} (\log(\frac{L_i T_i}{K_i L_i}) - \log(\frac{L_i^e T_i^e}{K_i^e L_i^e}) \right)$$

$$= -(\delta_i + \eta_i) + s_i \left(e^{y_i^e - k_i^e} + e^{y_i^e - k_i^e} (\log(\frac{L_i Y_i}{K_i L_i}) - \log(\frac{L_i^e Y_i^e}{K_i^e L_i^e}) \right) +$$

$$\sum_{j \neq i} s_{ij} e^{t(\eta_j - \eta_i)} \left(e^{y_j^e - k_i^e} + e^{y_j^e - k_i^e} (\log(\frac{L_i Y_j}{K_i L_j}) - \log(\frac{L_i^e Y_j^e}{K_i^e L_j^e}) \right) +$$

$$\tau_i \left(e^{t_i^e - k_i^e} + e^{t_i^e - k_i^e} (\log(\frac{L_i T_i}{K_i L_i}) - \log(\frac{L_i^e T_i^e}{K_i^e L_i^e}) \right)$$

$$= \tilde{s}_i \left(\log(\frac{L_i Y_i}{K_i L_i}) - \log(\frac{L_i^e Y_i^e}{K_i^e L_i^e}) \right) + \sum_{j \neq i} \tilde{s}_{ij} e^{t(\eta_j - \eta_i)} \left(\log(\frac{L_i Y_j}{K_i L_j}) - \log(\frac{L_i^e Y_j^e}{K_i^e L_j^e}) \right) +$$

$$\tilde{\tau}_i \left(\log(\frac{L_i T_i}{K_i L_i}) - \log(\frac{L_i^e T_i^e}{K_i^e L_i^e}) \right) + \underbrace{\left(-(\delta_i + \eta_i) + e^{-k_i^e} (s_i e^{y_i^e} + \sum_{j \neq i} s_{ij} e^{y_j^e + t(\eta_j - \eta_i)} + \tau_i e^{t_i^e}) \right)}_{= 0, \text{ see the conditions from which the equilibrium values are obtained.}}$$

$$= \tilde{s}_i \left(\log(\frac{Y_i}{L_i}) - \log(\frac{Y_i^e}{L_i^e}) - (\log(\frac{K_i}{L_i}) - \log(\frac{K_i^e}{L_i^e})) \right) +$$

$$\sum_{j \neq i} \tilde{s}_{ij} e^{t(\eta_j - \eta_i)} \left(\log(\frac{Y_j}{L_j}) - \log(\frac{Y_j^e}{L_j^e}) - (\log(\frac{K_i}{L_i}) - \log(\frac{K_i^e}{L_i^e})) \right) +$$

$$\tilde{\tau}_i \left(\log(\frac{T_i}{L_i}) - \log(\frac{T_i^e}{L_i^e}) - (\log(\frac{K_i}{L_i}) - \log(\frac{K_i^e}{L_i^e})) \right)$$

$$= \tilde{s}_i (y_{li} - k_{li}) + \sum_{j \neq i} \tilde{s}_{ij} e^{t(\eta_j - \eta_i)} (y_{lj} - k_{li}) + \tilde{\tau}_i (t_{li} - k_{li})$$

where, $\tilde{s}_i = s_i e^{y_i^e - k_i^e}$ and $\tilde{s}_{ij} = s_{ij} e^{y_j^e - k_i^e}$ and $\tilde{\tau}_i = \tau_i e^{t_i^e - k_i^e}$.

Similarly, using the MacLaurin series, expansion of $\log(x)$ around x^e again, we obtain next approximations.

- $\dot{t}_{li}(t) = \frac{\partial}{\partial t}(t_i(t) - t_i^e) = \dot{t}_i(t) = \frac{L_i(t)}{T_i(t)}\frac{\dot{T}_i(t)}{L_i(t)} - \eta_i = \frac{L_i(t)}{T_i(t)}\frac{1}{L_i(t)}\left(g_i T_i + \sum_{j\neq i} g_{ij} T_j + \epsilon_i K_i\right) - \eta_i$

$= -\eta_i + g_i + \sum_{j\neq i} g_{ij}\frac{L_i T_j L_j}{T_i L_j L_i} + \epsilon_i \frac{L_i K_i}{T_i L_i} = -\eta_i + g_i + \sum_{j\neq i} g_{ij} e^{t(\eta_j-\eta_i)}\frac{L_i T_j}{T_i L_j} + \epsilon_i \frac{L_i K_i}{T_i L_i}$

$\approx -\eta_i + g_i + \sum_{j\neq i} g_{ij} e^{t(\eta_j-\eta_i)}\left(\frac{L_i^e T_j^e}{T_i^e L_j^e} + \frac{L_i^e T_j^e}{T_i^e L_j^e}(\log(\frac{L_i T_j}{T_i L_j}) - \log(\frac{L_i^e T_j^e}{T_i^e L_j^e}))\right) +$

$\epsilon_i\left(\frac{L_i^e K_i^e}{T_i^e L_i^e} + \frac{L_i^e K_i^e}{T_i^e L_i^e}(\log(\frac{L_i K_i}{T_i L_i}) - \log(\frac{L_i^e K_i^e}{T_i^e L_i^e}))\right)$

$= -\eta_i + g_i + \sum_{j\neq i} g_{ij} e^{t(\eta_j-\eta_i)}\left(e^{t_j^e - t_i^e} + e^{t_j^e - t_i^e}(\log(\frac{L_i T_j}{T_i L_j}) - \log(\frac{L_i^e T_j^e}{T_i^e L_j^e}))\right) +$

$\epsilon_i\left(e^{k_i^e - t_i^e} + e^{k_i^e - t_i^e}(\log(\frac{L_i K_i}{T_i L_i}) - \log(\frac{L_i^e K_i^e}{T_i^e L_i^e}))\right)$

$= \sum_{j\neq i} \tilde{g}_{ij} e^{t(\eta_j-\eta_i)}\left(\log(\frac{L_i T_j}{T_i L_j}) - \log(\frac{L_i^e T_j^e}{T_i^e L_j^e})\right) +$

$\tilde{\epsilon}_i\left(\log(\frac{L_i K_i}{T_i L_i}) - \log(\frac{L_i^e K_i^e}{T_i^e L_i^e})\right) + \underbrace{\left(-\eta_i + g_i + e^{-t_i^e}(\sum_{j\neq i} g_{ij} e^{t_j^e + t(\eta_j-\eta_i)} + \epsilon_i e^{k_i^e})\right)}_{=0,\text{ see the conditions from which the equilibrium values are obtained.}}$

$= \sum_{j\neq i} \tilde{g}_{ij} e^{t(\eta_j-\eta_i)}\left(\log(\frac{T_j}{L_j}) - \log(\frac{T_j^e}{L_j^e}) - (\log(\frac{T_i}{L_i}) - \log(\frac{T_i^e}{L_i^e}))\right) +$

$\tilde{\epsilon}_i\left(\log(\frac{K_i}{L_i}) - \log(\frac{K_i^e}{L_i^e}) - (\log(\frac{T_i}{L_i}) - \log(\frac{T_i^e}{L_i^e}))\right)$

$= \sum_{j\neq i} \tilde{g}_{ij} e^{t(\eta_j-\eta_i)}(t_{lj} - t_{li}) + \tilde{\epsilon}_i(k_{li} - t_{li})$

where, $\tilde{g}_{ij} = g_{ij} e^{t_j^e - t_i^e}$ and $\tilde{\epsilon}_i = \epsilon_i e^{k_i^e - t_i^e}$.

- $\dot{e}_{li} = \frac{\partial}{\partial t}(e_i(t) - e_i^e) = \dot{e}_i(t) = \frac{L_i(t)}{E_i(t)}\frac{\dot{E}_i(t)}{L_i(t)} - \eta_i = \frac{L_i(t)}{E_i(t)}\frac{1}{L_i(t)}\left(\zeta_i F_i(t) + \sum_{j\neq i}\zeta_{ij} F_j(t) - \xi_i E_i(t)\right) - \eta_i$

$= -(\xi_i + \eta_i) + \zeta_i \frac{L_i F_i}{E_i L_i} + \sum_{j\neq i}\zeta_{ij}\frac{L_i F_j L_j}{E_i L_j L_i} = -(\xi_i + \eta_i) + \zeta_i \frac{L_i F_i}{E_i L_i} + \sum_{j\neq i}\zeta_{ij} e^{t(\eta_j-\eta_i)}\frac{L_i F_j}{E_i L_j}$

$\approx -(\xi_i + \eta_i) + \zeta_i\left(\frac{L_i^e F_i^e}{E_i^e L_i^e} + \frac{L_i^e F_i^e}{E_i^e L_i^e}(\log(\frac{L_i F_i}{E_i L_i}) - \log(\frac{L_i^e F_i^e}{E_i^e L_i^e}))\right) +$

$\sum_{j\neq i}\zeta_{ij} e^{t(\eta_j-\eta_i)}\left(\frac{L_i^e F_j^e}{E_i^e L_j^e} + \frac{L_i^e F_j^e}{E_i^e L_j^e}(\log(\frac{L_i F_j}{E_i L_j}) - \log(\frac{L_i^e F_j^e}{E_i^e L_j^e}))\right)$

$= -(\xi_i + \eta_i) + \zeta_i\left(e^{f_i^e - e_i^e} + e^{f_i^e - e_i^e}(\log(\frac{L_i F_i}{E_i L_i}) - \log(\frac{L_i^e F_i^e}{E_i^e L_i^e}))\right) +$

$\sum_{j\neq i}\zeta_{ij} e^{t(\eta_j-\eta_i)}\left(e^{f_j^e - e_i^e} + e^{f_j^e - e_i^e}(\log(\frac{L_i F_j}{E_i L_j}) - \log(\frac{L_i^e F_j^e}{E_i^e L_j^e}))\right)$

$= \tilde{\zeta}_i\left(\log(\frac{L_i F_i}{E_i L_i}) - \log(\frac{L_i^e F_i^e}{E_i^e L_i^e})\right) + \sum_{j\neq i}\tilde{\zeta}_{ij} e^{t(\eta_j-\eta_i)}\left(\log(\frac{L_i F_j}{E_i L_j}) - \log(\frac{L_i^e F_j^e}{E_i^e L_j^e})\right) +$

$\underbrace{\left(-(\xi_i + \eta_i) + e^{-e_i^e}(\zeta_i e^{f_i^e} + \sum_{j\neq i}\zeta_{ij} e^{f_j^e + t(\eta_j-\eta_i)})\right)}_{=0,\text{ see the conditions from which the equilibrium values are obtained.}}$

$= \tilde{\zeta}_i\left(\log(\frac{F_i}{L_i}) - \log(\frac{F_i^e}{L_i^e}) - (\log(\frac{E_i}{L_i}) - \log(\frac{E_i^e}{L_i^e}))\right) +$

$\sum_{j\neq i}\tilde{\zeta}_{ij} e^{t(\eta_j-\eta_i)}\left(\log(\frac{F_j}{L_j}) - \log(\frac{F_j^e}{L_j^e}) - (\log(\frac{E_i}{L_i}) - \log(\frac{E_i^e}{L_i^e}))\right)$

$= \tilde{\zeta}_i(f_{li} - e_{li}) + \sum_{j \neq i} \tilde{\zeta}_{ij} e^{t(\eta_j - \eta_i)}(f_{lj} - e_{li})$

where, $\tilde{\zeta}_i = \zeta_i e^{f_i^e - e_i^e}$ and $\tilde{\zeta}_{ij} = \zeta_{ij} e^{f_j^e - e_i^e}$.

Appendix D. Objectives Linearized Model

Under the assumption that the players reach an equilibrium in the nonlinear model (8) and (9), it follows that if this equilibrium is perturbed by a small disturbance, the dynamics of the corresponding disturbed system are obtained by linearizing the nonlinear system (8) around this equilibrium. Furthermore, without going into detail (see for more details see, e.g., in [33] (p. 177)), assuming this disturbance is measured by "ϵ" one can make a second-order Taylor expansion of the cost function $J(\epsilon)$ around $\epsilon = 0$, yielding

$$J_i(\epsilon) = J_i^* + \epsilon \frac{\partial J_i(\epsilon)}{\partial \epsilon}(0) + \frac{1}{2}\epsilon^2 \frac{\partial^2 J_i(\epsilon)}{\partial \epsilon^2}(0) + O(\epsilon^2).$$

As for $\epsilon = 0$ we are at the equilibrium of the optimized nonlinear model, it follows that $\frac{\partial J_i(\epsilon)}{\partial \epsilon}(0) = 0$. Therefore, if the system is out of equilibrium, the consistent optimal response of players is approximately obtained by minimizing $\frac{\partial^2 J_i(\epsilon)}{\partial \epsilon^2}(0)$ subject to the linearized model. Unfortunately, the involved quadratic form depends on the realization of the corresponding equilibrium paths of as well state, control and co-state variables. As a consequence, basically, the involved optimization problem reduces to a linear quadratic control problem with time-varying cost matrices. This time-dependency makes the problem non-standard and difficult to tackle. However, as our welfare functions w_i are assumed to be quadratic functions, we know that the part of $J_i(\epsilon)$ that does not depend on the co-state variables coincides with w_i. Furthermore, due to the convergence of the variables $e^{(\theta-2\eta)t}\lambda_{i,j}$, a rough inspection of Equation (5–7) of the algebraic equations defining the equilibrium indicates that the initial value of these variables at $t = 0$ will be close to zero. This implies that the contribution of the costate variables we obtain after twice differentiation of the Hamiltonian will be probably not that large for smaller values of t, whereas for larger values of t the impact of the discount factor makes these contributions small.

Note that, with $x^T(t) = [k_{l1}\, k_{l2}\, t_{l1}\, t_{l2}\, e_{l1}\, e_{l2}]$, $u^T(t) = [f_{l1}\, g_{l1}\, f_{l2}\, g_{l2}]$ and $z^T := [x^T\, u^T]$, the welfare function (9) equals

$$\int_0^\infty g(t, z(t))dt, \text{ where } g(t, z(t)) = e^{(2\eta_i - \theta_i)t}\left(-u^2(t) - \pi_i\left(e_i(t) + \sum_{j \neq i} p_i e_j(t)\right)^2 - \rho_i g_i^2(t)\right).$$

Based on above considerations, we approximate $\frac{\partial^2 J_i(\epsilon)}{\partial \epsilon^2}(0)$ by $\int_0^\infty z^T(t) H_i'' z(t)dt$, where $H_i'' = (\frac{\partial^2 g(z)}{\partial z_i \partial z_j})$ (with some abuse of notation). After some calculations we obtain the next matrices

$$H_1'' = \begin{bmatrix} 0.2074 & 0 & 0.0243 & 0 & -0.0189 & 0 & -0.6440 & -0.6440 & 0 & 0 \\ 0 & 0 & 0 & 0 & 0 & 0 & 0 & 0 & 0 & 0 \\ 0.0243 & 0 & 0.0029 & 0 & -0.0022 & 0 & -0.0756 & -0.0756 & 0 & 0 \\ 0 & 0 & 0 & 0 & 0 & 0 & 0 & 0 & 0 & 0 \\ -0.0189 & 0 & -0.0022 & 0 & 0.0053 & 0.0010 & 0.0588 & 0.0588 & 0 & 0 \\ 0 & 0 & 0 & 0 & 0.0010 & 0.0003 & 0 & 0 & 0 & 0 \\ -0.6440 & 0 & -0.0756 & 0 & 0.0588 & 0 & 2.0000 & 2.0000 & 0 & 0 \\ -0.6440 & 0 & -0.0756 & 0 & 0.0588 & 0 & 2.0000 & 2.0044 & 0 & 0 \\ 0 & 0 & 0 & 0 & 0 & 0 & 0 & 0 & 0 & 0 \\ 0 & 0 & 0 & 0 & 0 & 0 & 0 & 0 & 0 & 0 \end{bmatrix}$$

$$H_2'' = \begin{bmatrix} 0 & 0 & 0 & 0 & 0 & 0 & 0 & 0 & 0 & 0 \\ 0 & 0.5151 & 0 & 0.0162 & 0 & -0.0736 & 0 & 0 & -1.0150 & -1.0150 \\ 0 & 0 & 0 & 0 & 0 & 0 & 0 & 0 & 0 & 0 \\ 0 & 0.0162 & 0 & 0.0005 & 0 & -0.0023 & 0 & 0 & -0.0319 & -0.0319 \\ 0 & 0 & 0 & 0 & 0.0444 & 0.0121 & 0 & 0 & 0 & 0 \\ 0 & -0.0736 & 0 & -0.0023 & 0.0121 & 0.0138 & 0 & 0 & 0.1450 & 0.1450 \\ 0 & 0 & 0 & 0 & 0 & 0 & 0 & 0 & 0 & 0 \\ 0 & 0 & 0 & 0 & 0 & 0 & 0 & 0 & 0 & 0 \\ 0 & -1.0150 & 0 & -0.0319 & 0 & 0.1450 & 0 & 0 & 2.0000 & 2.0000 \\ 0 & -1.0150 & 0 & -0.0319 & 0 & 0.1450 & 0 & 0 & 2.0000 & 2.0047 \end{bmatrix}.$$

Appendix E. Simulations

Appendix E.1. Stochastic Parameter θ

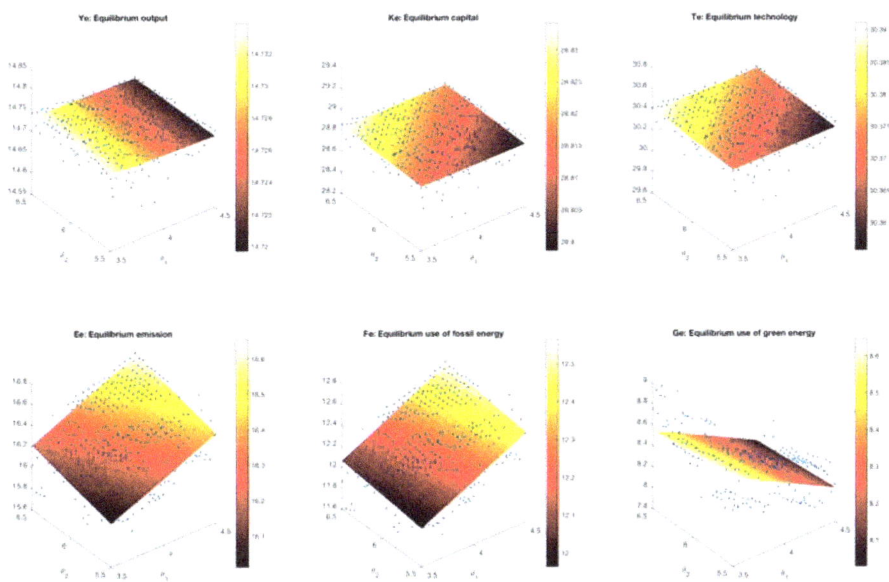

Figure A2. Equilibrium values on grid, OECD countries: fitted plane.

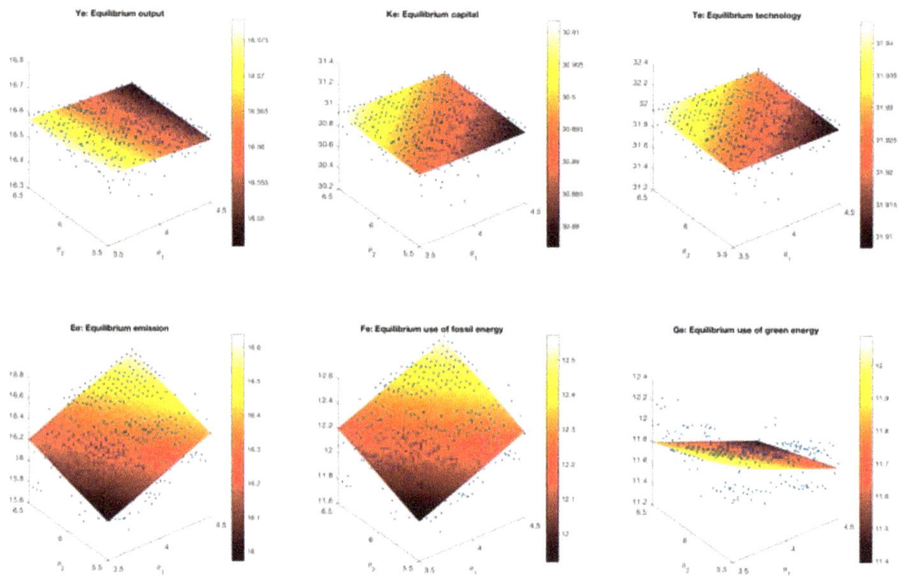

Figure A3. Equilibrium values on grid, non-OECD countries: fitted plane.

Appendix E.2. Stochastic Policy Parameter π

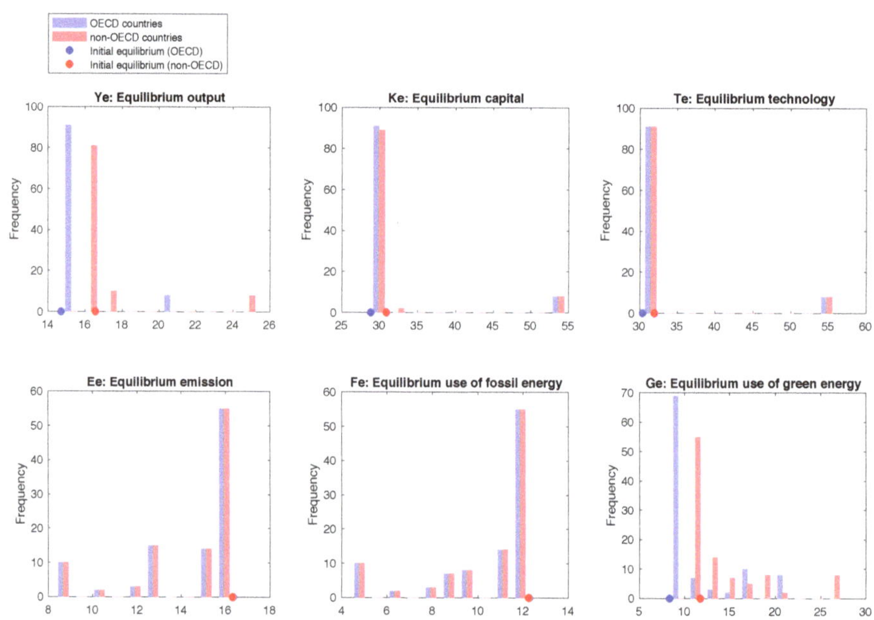

Figure A4. Equilibrium values with stochastic π_i.

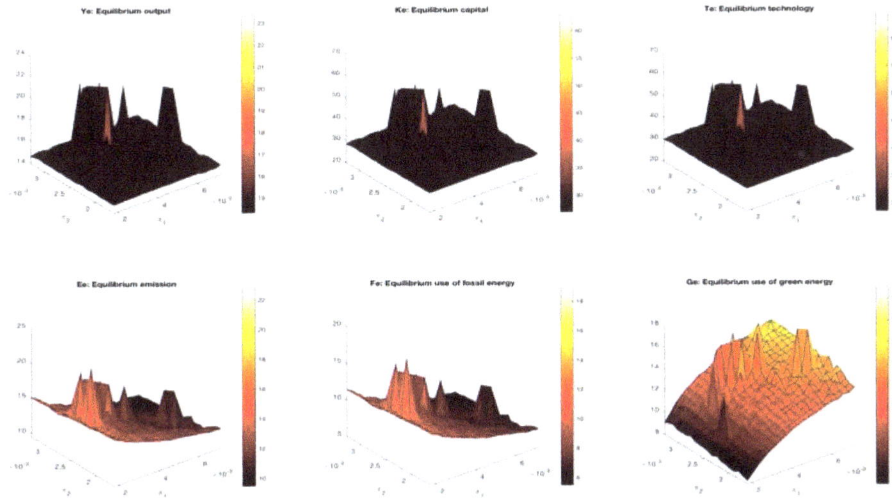

Figure A5. Equilibrium values OECD countries with different π_i.

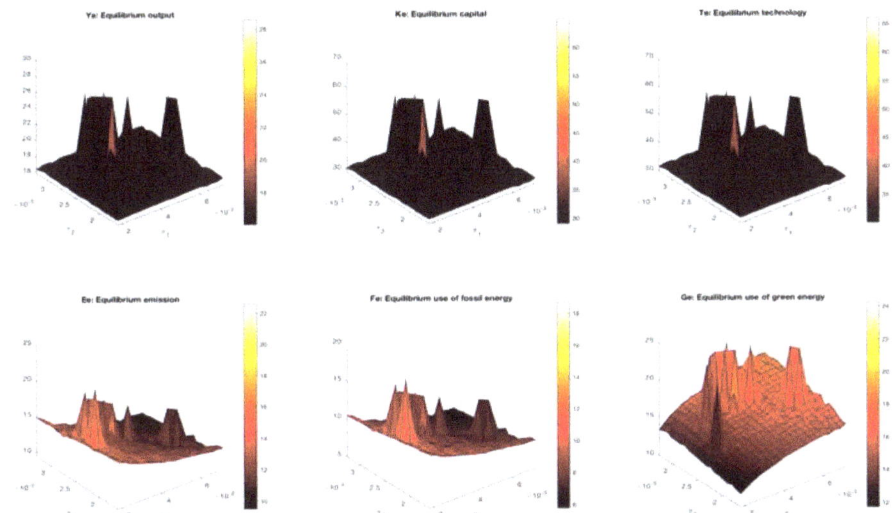

Figure A6. Equilibrium values non-OECD countries with different π_i.

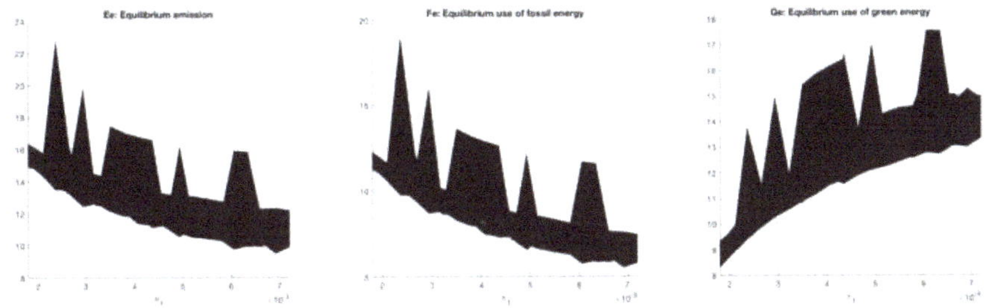

Figure A7. e, f and g for OECD countries.

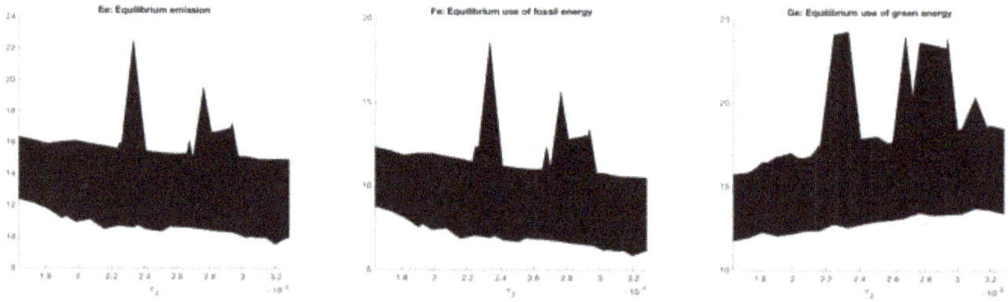

Figure A8. e, f and g for non-OECD countries.

Appendix E.3. Stochastic Relation $\dot{k}$

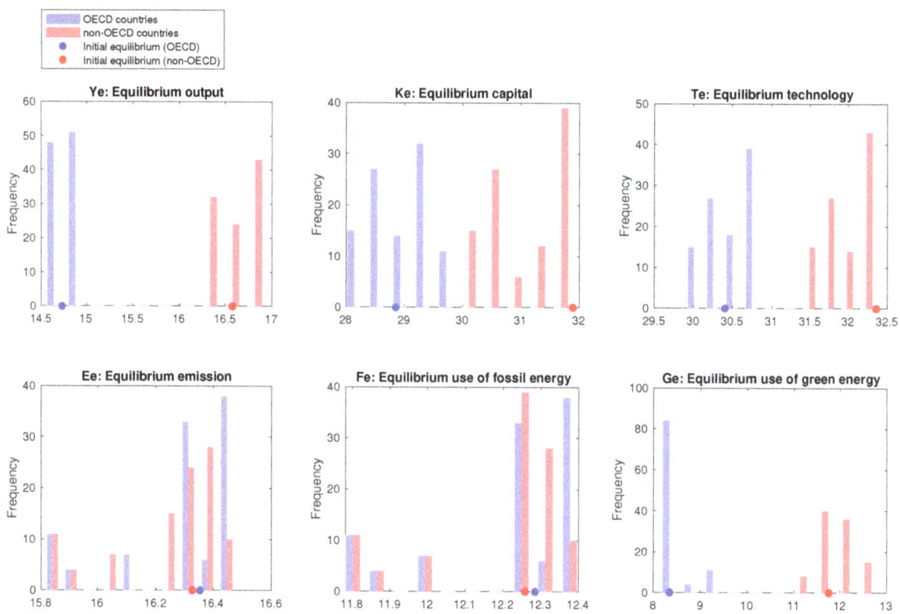

Figure A9. Simulation with $\dot{k}$, Λ stochastic: equilibrium values.

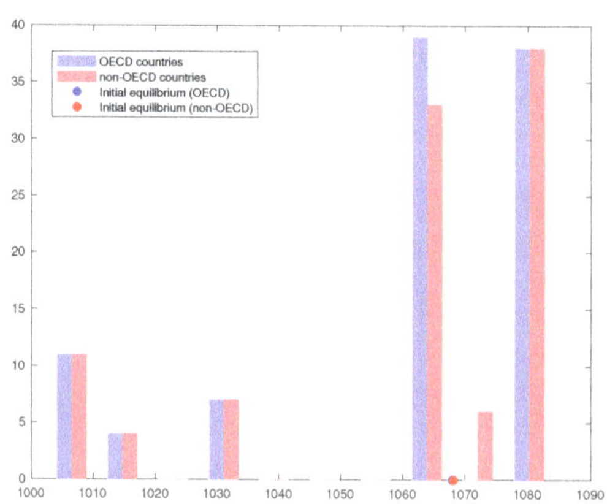

Figure A10. Simulation with $\dot{k}$, Λ stochastic: objective function values.

Appendix E.4. Scenario Analysis ρ, g

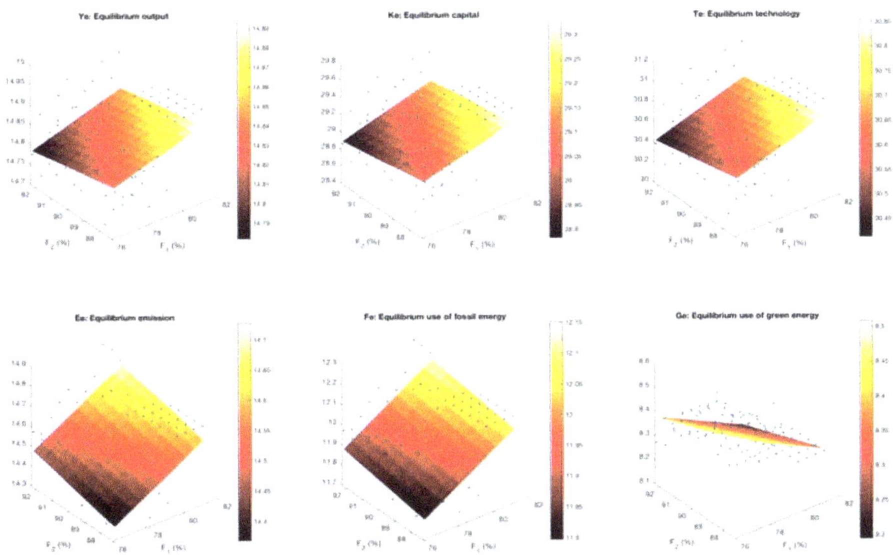

Figure A11. Equilibrium values on grid, OECD countries: fitted planes.

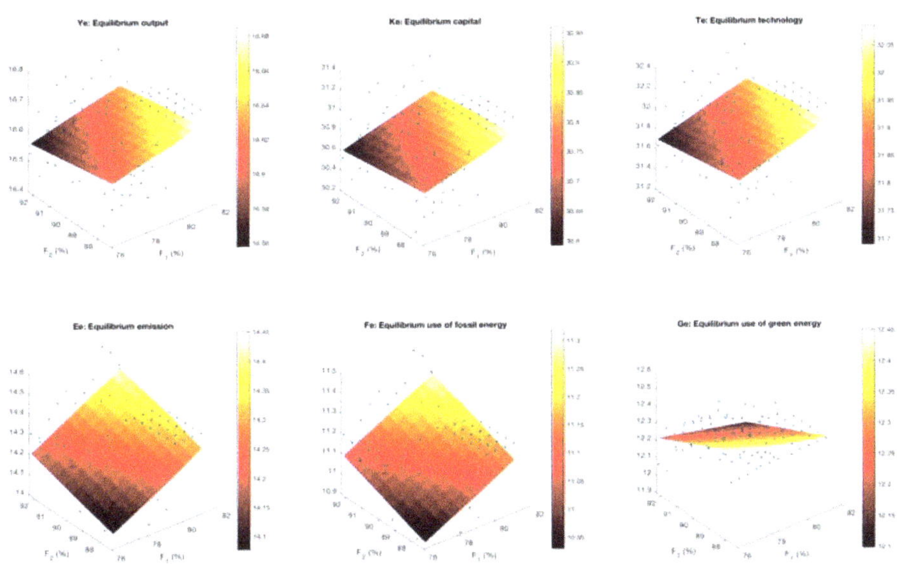

Figure A12. Equilibrium values on grid, non-OECD countries: fitted planes.

References

1. European Environment Agency (EEA). Available online: https://www.eea.europa.eu (accessed on 30 November 2015).
2. Intergovernmental Panel on Climate Change (IPCC). Available online: http://www.ipcc.ch (accessed on 30 November 2015).

3. National Centers for Environmental Information (NCEI). Available online: https://www.ncei.noaa.gov/ (accessed on 30 November 2015).
4. Kann, A.; Weyant, J.P. Approaches for performing uncertainty analysis in large-scale energy/economic policy models. *Environ. Model. Assess.* **1999**, *5*, 29–46. [CrossRef]
5. Pizer, W.A. The optimal choice of climate change policy in the presence of uncertainty. *Resour. Energy Econ.* **1999**, *21*, 255–287. [CrossRef]
6. Gillingham, K.; Nordhaus, W.; Anthoff, D.; Blanford, G.; Bosetti, V.; Christensen, P.; McJeon, H.; Reilly J.; Sztorc, P. Modeling uncertainty in climate change: A multi-model comparison. *J. Assoc. Environ. Resour. Econ.* **2018**, *5*, 791–826.
7. Kelly, D.L.; Kolstad, C.D. Integrated assessment models for climate change control. *Int. Yearb. Environ. Resour. Econ.* **1998**, *2000*, 171–197.
8. Bahn, O.; Drouet, L.; Edwards, N.R.; Haurie, A.; Knutti, R.; Kypreos, S.; Stocker T.; Vial, J.P. The coupling of optimal economic growth and climate dynamics. *Clim. Chang.* **2006**, *79*, 103–119. [CrossRef]
9. Blanford, G.J.; Kriegler, E.; Tavoni, M. Harmonization vs. fragmentation: Overview of climate policy scenarios in EMF27. *Clim. Chang.* **2014**, *123*, 383–396. [CrossRef]
10. Fragkos, P.; Kouvaritakis, N.; Capros, P. Incorporating uncertainty into world energy modelling: The PROMETHEUS Model. *Environ. Model. Assess.* **2015**, *20*, 549–569. [CrossRef]
11. Tol, R. Is the uncertainty about climate change too large for expected cost-benefit analysis? *Clim. Chang.* **2015**, *56*, 265–289. [CrossRef]
12. Jørgensen, S.; Martín-Herrán, G.; Zaccour, G. Dynamic games in the economics and management of pollution. *Environ. Model. Assess.* **2010**, *15*, 433–467. [CrossRef]
13. Van Long, N. Dynamic games in the economics of natural resources: A survey. *Dyn. Games Appl.* **2011**, *1*, 115–148. [CrossRef]
14. Bernard, L.; Semmler, W. *The Oxford Handbook of the Macroeconomics of Global Warming*; Oxford University Press: Oxford, UK, 2015.
15. Nordhaus, W.; Yang, Z. A regional dynamic general-equilibrium model of alternative climate-change strategies. *Am. Econ. Rev.* **1996**, *86*, 741–765.
16. Bosetti, V.; Carraro, C.; Galeotti, M.; Massetti, E.; Tavoni, M. WITCH a world induced technical change hybrid model. *Energy J.* **2006**, *26*, 13–37. [CrossRef]
17. Engwerda, J.C.; Michalak, T. Economic growth and choice of energy: A simplistic strategic approach. *Environ. Model. Assess.* **2014**, *20*, 321–342. [CrossRef]
18. Solow, R. A contribution to the theory of economic growth. *Q. J. Econ.* **1956**, *70*, 65–94. [CrossRef]
19. Michalak, T.; Engwerda, J.; Plasmans, J. A numerical toolbox to solve N-player affine LQ open-loop differential games. *Comput. Econ.* **2011**, *37*, 375–410. [CrossRef]
20. Başar, T.; Zhu, Q. Prices of anarchy, information, and cooperation in differential games. *Dyn. Games Appl.* **2011**, *1*, 50–73. [CrossRef]
21. Mäler, K.-G.; de Zeeuw, A.J. The acid rain differential game, *Environ. Resour. Econ.* **1998**, *12*, 167–184. [CrossRef]
22. Inman, M. Carbon is forever. *Nat. Rep. Clim. Chang.* **2008**, *2*, 156–158. [CrossRef]
23. Impact Data Source. Available online: http://www.impactdatasource.com/choosing-a-discount-rate (accessed on 30 November 2015).
24. Tol, R. The marginal damage costs of carbon dioxide emissions: An assessment of the uncertainties. *Energy Policy* **2005**, *33*, 2064–2074. [CrossRef]
25. WHO Regional Office for Europe (OECD). *Economic Cost of the Health Impact of Air Pollution in Europe: Clean Air, Health and Wealth*; WHO Regional Office for Europe: Copenhagen, Denmark, 2015.
26. World Bank. Available online: http://data.worldbank.org (accessed on 30 November 2015).
27. Tol, R. The damage costs of climate change toward more comprehensive calculations. *Environ. Resour. Econ.* **1995**, *5*, 353–374.
28. OECD Data Bank. Available online: http://data.oecd.org (accessed on 30 November 2015).
29. Oulton, N.; Srinivasan, S. Capital Stocks, Capital Services, and Depreciation: An Integrated Framework (Bank of England Working Paper No. 192). 2003. Available online: http://papers.ssrn.com/sol3/papers.cfm?abstract_id=492062 (accessed on 30 November 2015).

30. Dipert, B. It's Elementary [Blog Post]. 15 April 1999. Available online: http://www.edn.com/electronics-blogs/other/4361480/It-s-elementary (accessed on 30 November 2015).
31. Berlemann, M.; Wesselhöft J.E. Estimating aggregate capital stocks using the perpetual inventory method—New empirical evidence for 103 countries. *Rev. Econ.* **2012**, *65*, 1–34. [CrossRef]
32. International Institute for Environment and Development. Available online: http://www.iied.org/climate-change-group (accessed on 30 November 2015).
33. Engwerda, J.C. *LQ Dynamic Optimization and Differential Games*; John Wiley & Sons: Chichester, UK, 2005.

energies

MDPI

Article

Cost Allocation Mechanism Design for Urban Utility Tunnel Construction Based on Cooperative Game and Resource Dependence Theory

Zhi Zhang [1], Jiaorong Ren [2], Kaichao Xiao [1], Zhenzhi Lin [1,*], Jiayu Xu [2], Wei Wang [2] and Chuanxun Pei [2]

1 College of Electrical Engineering, Zhejiang University, Hangzhou 310027, China
2 Ningbo Electric Power Supply Company of State Grid Zhejiang Electric Power Co., Ningbo 315121, China
* Correspondence: linzhenzhi@zju.edu.cn; Tel.: +86-571-8795-1542

Received: 30 July 2019; Accepted: 26 August 2019; Published: 28 August 2019

Abstract: The urban utility tunnel presents solutions for the sustainable development of urban underground space, and is an important carrier of power distribution network and integrated energy systems. Considering the high investment cost of utility tunnels and the limitations of traditional cost sharing methods (i.e., spatial proportional method, direct-laying cost method and benefit-based proportional method), it is of great significance to establish a fair and practical cost sharing mechanism. First, an improved Shapley value-based spatial proportional method is proposed. A comprehensive decision-making mechanism for utility tunnel construction cost allocation is established by using the improved spatial proportion, the life-cycle direct-laying cost proportion, and the benefit proportion of pipeline companies as the cost allocation indexes. The resource dependence theory is introduced to quantify the bargaining power of each pipeline company in the negotiation of the cost allocation. The weights of the cost allocation indexes in the comprehensive decision-making model are optimized with the objective of maximizing the overall satisfaction of the pipeline companies. Simulations based on the data of utility tunnel pilots in China illustrate that the proposed cost allocation mechanism has the highest overall satisfaction and is more acceptable and more feasible than the traditional cost allocation methods. For power companies, the cost of laying power cables can be significantly reduced by utility tunnels, and laying 10 kV power cables has been shown to have higher economic benefits.

Keywords: urban utility tunnel; cost allocation; Shapley value; life cycle cost; resource dependence theory; economic benefit evaluation

---

1. Introduction

With the acceleration of urbanization, residents' demands for system reliability, power quality and living environment are increasing. Traditional overhead transmission networks are no longer suitable for future urban development due to environmental concerns and energy policies [1]. Given this background, underground power transmission, which presents solutions to some of the environmental and aesthetic problems involved with overhead transmission, is becoming a major trend. Underground power corridors are a precondition of realizing the construction of an urban underground power transmission network, and the urban utility tunnel (UUT), which not only functions as a power corridor, but also houses municipal engineering pipelines such as communication cables, water supply pipes, drainage pipes, heating pipes, and gas pipes, is becoming the carrier of the network of future urban underground integrated energy systems (IES) [2]. UUT permits the installation, maintenance, and removal of pipelines without the necessity of road excavations, which avoids wasting resources and causing inconvenience for society, improves the reliability of power supply, increases the value of developable land, and promotes total urban functionality. The laying of power pipelines into UUT has become an important part of the 13th five-year plan of electricity for the State Grid Corporation in China. However, UUT is more expensive to construct compared to classical solutions for urban utilities [3]. Relying solely on the financial support of the government, it is difficult to make up for the high construction costs of utility tunnels. In fact, the increase in network maintainability leads to a cost reduction of pipeline maintenance and renovation, making pipeline companies the main beneficiaries of UUT [4]. As a result, the Ministry of Housing and Urban-Rural Development of China, together with the National Development and Reform Commission, issued an order in 2015, which requires pipeline companies to pay usage charges to the construction and operation units of the tunnel to alleviate the financial pressure faced by local governments [5]. Although the value and importance of urban utility tunnels has been widely confirmed [6,7], the usage charges have put a heavier burden on pipeline companies, leading to resistance from these companies. Therefore, it is of great significance to establish a fair and practical cost sharing mechanism for improving the enthusiasm of pipeline companies such as power companies and communication companies towards participation in the construction of UUT.

The financing criteria are often the key issues when building a utility tunnel. The utility tunnels in Britain, France and other European countries are considered as communal facilities which are completely funded by the government. The government owns the UUT and leases it to pipeline companies in the form of a paid lease [8]. Japan adopts a diversified investment pattern, and the distribution ratios of pipeline companies as well as road management units is determined based on the specific conditions of each region according to the utility tunnel law. The diversified investment pattern is also employed in some UUT pilot projects in China. For example, Taiwan Province adopted the government-fallback-balance mechanism, which guarantees that the construction cost shared by pipeline companies is basically the same as the traditional pipeline laying cost, and the remaining cost (of about 33% of the total cost) is borne by the government. The construction costs of the UUT pilot projects in Haidong city of Qinghai province and Yinchuan city of Ningxia Hui Autonomous Region of China are allocated based on the proportion of pipeline laying costs, while the construction cost of the UUT in Guangzhou Higher Education Mega Center of Guangdong province, China is allocated according to the proportion of the pipeline cross-sectional area [9]. Indeed, there is no clear and instructive regulation on the cost-sharing mode for the construction cost of UUT under China's diversified investment pattern.

Traditional proportion-based methods, such as the spatial proportional method (SPM) and the direct-laying cost method (DCM), have been employed in most of the existing publications to allocate the construction cost of UUT [8–11]. The SPM takes the proportion of area occupied by pipelines as the allocation ratio of UUT construction cost, and the effective area of a pipeline includes the cross-sectional area and the necessary operating space for its installation and maintenance. The DCM allocates the UUT construction cost according to the proportion of pipeline laying cost, including the direct cost of pipeline laying and the indirect cost of road excavation. Moreover, a modified incremental method is used in reference [8] to allocate the marginal cost of operation and maintenance of UUT. A multiobjective programming model is established in reference [10] for UUT construction cost allocation. An improved proportion-based allocation method which combines SPM with DCM is proposed in reference [11] to obtain the ultimate cost distribution ratio of pipelines. Subjective group-decision [12], the Min-Max method [13] and cooperative game theory [14,15] are also used to solve allocation problems. However, the above-mentioned studies only focus on the cost sharing from a single point of view and consider more about the rationality of the allocation method in theory, while the cost allocation among pipeline companies is a multi-attribute decision-making problem that needs to be solved in practice. In addition, it is worth noting that China's UUT originates from underground power corridors, which is different from those in European countries. For example, the utility tunnel in Paris is derived from underground drainage systems. Therefore, in China, power cables are considered as the most important pipelines set up in UUT, and the economic benefits of power companies should receive more attention. Although the difference of construction cost, operation cost, and maintenance cost between the UUT-laying pipeline and traditional buried pipeline within the life cycle of utility tunnel is proposed in references [8,9,11], the additional benefits brought from power network expansion in the life cycle of UUT are not considered, which makes the evaluation results inaccurate.

Given this background, a comprehensive decision-making mechanism for UUT construction cost allocation is proposed to overcome the limitations of traditional cost sharing methods and maximize the overall satisfaction of pipeline companies, and the economic benefit of UUT-laying power cables is also evaluated on this basis. First, the traditional SPM is improved based on the Shapley value. The comprehensive decision-making model of UUT construction cost allocation is established by using the improved spatial proportion, the life-cycle direct-laying cost proportion and the benefit proportion of pipeline companies as the cost allocation indexes. The process of determining the cost allocation ratio of each pipeline company is considered to be a bargaining, and the resource dependence theory (RDT) is introduced to quantify the bargaining power of pipeline companies. The weights of the cost allocation indexes are optimized with the objective of maximizing the overall satisfaction of the pipeline companies. The impact of power network expansion on the economic benefit of power company within the life cycle of UUT is also considered in the case studies. Hence, the contributions of this paper can be highlighted as follows:

- An improved spatial proportional cost allocation method is proposed based on the Shapley value, which takes into account the cross-sectional area of the pipeline itself and the public operating space, making the cost allocation ratio of each pipeline company more reasonable than traditional SPM.
- A comprehensive decision-making mechanism of UUT cost allocation is designed based on the occupied area and the direct laying costs of pipelines, as well as the economic benefits of pipeline companies. The bargaining power of pipeline companies is analyzed based on the resource dependence theory, which makes the proposed cost allocation method more practical.
- The impacts of power network expansion are first considered in calculating the economic benefits of UUT-laying cables under various scenarios, which makes the economic benefit evaluation of UUT-laying power cables more comprehensive and accurate.

The rest of this paper is organized as follows. In Section 2, the improved SPM, life-cycle DCM and benefit proportion method for cost allocation are proposed. The comprehensive decision-making model of UUT construction cost allocation is presented in Section 3. Case studies and conclusions are presented in Sections 4 and 5, respectively.

2. Proportional Allocation Method for UUT Construction Cost

Construction and maintenance costs are the main costs of an urban utility tunnel. For local governments, the benefits of UUT are indirect and lie within a long-term perspective, involving special public benefits such as no disruption of the highway, no traffic interruption caused by trench digging, and the increase of the developable surface land area. However, the relevant data is difficult to obtain, since they cannot be easily observed and measured. In this paper, the government is assumed to bear part of the construction cost of UUT and the total maintenance cost. The concern of pipeline companies is the allocation of the remaining part of the construction cost, which is much higher than the cost of traditional buried pipelines, and it is easy to hinder pipeline companies from participating in UUT construction.

According to reference [5], the main factors to be considered in cost allocation among pipeline companies include: (1) the proportion of area occupied by pipelines, (2) the cost of traditional buried pipeline laying, (3) the economic benefits such as the reduction of maintenance cost of UUT-laying pipeline (compared with the traditional buried pipeline) and the reasonable return on investment in UUT construction. Therefore, the following three proportional methods are employed to establish the comprehensive decision-making model for the cost allocation among pipeline companies.

2.1. Improved Spatial Proportional Method Based on Shapley Value

Generally, the more pipelines are laid in the UUT, the more the underground space is needed and the higher the construction cost becomes. As a result, the SPM is proposed to allocate the construction cost based on the proportion of area occupied by the pipelines, which includes the cross-sectional area of the pipeline itself and the required operating space for its installation and maintenance. The cross-sectional structure of UUT laying power cables and communication cables is shown in Figure 1. Apart from the space occupied by systems such as ventilation, lighting, and firefighting, the space of the UUT is mainly composed of the power cables, communication cables, and their public operation space (indicated by blue shadows).

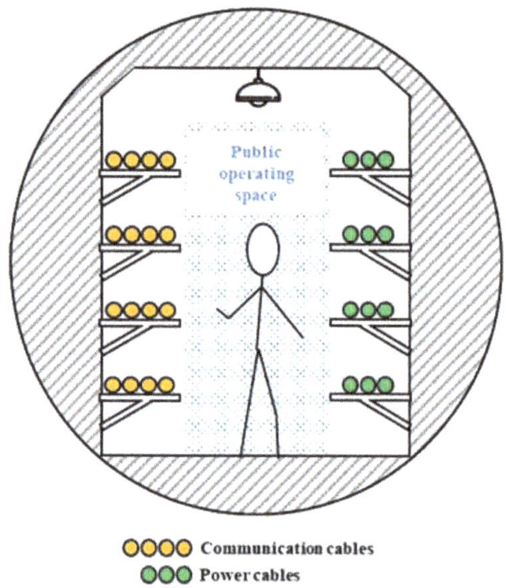

Figure 1. Cross-sectional view of the UUT laying power cables and communication cables.

Pipeline companies should not only bear the construction costs related to their own cross-sectional area, but also share the construction costs of public operating space. The traditional SPM allocates the public operating space (in the form of cross-sectional area) to each pipeline company according to the proportion of cross-sectional area of pipelines. However, there is no clear quantitative relationship between the operating space and the cross-sectional area of pipelines, which makes the traditional SPM unreasonable. Indeed, compared with multiple single cabins, when pipelines are laid in an integrated cabin, the total operating space required is reduced, which reduces the construction cost of the UUT, thus resulting in cooperation benefits. To this context, cooperative game theory could be used to improve the traditional SPM, and the Shapley value is used to solve the improved spatial proportional allocation index. The expected incremental cost that pipeline produces when laid into utility tunnel is the basis for the Shapley value-based cost allocation method, ensuring that the allocation solution is fair and desirable [16]. Since the order that the pipelines laid into the utility tunnel affects the incremental cost produced, the Shapley value considers all orders equally and gives them equal weights. Compared with other cooperative game methods such as Core and Nucleolus, the Shapley value solution is unique and more acceptable, and has a clear calculation process, which makes it more feasible in cost allocation [17]. Let N represent the set of n types of pipelines in the utility tunnel, and the improved spatial proportion cost allocation index of pipeline i can be expressed as:

$$\omega_i^S = S_i / \sum_{i \in N} S_i \tag{1}$$

$$S_i = \sum_{M \subset N} \frac{(m-1)!(n-m)!}{n!}(S_M - S_{M-i}) \tag{2}$$

where ω_i^S represents the cost allocation index of pipeline i under Shapley value-based SPM. S_i is the occupied area of pipeline i, including the cross-sectional area of pipeline and its allocated public operating space. M is a virtual set of m types of pipelines including pipeline i, and S_M is defined as the cross-sectional area of the utility tunnel containing these m types of pipelines. The term $(S_M - S_{M-i})$ models the incremental contribution that pipeline i makes to the cross-sectional area of the utility tunnel. Although the time complexity of Shapley value is $O(n!)$, it is generally acceptable to the computer because of the small number of pipeline types. The larger the index ω_i^S is, the higher the cost allocation ratio of pipeline i should be.

2.2. The Life Cycle Direct-Laying Cost Method

The direct-laying cost method (DCM) determines the cost allocation ratio of each pipeline company according to the proportion of traditional buried pipeline laying cost (referred to as direct-laying cost). The pipeline company with higher direct-laying cost should bear more allocated construction cost. The UUT-laying pipelines and traditional buried pipelines vary in service life and breakage rate, which leads to different maintenance cost and repeated laying frequency of pipelines. Therefore, when DCM is used for UUT construction cost allocation, the direct-laying cost of pipeline should be calculated in a specific time period. In this paper, the time period is set to the life cycle of UUT for the reason that life cycle cost theory considers all stages of UUT from planning, construction, operation to retirement, avoiding decision-making being limited to a certain period of time [18]. After discounting the direct-laying cost of pipelines in the life cycle to the present value, the DCM based cost allocation index of pipeline i can be expressed as:

$$\omega_i^C = (C_{i,T_g}) / \sum_{i \in N} C_{i,T_g} \tag{3}$$

$$C_{i,T_g} = \sum_{t=1}^{n_i^z} c_{i,t}^z (1+r)^{-(t-1)T_i^z} + \sum_{h=1}^{T_g} c_{i,h}^{zm} (1+r)^{-h} \tag{4}$$

$$n_i^Z = \lceil T_g / T_i^Z \rceil \tag{5}$$

where ω_i^C represents the cost allocation index of pipeline i. T_g is the life cycle of UUT. C_{i,T_g} is the total direct-laying cost of pipeline i in T_g. T_i^z is the service life of traditional buried pipeline i and n_i^z denotes the number of its repeated laying times. $c_{i,t}^z$ is the direct-laying cost of pipeline i at the t-th re-laying and $c_{i,h}^{zm}$ is the maintenance cost at the h-th year. r is the annual interest rate. The larger the index ω_i^C is, the higher the cost allocation ratio of pipeline i should be.

2.3. The Benefit-Based Proportional Allocation Method

According to the benefit principle [19], the benefit-based proportional method (BPM) determines the cost allocation ratio of each pipeline company according to the economic benefit that the pipeline produces when laid into a utility tunnel. The economic benefit of each pipeline company (also known as the investment income) mainly includes the construction cost reduction caused by the decrease of repeated laying times of the pipeline, and the maintenance cost reduction caused by the decrease of pipeline breakage rate and leakage rate within the life cycle of UUT. After being discounted to the present value, the cost allocation index of pipeline i under BPM can be expressed as:

$$\omega_i^R = \left(R_{i,T_g}^b + R_{i,T_g}^m\right) / \sum_{i \in \mathrm{N}} \left(R_{i,T_g}^b + R_{i,T_g}^m\right) \tag{6}$$

$$R_{i,T_g}^b = \sum_{t=1}^{n_i^z} c_{i,t}^z (1+r)^{-(t-1)T_i^z} - \sum_{t=1}^{n_i^g} c_{i,t}^g (1+r)^{-(t-1)T_i^g} \tag{7}$$

$$n_i^g = \left\lceil T_g / T_i^g \right\rceil \tag{8}$$

$$R_{i,T_g}^m = \sum_{h=1}^{T_g} \left(c_{i,h}^{zm} - c_{i,h}^{gm}\right)(1+r)^{-h} \tag{9}$$

where ω_i^R represents the cost allocation index of pipeline i under BPM. R_{i,T_g}^b and R_{i,T_g}^m denote the construction cost reduction and the maintenance cost reduction, respectively. T_i^g is the service life of UUT-laying pipeline i and n_i^g denotes the number of repeated laying times of pipeline i. $c_{i,t}^g$ is the laying cost of pipeline i in UUT at the t-th repeat and $c_{i,h}^{gm}$ is its maintenance cost at the h-th year. Similarly, the larger the index ω_i^R is, the higher the cost allocation ratio of pipeline i should be.

3. Comprehensive Decision-Making Mechanism of UUT Cost Allocation Considering the Bargaining Power of Pipeline Companies

Considering that the cost allocation ratio of each pipeline company varies greatly under the above-mentioned three allocation indexes, it is difficult to satisfy the wishes of all pipeline companies at the same time by allocating UUT construction cost with a single index. For example, the water supply companies will bear a large proportion of UUT construction cost under SPM, while the proportion under DCM is much smaller because the direct-laying cost of water supply pipelines is much lower than that of communication and power cables. All pipeline companies want to adopt cost allocation indexes that are beneficial to themselves, and it is difficult to reach an agreement. To solve this problem, the cost allocation among pipeline companies is regarded as a process of bargaining in this paper, and a comprehensive decision-making mechanism is established for the weight determination of the above-mentioned allocation indexes.

Negotiations are held among pipeline companies on their respective cost allocation ratios, and the pipeline company's satisfaction u_i with the outcome of the negotiations is defined as:

$$u_i = \frac{\omega_{i\max} - \omega_i}{\omega_{i\max} - \omega_{i\min}} \tag{10}$$

where ω_i is the comprehensive decision result (CDR) of cost allocation ratio of pipeline i. $\omega_{i\max}$ and $\omega_{i\min}$ denote the maximum and minimum cost allocation ratios of pipeline i under different cost allocation indexes, respectively. If the satisfaction of any pipeline company is too low, the traditional buried pipeline would replace the UUT-laying pipeline, thus hindering the construction of the utility tunnel. Thus, the optimization problem for the comprehensive decision-making mechanism of the cost allocation among pipeline companies is expressed as:

$$\max U = \sum_{i \in N} \alpha_i u_i \tag{11}$$

$$s.t. \quad \omega_i = \lambda_S \omega_i^S + \lambda_C \omega_i^C + \lambda_R \omega_i^R \tag{12}$$

$$\lambda_S + \lambda_C + \lambda_R = 1 \tag{13}$$

$$\omega_i(1-\lambda_G)C_g \leq R^b_{i,T_g} + R^m_{i,T_g} \tag{14}$$

$$0 \leq \omega_i \leq \beta_i \omega_{imax} \tag{15}$$

$$\sum_{i \in N} \beta_i \omega_{imax} \geq 1 \tag{16}$$

where U is the overall satisfaction of pipeline companies, and α_i is the coefficient indicating the bargaining power of the i-th pipeline company. λ_S, λ_C, and λ_R represent the weights of the improved SPM, DCM, and BPM within the CDR, respectively. C_g is the UUT construction cost and λ_G is the proportion of C_g shared by the government. β_i is the biggest discount that the pipeline company i can accept. Inequality constraint Equation (14) is the individual rational constraint, which ensures that the cost shared by the pipeline company is lower than its net income from laying the pipeline into a utility tunnel. Inequality constraint Equation (15) limits the cost allocation proportions of pipeline companies so as to ensure that UUT-laying pipelines will not be replaced by traditional buried pipelines. Inequality constraint Equation (16) is the group rational constraint, which ensures that the UUT construction cost can be fully shared by pipeline companies.

It is believed that there is an interdependent relationship between pipeline companies based on the two facts: (1) the construction of UUT requires mutual cooperation between pipeline companies; (2) the government shares part of the construction cost of UUT to ensure that pipeline companies can obtain profits. However, the differences in cost-sharing, the cost-benefit ratio, and payback period of pipelines companies lead to their different preferences for cooperation, which is reflected in the differences in bargaining power in alliance negotiations. It can be seen from Equation (11) that the overall satisfaction of pipeline companies is related to the bargaining power of each pipeline company. The greater the bargaining power of the i-th pipeline company in the cost allocation negotiation, the greater its impact on the overall satisfaction of pipeline companies, and the lower the cost allocation proportion it obtained through optimization. Therefore, the relationship between pipeline companies is interdependent and competitive. The bargaining power of pipeline companies is analyzed based on the resource dependence theory (RDT) [20] in this paper.

The RDT focuses on the fact that enterprises require resources from others for survival and operation, and suggests that resource dependence is an essential part of enterprise relationships and a reason for the bargaining power imbalance in negotiations [21]. The level of the interdependence between pipeline companies is mainly determined by two factors, i.e., the importance of external resources and the possibility of alternative suppliers. The construction of UUT requires the joint efforts of the government and pipeline companies, and any one of them is irreplaceable, i.e., there is no alternative suppliers. Thus, the importance of external resources, i.e., the economic benefits of pipeline companies when laying pipelines into utility tunnel, is the key to measure the level of interdependence between pipeline companies. The greater the benefit of the pipeline company, the stronger its dependence on others, the lower the bargaining power in negotiation, and the smaller α_i in Equation (11). The economic benefits of pipeline companies can be measured by indexes such as a net investment income, payback period, and cost-benefit ratio [22]. Then, the coefficient α_i indicating the bargaining power of pipeline company in the negotiation of UUT construction cost allocation can be expressed as:

$$\alpha_i = P_i \frac{\omega_i(1-\lambda_G)C_g}{\left(R^b_{i,T_g} + R^m_{i,T_g}\right)^2} \tag{17}$$

$$P_i = \max\left\{t \middle| R^b_{i,t} + R^m_{i,t} - \omega_i(1-\lambda_G)C_g = 0\right\} \tag{18}$$

where P_i is the payback period of pipeline company i. $R^b_{i,t}$ and $R^m_{i,t}$ represent construction cost reduction and the maintenance cost reduction within the time P_i, respectively. It can be seen from Equation (17) that α_i is proportional to the payback period P_i, and inversely proportional to the net investment income and the cost-benefit ratio (i.e., the ratio of the investment income to the allocated UUT construction cost). Equation (18) denotes that the payback period P_i is the maximum value of time t that the net investment income within t is equal to the allocated cost of pipeline company i.

Although RDT plays an important role in explaining the organization's behavior, it does have defects and limitations [23]. Considering the operational environment, relational networks as well as inter-organizational power dynamics, RDT cannot fully explain the relationship between organizations. Further research can be conducted on the combination of RDT and other theories, e.g., social network theory, game theory, and agency theory, to study the cost-sharing negotiations among pipeline companies, taking into account economic, policy, and social environmental factors.

In summary, the process of the proposed comprehensive decision-making mechanism for UUT construction cost allocation is shown in Figure 2. Based on the cost allocation mechanism, the economic benefits evaluation of UUT-laying power cables could also be obtained.

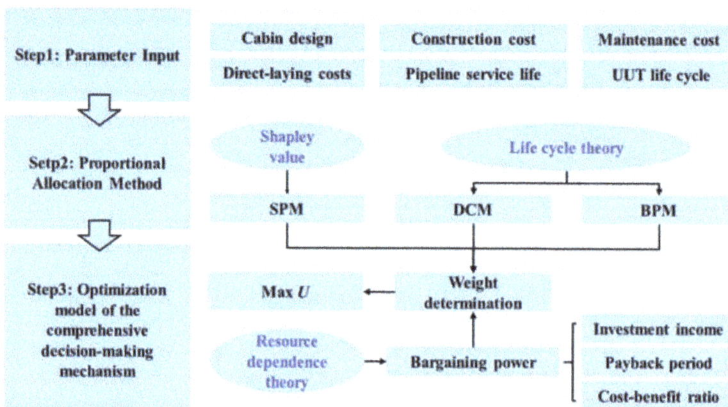

Figure 2. Flow chart of comprehensive decision-making mechanism for UUT construction cost allocation.

4. Case Studies

The data of the utility tunnel project of Shanghai Taopu Science and Technology Intelligence City in China is used for demonstrating the proposed comprehensive decision-making mechanism for UUT construction cost allocation. The utility tunnel consists of three cabins, in which 110 kV cables are laid in the high-voltage power cabin, gas pipelines are laid in the gas cabin, and 10 kV cables, communication cables and water supply pipelines are in the integrated cabin, as shown in Figure 3. The utility tunnel is 1000 m long, has a service life of 100 years, and has a total construction investment of 50.9 million CNY. Considering the reduction of civil engineering cost, the improvement of maintenance efficiency and the improvement of operation environment, reasonable assumptions are made on the laying cost, service life and maintenance cost of UUT-laying pipelines based on the data of traditional buried pipelines [24,25]. Assuming that the government shares 40% of the construction cost. The annual interest rate is 5%, and the largest discount that pipeline companies are willing to accept is 0.9. The construction cost of the underground power corridor is 5 million CNY/km [26]. Overall, the engineering parameters of the utility tunnel and pipelines are shown in Table 1, and the parameters of traditional buried pipelines are shown in Table 2.

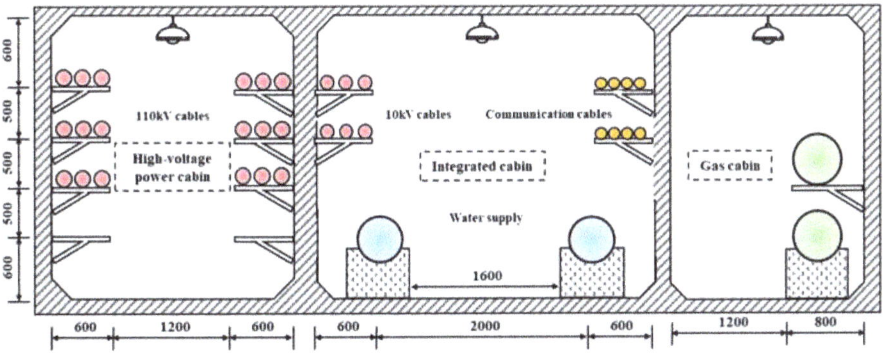

Figure 3. Cross-sectional view of utility tunnel in Taopu Science and Technology Intelligence City.

Table 1. Engineering parameters of UUT-laying pipelines.

	110 kV Cable	10 kV Cable	Water Supply	Communication	Gas
Occupied space (m^2)	6.48	2.26	2.26	4.12	5.40
Type and quantity	12	36	DN300*2	25	DN500*2
Laying cost (10^4 CNY/km*pipe)	35	20	30	15	70
Service life (years)	75	40	25	40	25
Maintenance cost (10^4 CNY/km)	0	0	0.2	0.2	0.2

Table 2. Engineering parameters of traditional buried pipelines.

	110 kV Cable	10 kV Cable	Water Supply	Communication	Gas
Laying cost (10^4 CNY/km*pipe)	40	25	60	20	100
Service life (years)	50	25	15	25	15
Maintenance cost (10^4 CNY/km)	0.1	0.1	1.5	0.2	1.0

4.1. Cost-Sharing Comparison of Pipeline Companies under Different Cost Allocation Methods

The weights of the improved SPM, DCM, and BPM in the comprehensive decision-making result (CDR) of the UUT cost allocation are 0.3398, 0.3027, and 0.3575, respectively. The allocated cost and proportion of each pipeline company are shown in Table 3 and Figure 4, respectively.

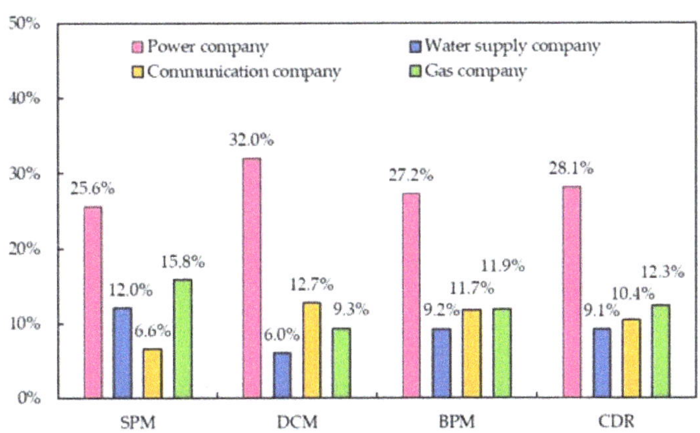

Figure 4. Cost allocation proportion of pipeline companies.

Table 3. Cost-sharing of pipeline companies (Unit: 10^4 CNY).

	Power	Water Supply	Communication	Gas
SPM	13.01	6.13	3.36	8.04
DCM	16.26	3.08	6.46	4.74
BPM	13.86	4.70	5.95	6.04
CDR	14.32	4.65	5.31	6.26

It can be seen from Table 3 and Figure 4 that the power company shares a higher proportion of the cost under different allocation methods, which makes it the most important pipeline company participating in the construction of a utility tunnel. Specifically, the differences in the cost allocation proportion of each pipeline company under different allocation methods are summarized as follows:

(1) Under SPM, the UUT construction cost is allocated based on the proportion of area occupied by the pipelines. For the sake of security, 110 kV cables and gas pipelines should be separated from other pipelines and need to be laid in their individual cabins, i.e., the high-voltage power cabin and the gas cabin, making the space occupied by power cables and gas pipelines larger (i.e., 8.75 m^2 and 5.40 m^2, respectively). The power company also needs to share part of the construction cost of the integrated cabin where 10 kV cables are laid, which makes it share the highest proportion of UUT construction cost (i.e., 25.6%).

(2) Under DCM, the proportion of direct-laying costs among pipelines is used to allocate the UUT construction cost. For power companies, the number of 110 kV and 10 kV cables to be laid is large, and underground power corridors need to be built for traditional buried cables, resulting in a higher total direct-laying cost than other pipelines. For water supply/gas companies, the laying cost of traditional buried pipeline and the frequency of repeated laying are relatively high. However, compared with the large number of power cables and communication cables laid, the laying of water supply and gas pipelines is not a high-cost business, which makes the cost shared by water supply companies and gas companies lower (i.e., 6.0% and 9.3%, respectively).

(3) Under BPM, the cost allocation proportion of each pipeline company is determined according to the economic benefit that the pipeline produces when laid into a utility tunnel. For power companies, the total cost of cable laying is reduced because the utility tunnel replaces the traditional underground power corridor that requires additional investment in construction, resulting in the highest economic benefits for the power companies. On this basis, the cost allocation proportion of power companies is the highest (i.e., 28.1%). For water supply and gas companies, compared with the traditional buried pipelines, UUT-laying pipelines have a lower laying cost, maintenance cost and repeated laying frequency, which increases their economic benefits. Therefore, the costs shared by water supply companies and gas companies under BPM are higher than those shared under other methods.

(4) Under the proposed comprehensive decision-making mechanism, the cost-sharing proportion of each pipeline company is the optimization result of the above-mentioned methods, which maximizes the overall satisfaction of pipeline companies. As shown in Table 4, the overall satisfaction of the pipeline companies under SPM, DCM and BPM is 54.8%, 72.5%, and 92.8% of the maximum overall satisfaction, respectively, which indicates that the proposed comprehensive decision-making mechanism of UUT cost allocation is more acceptable and more feasible.

Table 4. Economic benefits and overall satisfaction of pipeline companies.

	Economic Benefits (10^6 CNY)				Overall Satisfaction
	Power	Water Supply	Communication	Gas	
SPM	7.75	0.91	5.55	1.00	54.8%
DCM	4.50	3.97	2.45	4.30	72.5%
BPM	6.90	2.34	2.96	3.01	92.8%
CDR	6.44	2.39	3.60	2.78	100.0%

4.2. Economic Benefit Evaluation of UUT-Laying Power Cable

Under the comprehensive decision-making allocation mechanism, the power cable laying cost of the power company reduced by 12.5% (i.e., 6.44 million CNY) in the life cycle of the UUT. The payback period (i.e., the time when the cumulative laying cost of UUT-laying power cables is lower than that of traditional buried power cables) of the power company is 51 years, as shown in Figure 5.

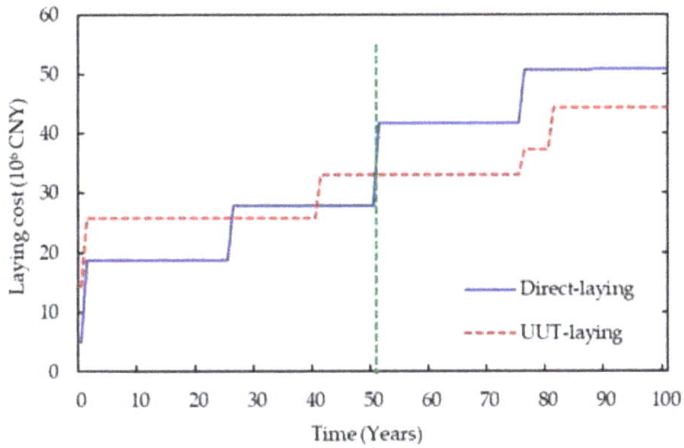

Figure 5. Comparison of power cable laying costs.

It is worth noting that the service life of traditional buried cables and UUT-laying cables are different, which makes the time point of cable re-laying different. Therefore, the cumulative laying cost of UUT-laying cables may be lower than that of the traditional buried cables during a certain period, but the power company cannot recover the allocated UUT construction cost. Take a 10 kV power cable as an example, due to the improvement of environmental conditions, the average service life of the UUT-laying cable is 40 years, 15 years longer than that of the traditional buried cables. It can be seen from Figure 5 that in the first 25 years, the UUT-laying cost curve is always higher than the direct-laying cost curve. Within the payback period, the traditional buried 10 kV cables need to be re-laid in the 26th and 51st years at a cost of 9 million CNY per time, while the UUT-laying 10 kV cables only need to be re-laid in the 41st year at a cost of 7.2 million CNY. As a result, in the 15 years period from the 26th to the 40th year, the UUT-laying cost curve is lower than the direct-laying cost curve, but in fact the cost shared by the power company is not offset. After the 51st year, the UUT-laying cost curve is always lower than the direct-laying cost curve, which shows that the power company has completed cost recovery.

It is important to assume that due to the increase of load in the UUT pilot area, the distribution network needs to double its capacity in the t_1-th and the t_2-th year after cable laying. The capacity of other types of pipelines has also increased at the same time, and cost allocation proportion of each pipeline company remains unchanged. Then, considering the expansion of distribution network, the cost comparison of traditional buried cables and UUT-laying cables under two scenarios is shown in Figures 6 and 7, respectively.

It can be seen from Figures 6 and 7 that when considering the expansion of distribution network, the payback period of the power company is shortened from 51 years to 26 years under Scenario A, and 46 years in the case of Scenario B. The power cable laying cost reduction of the power company within the life cycle of UUT is increased from 6.44 million CNY to 40.22 million CNY (under Scenario A) and 47.37 million CNY (under Scenario B), respectively. The earlier the distribution network is expanded or rebuilt, the shorter the payback period is, and the higher the economic benefit of UUT-laying power cables is. Generally, in the life cycle of UUT (usually 100 years), the expansion of distribution network and the transformation of power pipelines caused by the growth of regional energy consumption and the development of distributed generation are inevitable [27]. Therefore, it can be concluded that the use of UUT-laying power cables in place of traditional buried cables in areas with rapid load growth has great economic benefits.

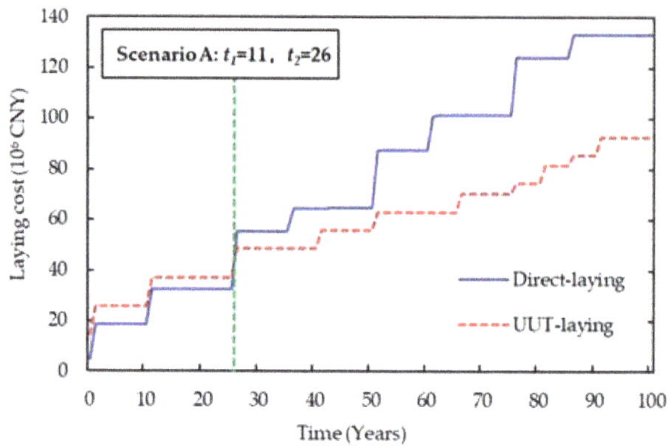

Figure 6. Comparison of power pipeline laying costs under Scenario A.

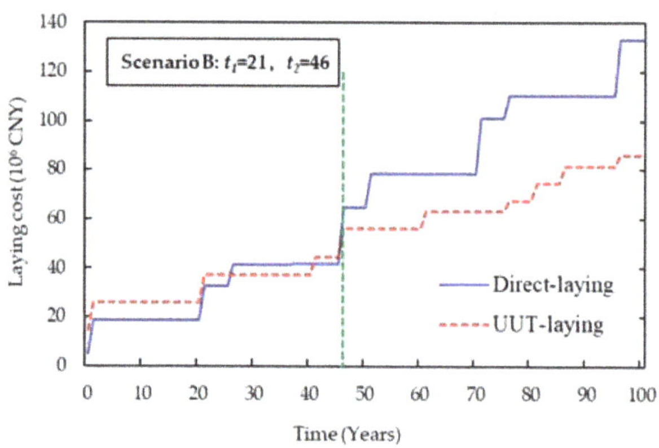

Figure 7. Comparison of power pipeline laying costs under Scenario B.

In addition, the relationship between the number of cables of different voltage levels laid in the utility tunnel and the economic benefits of the power company is shown in Figure 8. The upper limit of the number of 110 kV cables laid in the high-voltage power cabin is 30, and the upper limit of the number of 10 kV cables laid in the integrated cabin is 60.

It can be seen from Figure 8 that it is not economical for power companies to lay only 110 kV cables in a utility tunnel. With the increase of the number of 10 kV cables laid in the integrated cabin, the economic benefits of the power company have been continuously improved. When the number of 110 kV and 10 kV cables reaches their upper limits, the maximum net income (i.e., 15.31 million CNY) of the power company without considering the expansion of distribution network can be obtained, and the cost shared by the power company is about 17.56 million CNY, accounting for 34.5% of the total UUT construction cost. Compared with traditional buried cables, 10 kV UUT-laying cables have the advantages of lower laying cost, lower maintenance cost, and longer service life (i.e., lower repeated laying frequency), which is the main source of economic benefits of laying cables in UUT. Therefore, power companies should pay attention to the coordination of 110 kV and 10 kV UUT-laying cables in the distribution network planning to maximize their economic benefits.

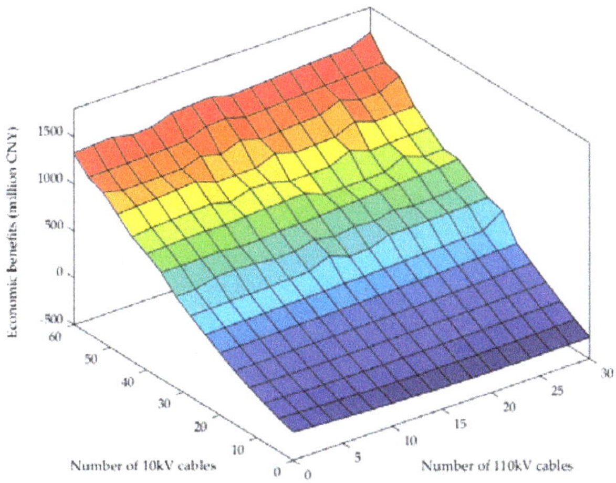

Figure 8. The relationship between the number of cables and the economic benefits.

4.3. Cost-Sharing Comparison of Single Integrated Cabin under Different Cost Allocation Methods

It is worth noting that the construction costs of UUT projects in different regions can vary hugely, which may be caused by the difference of cabin types and regional economic development [11]. To further prove the effectiveness of the proposed comprehensive decision-making cost allocation mechanism, the UUT project in Ningbo City, Zhejiang Province, China, is also used for demonstration [28]. The cross-sectional view of utility tunnel of Yincounty Avenue in Ningbo is shown in Figure 9. It is an integrated cabin in which 10 kV power cables, communication cables, and water supply pipelines are laid, while 110 kV cable and gas pipeline need not be laid. The total investment in construction is 60 million CNY per kilometer, 40% of which is funded by the government. The engineering parameters of the tunnel and pipelines are shown in Table 5.

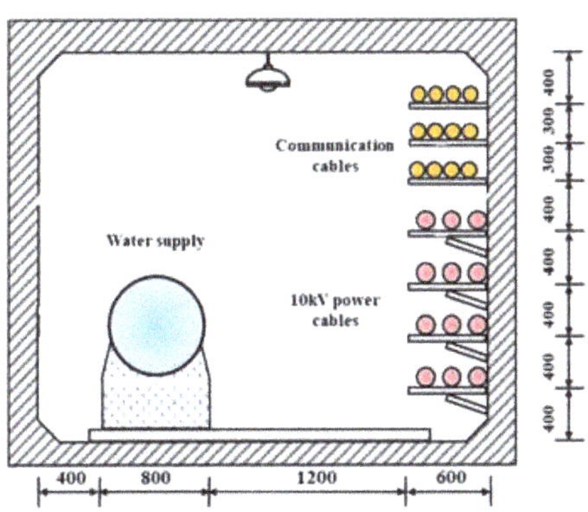

Figure 9. Cross-sectional view of the Yincounty Avenue utility tunnel.

Table 5. Engineering parameters of pipelines laid in the Yincounty Avenue utility tunnel.

	10 kV Cable	Water Supply Pipeline	Communication Cable
Occupied space (m^2)	2.4	4.8	1.8
Type and quantity	12	DN300	12

The optimization results of the proposed comprehensive decision-making mechanism are shown in Table 6. The weights of the improved SPM, DCM, and BPM in the comprehensive decision-making result (CDR) of the UUT cost allocation are 0.5361, 0, and 0.4639, respectively. As shown in Table 6, under DCM, the cost-sharing proportion of power company and telecommunications company reaches the maximum at the same time, i.e., 33.1% and 18.7%, respectively. It is believed that the two companies will oppose the DCM, and as expected, DCM has the lowest overall satisfaction among the four methods. The overall satisfaction of the pipeline companies under DCM and BPM is 69.2% and 72.0% of CDR, respectively. It can be concluded that although the data from different UUT projects may have an impact on the simulation results, under the proposed comprehensive decision-making mechanism, the cost-sharing ratio that achieves the maximum overall satisfaction can always be obtained.

Table 6. Cost allocation proportion and overall satisfaction of pipeline companies.

	Cost Allocation Proportion			Overall Satisfaction
	Power	Water Supply	Communication	
SPM	16.0%	32.0%	12.0%	69.2%
DCM	33.1%	8.2%	18.7%	30.4%
BPM	32.4%	13.3%	14.3%	72.0%
CDR	23.6%	23.3%	13.1%	100%

For power companies, the cost-sharing proportion under SPM (i.e., 16.0%) is significantly lower than that in Figure 4 (i.e., 25.6%). This is because when only 10 kV cables are laid, there is no need to build an individual high-voltage power cabin, which reduces the proportion of area occupied by the power cables. Accordingly, under the comprehensive decision-making mechanism, the cost allocation ratio of the power company is also reduced. Overall, laying 10 kV power cables in UUT has higher economic benefits than 110 kV cables.

5. Conclusions

The urban utility tunnel is an important infrastructure of urban multi-energy supply system, which presents solutions for the sustainable development of urban underground space, and has great economic and social benefits. To overcome the shortcomings of UUT construction cost allocation methods in China, a comprehensive decision-making mechanism for UUT cost allocation is designed in this paper. Firstly, the traditional spatial proportion method is improved based on the Shapley value, and the improved spatial proportion, the direct-laying cost proportion, and the benefit proportion of pipeline companies are taken as UUT cost allocation indexes. The bargaining power of pipeline companies is considered in the decision-making and analyzed based on the resource dependence theory. The weights of the cost allocation indexes are optimized with the objective of maximizing the overall satisfaction of the pipeline companies. Simulation results show that the proposed comprehensive decision-making mechanism for UUT construction cost allocation is more acceptable and more feasible than traditional cost-sharing methods. For power companies, compared with traditional buried pipelines, the cost of laying power cables can be significantly reduced by UUT, and the laying of 10 kV power cables has been shown to have higher economic benefits than laying 110 kV cables.

Author Contributions: Z.Z. and Z.L. conceptualized the study; Z.Z., J.R., Z.L. and J.X. performed the analysis; K.X., W.W. and C.P. performed investigations; J.R., Z.L. and J.X. acquired resources; J.R., W.W. and C.P. acquired funding; Z.Z. wrote the original draft; J.R., Z.L. and K.X. reviewed and edited the manuscript.

Funding: This work was supported by Science and Technology project of State Grid Corporation of China (52110418000V), National Natural Science Foundation of China (51777185) and Natural Science Foundation of Zhejiang Province (LY17E070003).

Conflicts of Interest: The authors declare no conflict of interest.

References

1. He, J.; Liu, L.; Ding, F.; Li, C.; Zhang, D. A new coordinated backup protection scheme for distribution network containing distributed generation. *Prot. Control Mod. Power Syst.* **2017**, *2*, 10. [CrossRef]
2. Ye, K.; Zhou, X.; Yang, L.; Tang, X.; Zheng, Y.; Cao, B.; Peng, Y.; Liu, H.; Ni, Y. A multi-scale analysis of the fire problems in an urban utility tunnel. *Energies* **2019**, *12*, 1976. [CrossRef]
3. Canto-Perello, J.; Curiel-Esparza, J. Assessing governance issues of urban utility tunnels. *Tunn. Undergr. Space Technol.* **2013**, *33*, 82–87. [CrossRef]
4. Legrand, L.; Blanpain, O.; Buyle-Bodin, F. Promoting the urban utilities tunnel technique using a decision-making approach. *Tunn. Undergr. Space Technol.* **2016**, *19*, 79–83. [CrossRef]
5. Guidance Opinions on Implementing the Paid Use System of Urban Utility Tunnel. Available online: http://www.ndrc.gov.cn/fzgggz/jggl/zcfg/201512/t20151209_761905.html (accessed on 14 July 2019).
6. Sun, F.; Liu, C.; Zhou, X.G. Utilities tunnel's finance design for the process of construction and operation. *Tunn. Undergr. Space Technol.* **2017**, *69*, 182–186. [CrossRef]
7. Hunt, D.V.L.; Nash, D.; Rogers, C.D.F. Sustainable utility placement via multi-utility tunnels. *Tunn. Undergr. Space Technol.* **2014**, *39*, 15–26. [CrossRef]
8. Chen, S. The Research on Investment Mode and Fare Apportion of Utility Tunnel. Master's Thesis, Tongji University, Shanghai, China, 2005.
9. Wang, T.; Tan, L.; Xie, S.; Ma, B. Development and applications of common utility tunnels in China. *Tunn. Undergr. Space Technol.* **2018**, *76*, 92–106. [CrossRef]
10. Cui, Q. Study on the Urban Utility Tunnel Toll Pricing under PPP Model. Master's Thesis, Beijing University of Civil Engineering and Architecture, Beijing, China, 2017.
11. Meng, X. Study on the Cost Sharing of the Underground Integrated Pipe Gallery. Master's Thesis, Qingdao University of Technology, Qingdao, China, 2016.
12. Sanayei, A.; Farid Mousavi, S.; Abdi, M.R.; Mohaghar, A. An integrated group decision-making process for supplier selection and order allocation using multi-attribute utility theory and linear programming. *J. Frankl. Inst.* **2008**, *7*, 731–747. [CrossRef]
13. Rao, M.S.S.; Soman, S.A.; Chitkara, P.; Gajbhiye, R.K.; Hemachandra, N.; Menezes, B.L. Min-max fair power flow tracing for transmission system usage cost allocation: A large system perspective. *IEEE Trans. Power Syst.* **2010**, *25*, 1457–1468. [CrossRef]
14. Bhakar, R.; Sriram, V.S.; Padhy, N.P.; Gupta, H.O. Probabilistic game approaches for network cost allocation. *IEEE Trans. Power Syst.* **2010**, *25*, 51–58. [CrossRef]
15. Molina, Y.P.; Prada, R.B.; Saavedra, O.R. Complex losses allocation to generators and loads based on circuit theory and Aumann-Shapley method. *IEEE Trans. Power Syst.* **2010**, *25*, 1928–1936. [CrossRef]
16. Zeng, L.; Zhao, L.; Wang, Q.; Wang, B.; Ma, Y.; Cui, W.; Xie, Y. Modeling interprovincial cooperative energy saving in China: An electricity utilization perspective. *Energies* **2018**, *11*, 241. [CrossRef]
17. Zolezzi, J.M.; Rudnick, H. Transmission cost allocation by cooperative games and coalition formation. *IEEE Trans. Power Syst.* **2002**, *17*, 1008–1015. [CrossRef]
18. Xu, T.; Meng, H.; Zhu, J.; Wei, W.; Zhao, H.; Yang, H.; Li, Z.; Ren, Y. Considering the life-cycle cost of distributed energy-storage planning in distribution grids. *Appl. Sci.* **2018**, *8*, 2615. [CrossRef]
19. Liu, H.; Fan, X. Value-added-based accounting of CO2 emissions: A multi-regional input-output approach. *Sustainability* **2017**, *9*, 2220. [CrossRef]
20. Biermann, R.; Harsch, M. Resource Dependence Theory. In *Palgrave Handbook of Inter-Organizational Relations in World Politics*; Koops, J.A., Biermann, R., Eds.; Palgrave Macmillan: London, UK, 2017; pp. 135–155. [CrossRef]
21. Shin, N.; Park, S.H.; Park, S. Partnership-based supply chain collaboration: Impact on commitment, innovation, and firm performance. *Sustainability* **2019**, *11*, 449. [CrossRef]
22. Chen, L.; Wu, T.; Xu, X. Optimal configuration of different energy storage batteries for providing auxiliary service and economic revenue. *Appl. Sci.* **2018**, *8*, 2633. [CrossRef]
23. Werner Nienhüser. Resource dependence theory: How well does it explain behavior of organizations? *Manag. Rev.* **2008**, *19*, 9–32.
24. Liu, L.; Zhu, X.; Liu, S. Study on technical and economic performance of power pipeline laying into Urban utility tunnel. *Eng. Constr.* **2017**, *4*, 434–441.

25. Wei, H.; Liu, W. Incentive mechanism study on the pricing control of underground pipe gallery under PPP model. *Chin. J. Undergr. Space Eng.* **2018**, *14*, 585–594.
26. Wang, Y.; Fang, R.; Zhou, Y. Study on calculation method of fees for pipe-gallery power cable and cost adjustment strategy. *Hubei Electr. Power* **2018**, *2*, 41–50.
27. Hong, H.; Hu, Z.; Guo, R.; Ma, J.; Tian, J. Directed graph-based distribution network reconfiguration for operation mode adjustment and service restoration considering distributed generation. *J. Mod. Power Syst. Clean Energy* **2017**, *5*, 142–149. [CrossRef]
28. Ningbo Bureau of Natural Resources and Planning; Ningbo Housing and Urban-rural Development Bureau. *The Special Plan of Urban Utility Tunnel of Ningbo City (2016–2020)*; Ningbo Bureau of Natural Resources and Planning; Ningbo Housing and Urban-rural Development Bureau: Ningbo, China, 2016.

energies

MDPI

Article

Study on Global Industrialization and Industry Emission to Achieve the 2 °C Goal Based on MESSAGE Model and LMDI Approach

Shining Zhang [1], Fang Yang [1], Changyi Liu [1,*], Xing Chen [1], Xin Tan [1], Yuanbing Zhou [1], Fei Guo [2] and Weiyi Jiang [3]

1 Global Energy Interconnection Development and Cooperation Organization, Xicheng District, Beijing 100031, China; shining-zhang@geidco.org (S.Z.); fang-yang1@geidco.org (F.Y.); xing-chen@geidco.org (X.C.); xin-tan@geidco.org (X.T.); zhouyuanbing@geidco.org (Y.Z.)

2 International Institute for Applied Systems Analysis (IIASA), Schlossplatz 1-A, 2361 Laxenburg, Austria; guof@iiasa.ac.at

3 Faculty of science, Camperdown campus, University of Sydney, Camperdown, Sydney 2006, Australia; wjia5733@uni.sydney.edu.au

* Correspondence: changyi-liu@geidco.org; Tel.: +86-10-6341-1703

Received: 2 January 2020; Accepted: 6 February 2020; Published: 13 February 2020

Abstract: The industrial sector dominates the global energy consumption and carbon emissions in end use sectors, and it faces challenges in emission reductions to reach the Paris Agreement goals. This paper analyzes and quantifies the relationship between industrialization, energy systems, and carbon emissions. Firstly, it forecasts the global and regional industrialization trends under Representative Concentration Pathway (RCP) and Shared Socioeconomic Pathway2 (SSP2) scenarios. Then, it projects the global and regional energy consumption that aligns with the industrialization trend, and optimizes the global energy supply system using the Model for Energy Supply Strategy Alternatives and their General Environmental Impact (MESSAGE) model for the industrial sector. Moreover, it develops an expanded Kaya identity to comprehensively investigate the drivers of industrial carbon emissions. In addition, it employs a Logarithmic Mean Divisia Index (LMDI) approach to track the historical contributions of various drivers of carbon emissions, as well as predictions into the future. This paper finds that economic development and population growth are the two largest drivers for historical industrial CO_2 emissions, and that carbon intensity and industry energy intensity are the top two drivers for the decrease of future industrial CO_2 emissions. Finally, it proposes three modes, i.e., clean supply, electrification, and energy efficiency for industrial emission reduction.

Keywords: industrialization; industrial CO_2 emission; MESSAGE model; Kaya identity; LMDI approach

1. Introduction

The industrial sector is the largest sector of energy consumption with largest CO_2 emission among the final sectors [1]. In 2017, the total energy consumption of the industrial sector accounted for 29% of the end-use energy consumption and 24% of the total CO_2 emissions. Considering the indirect energy consumption and CO_2 emissions from industrial electricity and heat, the percentage of industrial energy consumption and emissions are 43% and 42% respectively [2]. In order to understand the role of the industrial sector in the future energy consumption and emission pathways, especially under the rapid development of renewable energy and electrification, there is a need to analyze the global industrialization process, study the energy system and the drivers, and explore a future emission path for the industrial sector that aligns with the 2 °C global temperature control goal of the Paris Agreement.

There is much existing literature evaluating industrialization from the economic perspective. Regarding industrialization, Chang Pei-Kang studied the industrialization process and its relationship with the agriculture sector [3]. In this paper, we define the "industrialization" stage as the proportion of the manufacturing industries' output in the GDP structure. The manufacturing industries include iron and steel, cement, chemicals, pulp and paper, non-ferrous metals, food processing, textiles, leather, and mining etc. Economists have divided economic development into three stages: pre-industrialization, industrialization, and post-industrialization [4]. The industrialization process of a country is closely related to the stage of economic growth [3,4]. The basic characteristics of the industrialization are shown in the following aspects: (a) The increase in the proportion of manufacturing activities in national income structure. (b) The increase in the proportion of the labor population in the manufacturing industry. According to various standards, e.g., income level of GDP per capita, national income structure (percent of the first, second, and third industry respectively), employment structure, urbanization level, etc. Therefore, the industrialization stage can be further divided into three sub-stages: initial stage, intermediate stage, and late stage (Table 1).

Table 1. Different industrialization stages and criterions.

Criterions	Pre-Industrialization (I)	Industrialization Stages			Post-Industrialization Stage (V)
		Early Stage of Industrialization (II)	Intermediate Stage of Industrialization (III)	Late Stage of Industrialization (IV)	
(1) GDP per capita (2015 USD, in PPP)	<1000	1000~5000	5000~18,000	18,000~30,000	>30,000
(2) Economic structure	A > I	A > 20% and A > I	A < 20% and A > S	A < 10% and I > S	A < 10% and I < S
(3) Urbanization rate by population	Below 30%	30~50%	50~60%	60~75%	Above 75%

Note: (1) criterions refers to Chen et al. [5], the standard of per capita GDP levels be updated by the authors; (2) *A* denotes agriculture, *I* denotes second industry, *S* denotes the services sector.

The manufacturing industries represent the industrialization process and dominate energy consumption and emission in the end-use sectors. Along with the industrialization process, the proportion of output value of the manufacturing industries in the national economy will experience an inverse-U shaped process which is also known as the "Kuznets curve." In general, the proportion of the second industry in GDP rises first, then reaches peak, and falls afterwards. The internal structure of the manufacturing industry will experience the same process as well. During the early stage of industrialization, the dominant industries are labor-intensive light industries such as the textile and food sectors, whose proportion will first increase and peak, and then begin to decline at the end of the early stage. While in the intermediate stage of industrialization, the dominant industries are capital-intensive heavy industries, such as the iron and steel, cement, electricity, and other energy and raw material-related industries, in the late stage of industrialization, the dominant industries involve technology-intensive high-value machining manufacturing, such as equipment manufacturing sectors. When entering the post-industrialization stage, the proportion of secondary industry output falls, while the proportion of third industry rises [6]. According to the experience of the major developed countries, the proportion of industrial output in the total output will keep stable between 20%~30%. This result is driven by both the international industry transfer and domestic industry upgrade trends [7].

There is also extensive literature regarding industrial energy and emissions. For example, Caraiani et al. [8] study the causality relationship between energy consumption and economic growth in the context of emerging European countries, while Ntanos et al. [9] study renewable energy and economic growth based on European countries. Taeyoung and Jinsoo [10] study the relationship between coal consumption and economic growth based on the OECD and non-OECD countries. Chen et al. [3] study the relationship between industrialization and industrial CO_2 emissions for China. Wang and Chen [11] study the decarbonization pathways of industrial energy consumption under 2 °C scenario for a comparison of China, India and Western European countries. Van Ruijven et al. [12] study the energy use and CO_2 emissions from the global steel and cement industries based on model projections.

International Energy Agency (IEA) studies the industrial energy efficiency and CO_2 emissions [13]. However, there are few papers attempting to link the industrialization process with the industrial energy consumption and CO_2 emissions together, especially at the global level [12]. This is one of the main innovative contributions of this paper: to establish and quantify the relationship between industrialization, energy system, and carbon emissions.

When identifying drivers of CO_2 emissions, Kaya identity [14] is commonly used in Intergovernmental Panel on Climate Change (IPCC) reports [15] and IEA reports for tracking the trends of key drivers [16]. As indicated from IEA statistics, energy intensity has decreased by 34% globally compared with the reference year of 1990, while population and GDP per capita has increased by 41% and 68% respectively [16]. In order to further study the impact of these drivers, the Logarithmic Mean Divisia Index (LMDI) approach [17,18] is proposed and widely used to decompose the drivers for national CO_2 emission. For example, Freitas and Kaneko [19] used the LMDI approach to evaluate the changes in CO_2 emissions from energy consumption in Brazil for the period 1970–2009 and the results demonstrated that economic activity and demographic pressure are the leading forces explaining emissions increase. Fatima et al. [20] applied LMDI method to study energy-related CO_2 emission in China's industrial sector and found that CO_2 emission experienced a significant increase from 1991 to 2013, and started to decrease in 2016. They also identified that the income effect and labor effect are the top two contributors to emissions. Arsalan et al. [21] used the LMDI approach to decompose the changes in CO_2 emissions in Pakistan for the time periods of 1990–2017 and found that activity effect, structural effect and intensity effect were identified as the three major factors responsible for changes in overall CO_2 emissions in the country. Besides Brazil [19], China [20,22,23], and Pakistan [21], the LMDI approach is also applied in the study for Iran [24], Philippines [25], Spain [26], Portugal [27], Iran [28], Turkey [29], and Greece [30] etc. In terms of regional study, Wang and Chen [11] applied a 14-region energy system model (Global TIMES) to analyze the transition pathways of the industrial sector using the LMDI approach and found that the changes in socio-economic development pattern could slow the emission growth. Moutinho et al. [31] identified the relevant factors that have influenced the changes in the level of CO_2 emissions among four groups (eastern, western, northern and southern) of European countries and found that energy mix, switching to cleaner fuels for end-user energy production contributes significantly to emission reduction. To sum up, most of the previous studies focused on analyzing national historical data to identify the contribution of drivers in CO_2 emission using the LMDI approach, while the key drivers' potential in future CO_2 emission reductions are yet to be discussed [17,19–31]. Furthermore, a traditional Kaya identity equation consisting of energy intensity, CO_2 intensity, economy, and population is used in the decomposition analysis, while more detailed factors are overlooked for better understanding the source of those drivers [11]. Another aim of this paper is to fill these gaps.

The aim of this paper is to first explore the global industrialization trend, the energy demand, and emission trend for the world and 11 regions achieving the temperature control goal of 2 °C of the Paris Agreement, and then to expand the traditional Kaya identity to study the drivers of CO_2 emissions from historical data and a future scenario using the LMDI approach. Section 2 provides a description of the method and data, including the industrialization regression and projection method; energy and emission prediction and optimization software by the Model for Energy Supply Strategy Alternatives and their General Environmental Impact (MESSAGE) model and Representative Concentration Pathways (RCPs) and Shared Socioeconomic Pathways (SSPs) scenarios database, the decomposition method of Kaya identity, and contribution analysis method of the LMDI approach. Section 3 is the analysis and results, it assesses the global industry development and industrialization trend in the 21st century, the global energy demand and emissions from the industrial sector under the 2 °C scenario, and the decomposition and contribution analysis using the LMDI approach. Section 4 summarizes the modes for industrial emission mitigation. Section 5 presents the conclusions and future works.

2. Method and Data

2.1. Industrialization Projection Method

This paper analyzes the status quo of the current industrialization stages globally, for 11 regions. The world average industrialization rate is 24% in 2015. Based on the industrialization rate and GDP per capita levels for 11 regions, from Figure 1, it clearly shows the inverse-U shape of the industrialization process. The least developing region, i.e., Sub-Saharan Africa (AFR) is still at the first stage of industrialization or Before Industrialization stage. South Asia (SAS) is at the Early stage of Industrialization. These two regions are still experiencing an increasing trend in terms of their industrialization rate. According to the criterions from Table 1, and based on the historical data for the industrialization rate and GDP per capita levels for these six regions, these regions are in different sub-stages of industrialization. The Central and Eastern Europe (EEU), Former Soviet Union (FSU) and Latin America and the Caribbean (LAC) are in the late stage of industrialization, while the Centrally planned Asia and China (CPA), Middle East and North Africa (MEA) and Other Pacific Asia (PAS) are in the intermediate stage of industrialization. The three most developed regions, i.e., North America (NAM), Western Europe (WEU), Pacific OECD (PAO) are in the post-industrialization stage.

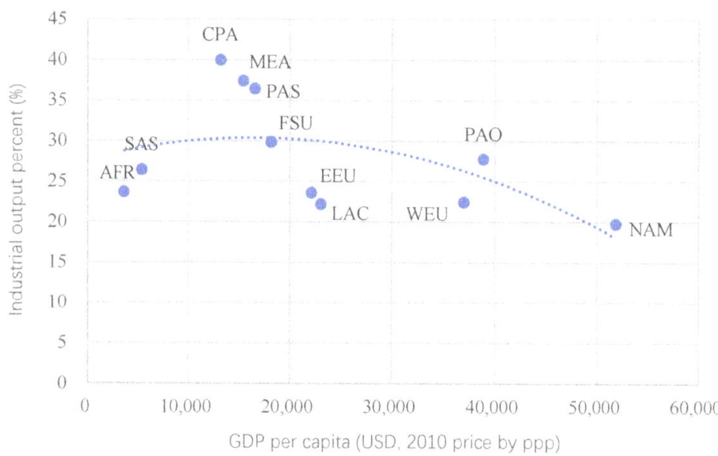

Figure 1. The status of industrialization globally and across 11 regions in 2015. Note: (1) 11 regions are defined by International Institute for Applied Systems Analysis (IIASA) [32]. (2) Data of GDP per capita and industrial output percent are retrieved from the World Bank Database [33]. (3) Dashed line is the world average; dotted line is the regression line from the 11 regions GDP per capita and industrialization data.

Based on the historical data of industrialization for the 11 regions across the globe, we develop a regression method to predict the future industrialization for those regions. The country-level historical annual data such as the GDP, population and GDP per capita (denoted as *gdppc*), from 1990 to 2017 are collected from the World Bank Database [31]. Then, country-level data are aggregated to 11 regions defined by IIASA [32]. We collect the annual GDP per capita (*gdppc*) along with its quadratic and cubic forms as the main explanatory variables. The explained variable is the industrialization level (*ind*). The basic regression equation is defined as following:

$$ind_{it} = constant + a{\cdot}gdppc_{it} + b{\cdot}{gdppc_{it}}^2 + c{\cdot}{gdppc_{it}}^3. \quad (1)$$

where i denotes 11 regions, t denotes different years, a, b, c are parameters.

When we arrive at the regression results for 11 regions, we apply the future *gdppc* data derived from the second Shared Social-economic Pathways (SSP2) [34] to predict the future industrialization level for 11 regions. Two caveats arise here: the first is that there is only one explanatory variable, *gdppc*, because we try to catch the relationship between income and industrialization level; the second is that the prediction period of 2018–2100 is much longer than the historical period in order to match the "S-curve" method for energy projection and the Kaya and LMDI methods.

2.2. Industrial Energy Demand Projection Method

A hump-shaped function method is used to project the industrial energy demand in each region. Historical data show that the relationship between the industrial energy consumption per capita and GDP per capita follows the "S-Curve": with the increase of GDP per capita, the industrial energy consumption per capita first increases, and then peaks then decreases along with the industrialization process [3]. Studies reveal that the "S-curve" method can capture the relationship between industrial energy consumption per capita and GDP per capita. Based on the industrialization projection method in Section 2.1, here we apply a simple top-down method, i.e., the "S-curve" method to project the global energy consumption for the industry sector.

The mathematic equation to describe the S-curve relationship [35] between the industrial energy consumption per capita and GDP per capita is:

$$E - E_i = A\frac{\exp(\alpha_1(G - G_i)) - \exp(-\alpha_3(G - G_i))}{2\cosh(\alpha_3(G - G_i))} \tag{2}$$

where i is the turning point of the S-curve; G_i is the GDP per capita at the turning point; E_i is the industrial energy consumption per capita at the turning point; α_1, α_2, α_3, and A are country specific parameters which can be estimated from regression results.

According to the time of industrialization process, industrial structure, duration of industrialization as studied in Section 2.1 and urbanization development in each country or region, the S-curve could be classified into three main types: high S type, middle S type, and low S type [35]. Countries categorized with high S type consumed more energy intensive products than middle and low type, resulting in a larger magnitude of turning point. Furthermore, the timing of the high S type turning point occurrence lags middle and low S type, resulting in high GDP per capita when the turning point occurred. Threshold GDP per capita for different S-curves is summarized in Table 2. Major boundary conditions and assumptions in the MESSAGE model are described in Table 3.

Table 2. Threshold GDP per capita levels for different S-curve.

Type	Take-Off Point	G_i and E_i at Turning Point	G_i and E_i at Zero Growth Point	Representative Countries
High "S" type	2500~3000	10,000~12,000 1.6~1.8	15,000~17,000 2~2.5	U.S., Canada
Mid "S" type		10,000~12,000 1.4~1.5	15,000~17,000 1.7~1.8	Sweden, Belgium
Low "S" type		7000~9000 0.6~0.8	10,000~12,000 0.7~1.2	England, France, Germany, Japan, Italy, etc.

Table 3. Major boundary conditions and assumptions.

Items		Description
Model related	Socio-economic	SSP2
	Industry energy demand	S-Curve
	Resource potential	Resource curves for each region
	Solve	Global optimization
Scenario related	2 °C target	Carbon budget for 2018–2100: 1280 Gt CO_2 [1]
	Reference scenario	No carbon limit, NPi_v4 from IIASA
	Technology	Exogenous technological progress

In the present study, traditional Kaya identity [14] is further expanded to decompose the industrial carbon emission and a LMDI approach is used to investigate the contribution of each drivers in carbon emission based on historical data. Furthermore, on the basis of MESSAGE optimized energy system results under conditions of the 2 °C target, decomposition analysis for projected industrial carbon emissions is carried out. The analysis framework is shown in Figure 2.

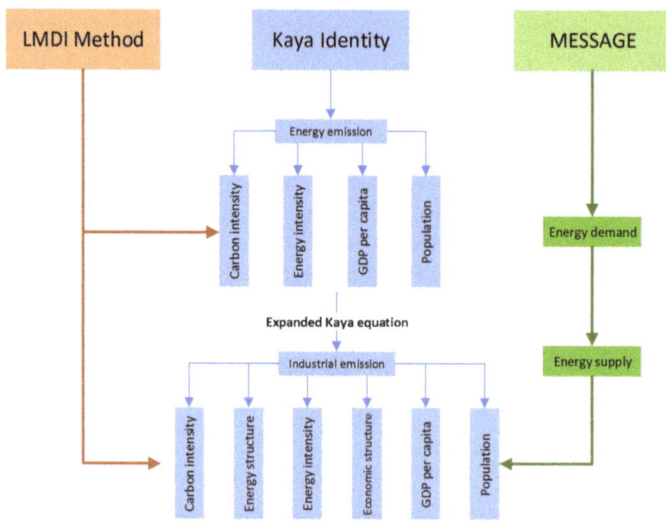

Figure 2. The framework of present and projection analysis on industrial emission reduction based on MESSAGE optimization and expanded Kaya equation with LMDI decomposition.

2.3. Expanded Kaya Equation and Contribution Decomposition

2.3.1. Expanded Kaya Equation

Traditional Kaya identity [14] is a commonly used way to decompose carbon dioxide emissions, which is expressed as the product of four factors: population, GDP per capita, energy intensity and carbon intensity (see Equation (3)).

$$CO_2 = POP \times \frac{GDP}{POP} \times \frac{E}{GDP} \times \frac{CO_2}{E}. \quad (3)$$

where CO_2 is the total CO_2 emissions, POP is total population, GDP is economic output, E. is total primary energy consumption, while GDP/POP denotes GDP per capita indicating income level, E/GDP represents energy intensity which indicates energy efficiency, CO_2/E stands for carbon intensity, reflecting the effect from energy structure changes.

Traditional Kaya identity is widely used to decompose carbon emissions [16], but it fails to consider the impact of industrialization and electrification on carbon reduction. An expanded Kaya identity is re-written as follows,

$$CO_2 = \frac{CO2}{FOF} \times \frac{FOF}{TOE} \times \frac{TOE}{IND} \times \frac{IND}{GDP} \times \frac{GDP}{POP} \times POP \tag{4}$$

Here, the CO_2 is the total industrial CO_2 emissions, *FOF* is total industrial fossil fuels consumption, *TOE* is total industrial energy consumption, *IND* is industrial output, *GDP* is economic output, *POP* is total population. Therefore, the industrial CO_2 emissions is decomposed into six drivers, while the CO_2/FOF denotes CO_2 emission intensity of fossil fuels, *FOF/TOE* denotes the energy structure of fossil fuel among total energy consumption, *TOE/IND* denotes energy intensity of industrial output, *IND/GDP* denotes the industrialization level, *GDP/POP* denotes GDP per capita. This paper applies the Kaya method, which is in essence a top-down method, while combined with a bottom-up study on specific industries based on local endowments of different regions.

2.3.2. Contribution Analysis Based on LMDI Approach

When taking all industry sectors into consideration, the total industrial emissions could be expressed in the following way as in Equation (5).

$$\begin{aligned} CO_2 = \sum_{ij} CO_{2ij} = \sum_{ij} \frac{CO_{2ij}}{FOF_{ij}} \times \frac{FOF_{ij}}{TOE_i} \times \frac{TOE_i}{IND_i} \times \frac{IND_i}{GDP} \times \frac{GDP}{POP} \times POP \\ = \sum_{ij} C_{ij} \times S_{ij} \times I_i \times Q_i \times G \times P \end{aligned} \tag{5}$$

where CO_{2ij} is the CO_2 emissions arising from fuel j in industrial sector i, FOF_{ij} is the consumption of fuel j in industrial sector i, TOE_i is total energy consumption in industrial sector i, IND_i is total industrial output in industrial sector i; C_{ij},S_{ij},I_i,Q_i,G and P represent the drivers of carbon emission from carbon intensity, energy structure, industrial energy intensity, economic structure, economic development, and population, respectively.

According to Ang [18] and present CO_2 decomposition, changes in CO_2 emission from industry could be expressed in an additive decomposition way as follows,

$$\Delta C_{tot} = C^T - C^0 = \Delta C_{cei} + \Delta C_{str} + \Delta C_{iei} + \Delta C_{estr} + \Delta C_{ed} + \Delta C_{pop} \tag{6}$$

where,

$$\begin{aligned} \Delta C_{cei} &= \sum_{ij} \frac{C_{ij}^T - C_{ij}^0}{\ln C_{ij}^T - \ln C_{ij}^0} \ln\left(\frac{C^T}{C^0}\right) \\ \Delta C_{str} &= \sum_{ij} \frac{C_{ij}^T - C_{ij}^0}{\ln C_{ij}^T - \ln C_{ij}^0} \ln\left(\frac{C^T}{C^0}\right) \\ \Delta C_{iei} &= \sum_{ij} \frac{C_{ij}^T - C_{ij}^0}{\ln C_{ij}^T - \ln C_{ij}^0} \ln\left(\frac{S^T}{S^0}\right) \\ \Delta C_{estr} &= \sum_{ij} \frac{C_{ij}^T - C_{ij}^0}{\ln C_{ij}^T - \ln C_{ij}^0} \ln\left(\frac{I^T}{I^0}\right) \\ \Delta C_{ed} &= \sum_{ij} \frac{C_{ij}^T - C_{ij}^0}{\ln C_{ij}^T - \ln C_{ij}^0} \ln\left(\frac{Q^T}{Q^0}\right) \\ \Delta C_{pop} &= \sum_{ij} \frac{C_{ij}^T - C_{ij}^0}{\ln C_{ij}^T - \ln C_{ij}^0} \ln\left(\frac{P^T}{P^0}\right) \end{aligned} \tag{7}$$

3. Results and Discussion

3.1. Industrialization and Its Projection

In this paper, we use the historical industrialization data to predict the future industrialization process for the 11 regions. The regression results for the 11 regions are as in Table 4. From the regression results we can find that there are significant downward trends for the most developed regions and also the post-industrialization regions such as the NAM, WEU, and PAO. With regard to the late stage of industrialization regions such as EEU and FSU, we can find that there is significant downward trend for EEU; there is an inverse-U shape for the FSU; there is a downward cube trend for the LAC region. With regard to the regions in the intermediate stage of industrialization, CPA shows that it has passed the peak and goes downward of industrialization; MEA and PAS show downward cubic trend. With regard to the AFR and SAS regions, they show an upward trend, i.e., they are on the left side of the inverse-U curve and still climbing up of their industrialization levels.

Table 4. The regression results for the 11 regions.

Regions	Constant	a	b	c	Adj-R^2	F-stat	*p*-Value for F
CPA	38 ***	2.22×10^{-3} ***	-1.63×10^{-7} ***	/	0.78	48.2	2.64×10^{-9}
EEU	98.04 ***	-1.46×10^{-2} **	1.02×10^{-6} **	-2.34×10^{-11} **	0.39	5.07	0.012
FSU	119 ***	-1.42×10^{-2} **	7.53×10^{-7} **	-1.32×10^{-11} **	0.58	9.57	7.4×10^{-4}
LAC	847.4 ***	−0.2 ***	1.615×10^{-5} ***	-4.34×10^{-10} ***	0.526	11	9.74×10^{-5}
MEA	368.2 *	−0.12 *	1.42×10^{-5} **	-5.34×10^{-10} **	0.69	21	6.78×10^{-7}
NAM	31.2 ***	-2.12×10^{-4} ***	/	/	0.75	82	1.54×10^{-9}
PAS	66.3 ***	-9.05×10^{-3} **	8.08×10^{-7} **	-2.33×10^{-11} **	0.55	12.2	4.84×10^{-5}
PAO	66.4 ***	1.02×10^{-3} ***	/	/	0.68	57.4	3.83×10^{-8}
SAS	16.4 ***	6.24×10^{-3} ***	-8×10^{-7} ***	/	0.62	23.2	1.99×10^{-6}
AFR	4.47	2.48×10^{-2} **	-4.61×10^{-6} **	/	0.45	12	2.22×10^{-4}
WEU	39.8 ***	-4.93×10^{-4} ***	/	/	0.91	282	1.78×10^{-15}

Note: (1) Significance levels: * denotes $p < 0.1$; ** denotes $p < 0.05$; *** denotes $p < 0.01$. Use R software for regression. (2) The EEU and FSU regions went through economic recessions in 1990s thus authors use historical data during period 1998–2017 for regressions.

Based on the SSP2 scenario, we can predict the future industrialization levels of the 11 regions. The prediction results are as in Figure 3. The figure on the left is for six developed regions; while the figure on the right is for the five developing regions.

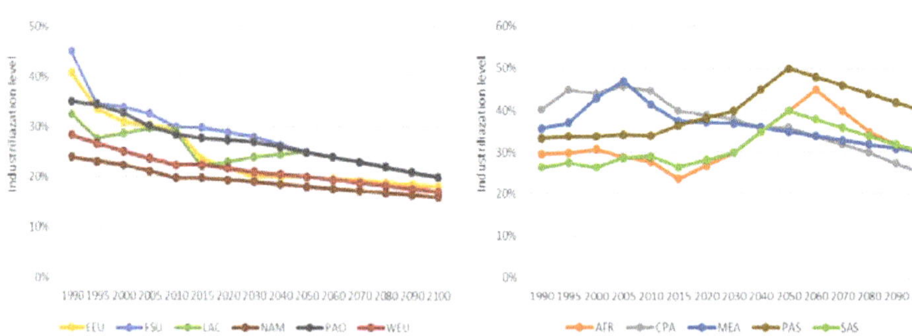

(a) Developed regions (b) Developing regions

Figure 3. The predicted industrialization results for 11 regions. Note: (**a**) Developed regions; (**b**) Developing regions.

3.2. Industrial Energy Consumption and Demand

In 2017, the direct energy consumption of the industrial sector was 2821 Mtoe, accounting for 29% of end-use energy consumption [2]. The industrial sector's direct emission is approximately

7.7 $GtCO_2$, which is about 24% of the world total emissions in 2016 [2]. Based on the historical emissions of the industrial sector, both direct and indirect emissions are increasing from 1970 to 2010, with a rapid increase after the 2000s, accompanying fast industrialization, especially the heavy industry development in China. It was also found that the indirect emissions from electricity and heat consumption in the industrial sector is increasing faster than direct emissions, implying that the electrification for the industrial sector plays an increasingly important role in mitigation options.

As introduced in Section 2.2, the total industry energy demand is firstly projected using an S-curve and then energy structure in industry sector is optimized by the MESSAGE model [36]. The identified parameters in S-curves for 11 regions are listed in Table 5.

Table 5. Major boundary conditions and assumptions.

Regions	"S" Type	E_i (toe/p)	A ($/capita)	α_1 (1/$)	α_2 (1/$)	α_3 (1/$)	G_i ($/capita)
CPA	Low	0.75	0.7	0.00002	0.00025	0.00009	9000
EEU	Mid	0.55	1	0.000015	0.00025	0.00009	7000
FSU	Mid	0.75	1.2	0.000015	0.00025	0.00009	6000
LAC	Mid	0.6	1	0.000015	0.00025	0.00009	10,000
MEA	High	0.55	1	0.00001	0.00035	0.0001	7000
NAM	High	0.75	1	0.000015	0.00025	0.00009	15,000
PAS	Low	0.55	1	0.000015	0.00025	0.00009	10,000
PAO	Low	0.7	1	0.000015	0.00025	0.00009	9000
SAS	Low	0.45	1	0.00002	0.00019	0.00011	10,000
AFR	Low	0.39	1	0.00002	0.00019	0.00011	7300
WEU	Mid	0.7	1	0.000015	0.00025	0.00009	10,000

The industrial energy consumption per capita estimation from the S-curve model for several specific regions is shown in Figure 4. Most of the countries in WEU are now in Post-industrialization stage and they concentrate more on technology-intensive products with high added value in industrial sectors. Furthermore, lots of efforts have been spent on energy efficiency improvement. For example, energy intensity in manufacturing sector from Ireland, Denmark, and United Kingdom and United States decreased by 46%, 26%, 20%, and 9%, respectively in the past five years. Consequently, industry energy consumption per capita in WEU will experience a decreasing trend in coming years. On the contrary, countries in SAS and AFR are all developing or undeveloped countries. Those countries are in the pre-industrialization stage or are experiencing industrialization, and they will pursue urbanization and economic development in the coming years. With more energy-intensive products such as steel, cement, chemicals and petrochemicals, and nonferrous metals produced in industrial sectors, the energy demand will firstly see a significant growing trend in coming years, but it will shift to a slow decrease as industrialization is completed and energy efficiency improves. In 2016, 34 of 53 countries in Africa are estimated to be in the pre-industrialization stage according to the method introduced by Chen et al. [5], while 14 countries are in the intermediate-industrialization stage and only five countries are estimated in the post-industrialization stage. Therefore, considering the relatively undeveloped situation, AFR is the last region to complete the industrialization process while energy consumption in AFR will keep increasing due to its upcoming booming economic development and increasing population.

With the acceleration of industrialization in Africa and Asia and the re-industrialization in Central and South America, the energy consumption in the industrial sector increases year by year. By 2050, the direct energy consumption in the industrial sector will increase by 54% to 4230 Mtoe, accounting for 40% of end-use energy consumption and surpassing the building sector to become the largest end-use energy consumption sector. In 2100, the industrial energy demand will increase to 5060 Mtoe, 22% more compared to 2050. As discussed in Section 2, a 2 °C scenario and a business-as-usual (BAU) scenario as the reference are used in this study, and the industrial sector is part of the whole global energy system in these scenarios [37]. Energy technologies in industry and other sectors are optimized to provide energy services in the MESSAGE model for both scenarios. As seen in Figure 5, more electricity is consumed in the 2 °C scenario, resulting in less energy consumption compared with the reference case. In 2050 and 2100, industry energy demand under 2 °C conditions are 5% and 20% respectively less than that in the reference scenario.

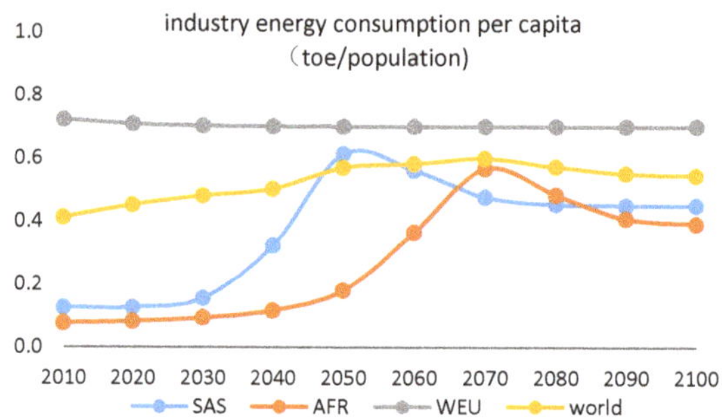

Figure 4. The evaluated regional and global industrial energy consumption per capita estimation from S-curve model.

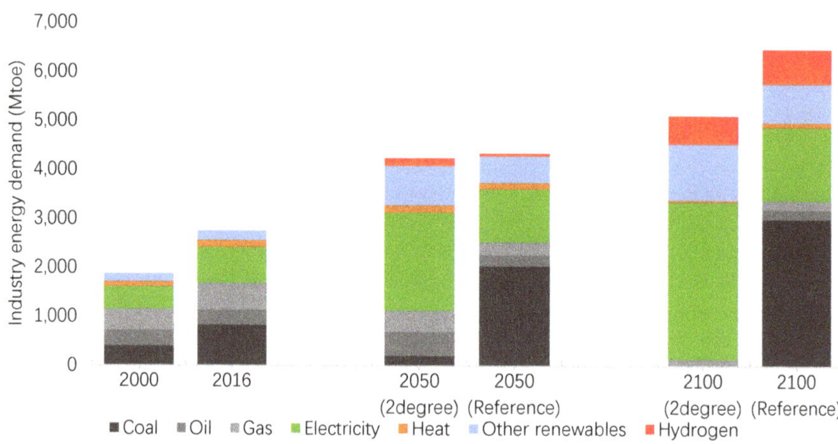

Figure 5. Industry energy demand and energy structure in 2050 and 2100 from 2 degree and reference scenarios.

The energy structure is optimized in the above two scenarios. In the 2 °C scenario, total fossil fuel consumption in the industrial sector fell to 1120 Mtoe in 2050, a decrease of 34% compared with 2017. The fossil fuel consumption reduction in the industrial sector is mainly due to increasing electricity consumption and the direct use of clean energy such as solar, bioenergy, and geothermal energy to provide heat within the low and middle range of temperature. Compared with 2017, the total electricity consumption in the industrial sector in 2050 increased by about 1240 Mtoe, accounting for 90% of the industrial energy increment. The share of electricity consumption in the industrial sector increases from 27.1% in 2016 to 48% in 2050 with an average annual growth rate of 0.6 percentage which is four times the growth from 2000–2017. Besides the increasing electricity consumption, the direct use of renewables in industry increases more than three times from 2016 to 2050, most of which are solar and bioenergy deployment. Solar energy utilization in the industrial sector has achieved extensive development from a few applications. With the increasing maturity of solar energy application technology, the cost of direct solar technology utilization in the industrial sector has dropped significantly. Solar water heaters, solar air heating systems, and solar collector systems are widely used in industrial processes for low-temperature heating and preheating in high-temperature demand. Furthermore, as the cost of clean energy power generation declines, hydrogen produced by electrolyzed water will gradually become economically competitive with fossil fuel energy hydrogen production, which is driving more hydrogen usage in industrial sector to provide high-temperature heat. 150 Mtoe hydrogen is expected to be used in 2050 while it increases to 570 Mtoe in 2100. In summary, more electricity and clean energy are used in 2 °C scenario compared with that in the reference scenario, as can be seen from leading to significant carbon reduction.

If the energy consumption for producing electricity, heat, and hydrogen are also allocated to consuming sectors, energy consumption in industrial sector increases to 4086 Mtoe in 2016 and 5270 Mtoe in 2050, increasing by 48% and 25% respectively compared with the case without electricity, heat and hydrogen energy consumption allocating to consuming sectors as shown in Figure 5. As seen in Figure 6, adjusted industry energy demand in 2 °C scenario is comparable to that in the reference scenario before 2050, while energy demand is smaller in the 2 °C scenario after 2050 due to massive electricity utilization and cleaner energy based power generation.

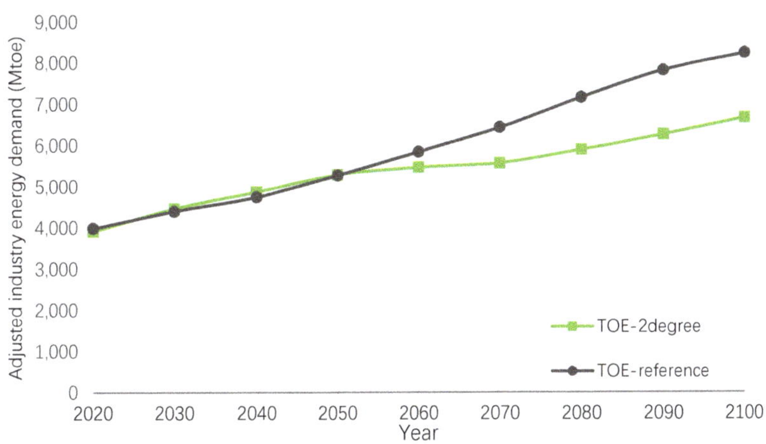

Figure 6. Adjusted industry energy demand with energy consumption in electricity, heat and hydrogen are allocated to consuming sectors in 2 degree and reference scenarios.

3.3. Industrial Emissions and Its Projection

It should be noted that CO_2 emissions from fuel combustion with electricity, heat, and hydrogen are allocated to consuming sectors in this research. Total direct and indirect industrial emission in 2016 is 13,537 Mt CO_2 accounting for 42% of total emissions. Industrial sector dominates the emissions in the end-use factors and the emissions differ a lot between countries. Nearly half (43%) of industrial carbon emissions come from China, while the second largest emission source (United States) accounts for 8.3% and the third largest emission source (India) accounts for 7.1%. In the case of reference scenario, carbon emission in 2100 reaches to 17,000 Mt CO_2 which is even 25% larger than that in 2016 (see Figure 7). Under conditions of 2 °C constraints, the industrial carbon emission in 2050 reduces to 3690 Mt CO_2 decreasing by 73% compared with that in 2016, and the carbon emission is further reduced to 350 Mt CO_2 in 2100 as shown in Figure 7. Carbon intensity improvement, energy intensity improvement, energy and economic structure optimization are believed to contribute to the carbon emission reduction in the 2 °C scenario [38].

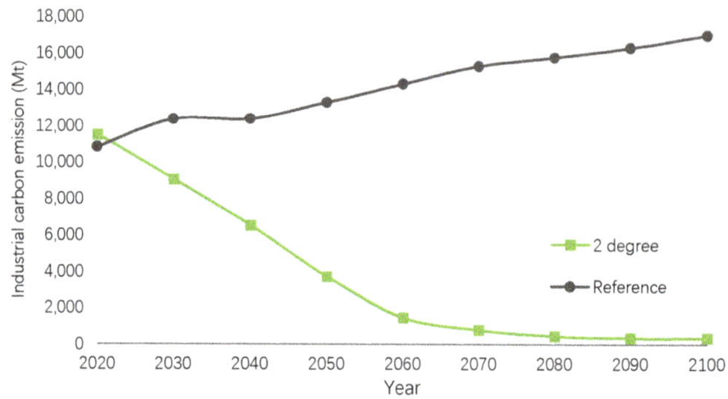

Figure 7. Projected industrial carbon emissions in 2-degree and reference scenarios.

3.4. Decomposition Analysis of Industrial Carbon Emission

The industrial carbon emission from 1995 to 2015 is decomposed according to the LMDI method introduced in Section 2. As can be seen from the decomposition results shown in Figure 8, every five-year CO_2 increment is positive from 1995 to 2015 and the increment peaks around 2010. The economic development and increasing population contribute the most to CO_2 incremental growth, while the economic structure contributes to the decreasing CO_2 emissions. Energy structure and industry energy intensity might contribute positively or negatively to emission reductions depending on the energy price, energy efficiency improvement, clean energy development rate, etc. Taking the case from 2010 to 2015 as an example, the contribution of carbon emission from economic development, population, carbon intensity, energy structure, industry energy intensity, and economic structure are 1510, 770, 40, −160, −700, and −760 Mt CO_2 respectively, resulting in a net 700 Mt CO_2 increment.

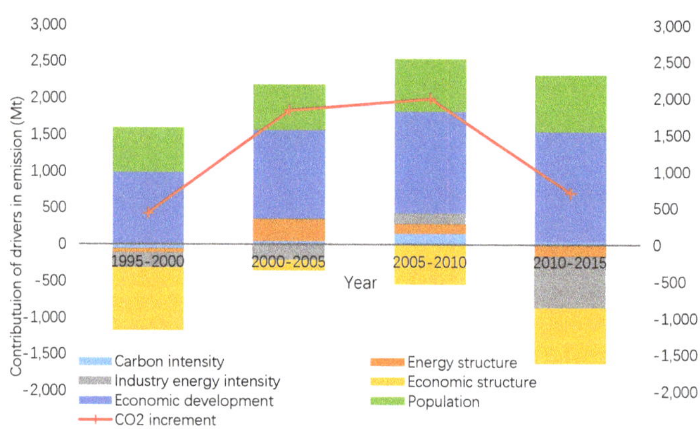

Figure 8. Decomposition of industrial carbon emission from 1995 to 2015.

Economic development is the biggest driving force of CO_2 emission increase, especially in the emerging economies like China, Brazil, India, Russia, and South Africa. Based on the World Bank Database, the averaged GDP growth rate from 2010–2015 in China and India was 8.3% and 7.4% respectively, while the world's averaged GDP growth is around 3% [7]. Globally, GDP per capita in 2015 reached to 14,825 USD/person, increasing by 12% compared with that in 2010. Economic development accounts for 65% of positive emissions. Increasing population is the second largest driver for CO_2 emission as industrial energy demand increases. Average population growth in Sub-Saharan Africa from 2010 to 2015 was 2.76%, more than twice the world's average growth. Increasing populations account for 33% of positive emissions. Industrial carbon intensity contributes to positive CO_2 emission, but its contribution decreases significantly due to much cleaner gas utilization in end-use sectors, and the use of increasing renewables power generation in power & heat generation sector. The share of gas consumption in fossil fuels is 30% and 30.4% in 2005 and 2010 respectively. In 2015, the share of gas further increases to 31.4%. As the share of electricity consumption increases, fossil fuel consumption in the industrial sector is limited, and therefore industrial carbon emission is restricted. Carbon intensity only accounts for 2% of positive emissions from 2010 to 2015. As the energy structure becomes cleaner, the role of energy structure adjustment in 2010–2015 contributes negatively to CO_2 emission. In 2010, the share of electricity consumption in the industrial sector was 24.3% with only a 0.2 percentage point increment from 2005. Correspondingly, the share of electricity consumption increased to 26.5% in 2015 and the increment from 2010 to 2015 was 2.2 percentage points which is 11 times the increment from 2005 to 2010. The share of fossil fuel consumption decreased by 1.2% from 2010 to 2015 as electricity consumption increased. As a result, energy structure contributed 10% of decreasing emissions. Industry energy intensity and economic structure contribute for 43% and 47% respectively in the decreasing emissions. Regarding energy intensity, it decreased from 1.53 to 1.45 toe/thousand USD with a decline of 5.2%. Declining energy intensity is mainly attributed to continuous energy efficiency enhancement and production of more high value-added products. Take United States as an example, energy saving reached to 414 PJ in manufacturing sector from 2010 to 2015. The saved energy is even more than the energy consumed in manufacturing sector from Austria and Czech Republic. As the statistics data from IEA members, manufacturing sector saves 1410 PJ from 2010 to 2015, while chemicals and chemical products sector, paper pulp and printing sector, and non-metallic minerals sector save 460 PJ, 390 PJ, and 77 PJ, respectively. The manufacturing sector contributes the most energy savings in industry sector. Due to greater production of high value-added products, 29 of 33 IEA members saw energy efficiency enhancement in the manufacturing sector. Ireland improved its manufacturing energy efficiency by 46% from 2010 to 2015 [39].

Among the driving factors, economic structure adjustment contributes the greatest emission reduction. Globally, economic structure transforms from industry driven economy to service driven economy. The share of industry value added in GDP declined in more than 160 countries, while service value added in GDP increased from 2010 to 2015. As global statistics data indicate, the share of industry value added in GDP decreased by 1.5 percentage points, while the share of service value added in GDP increased by 2.7 percentage point from 2010 to 2015 [7].

As can be seen from the annual industrial emission reduction shown in Figure 9, the industrial carbon emission path has three stages of characteristics: medium speed decline, high speed decline, and low speed decline. Maximum annual emission reduction peaks around 2040 to 2050 as −280 Mt CO_2/a. To ensure global temperature rises well below 2 °C at the end of this century, the industrial sector needs to reduce emissions substantially before 2060.

Figure 10 exhibits decomposition of industrial carbon emission from 2015 to 2060. Net negative emission is required to meet the global 2 °C temperature control goal. Contrary to the situation from 2010 to 2015, carbon intensity and industry energy intensity will contribute most of the decreasing emissions after 2015 rather than the economic structure factor. Carbon intensity could reduce by 28% from 2015 to 2030 as fossil fuel consumption decreases. The share of fossil fuels in the industrial sector decreases from 62% in 2015 to 46% in 2030 with an average annual drop of one percentage point. Compared with the emission in 2015, declining carbon intensity contributes 3600 Mt CO_2 decreasing emissions in 2030. Industry energy intensity is the second largest factor contributing to the decreasing emissions. Compared with industrial carbon emissions in 2015, the declining energy intensity has reduced carbon emissions by 2800 Mt CO_2. Energy intensity is expected to decrease to 22% from 2015 to 2030 due to increased energy efficiency through equipment updates and more digital equipment applications. Energy structure adjustment also plays an important role in future emission reduction. Share of fossil fuels reduces by 15% from 2015 to 2030 as electrification rate increases and more renewables are directly used in the industrial sector. As indicated by the MESSAGE result, the electrification rate could increase from 26.5% in 2015 to 35.7% in 2030 with an increase of nine percentage points, which is five times the increase during same time range from 2010 to 2015. Besides the contributions from more electricity consumption, the direct use of renewables like solar, modern bioenergy, and hydrogen in the industrial sector are beneficial for energy intensity improvement. The direct use of renewables increases by 125% from 2015 to 2030 with more solar and hydrogen energy applications in the industrial sector. In 2030, the share of solar direct use could reach 37% in terms of renewables and the share of hydrogen could hit 6% from zero in 2015.

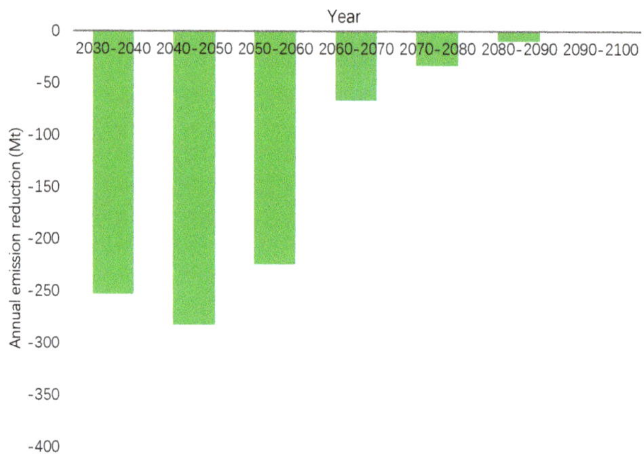

Figure 9. Annual emission reductions in industry sector from 2030 to 2100.

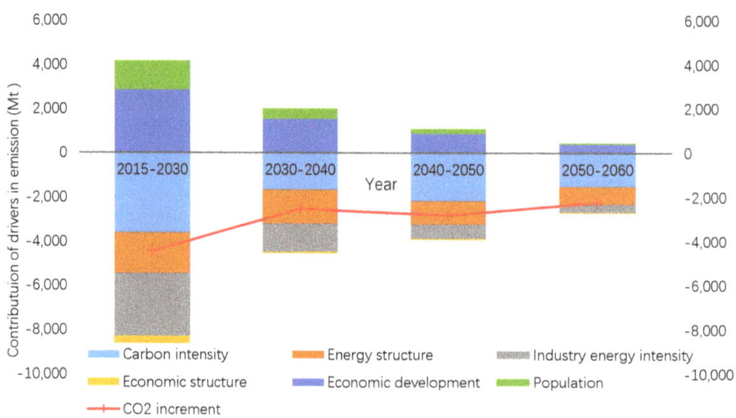

Figure 10. Decomposition of industrial carbon emission from 2015 to 2060.

As can be found from Figure 10, the contribution of population growth and economic development in driving an increase of emissions declines over time due to population saturation. The world's population will peak around 2070. Moreover, economy increase rate decreases over the next several decades, limiting the positive emission. GDP per capita increases by 30% in 2030 compared with that in 2020, whereas it only increases by 18% from 2050 to 2060. Totally, compared with emission in 2050, population and economic development drive an additional 460 Mt CO_2 emission in 2060. However, the role of energy structure transition will become more and more important after 2030, while carbon intensity is still the biggest contributor to decreasing emission before 2060.

As can be seen from Figure 7, the industrial emission is well controlled after 2060. Even the emission reduction decreases after 2060 as indicated in Figure 11, it will become increasingly difficult to reduce emissions more. Consequently, marginal costs of emissions reductions in the industrial sector increase, which is also the reason why the government should seize the opportunity to control emissions in the industrial sector in the earlier stages. From 2060 to 2070, energy structure transition accounts for 50% of total decreasing emissions and the main contribution of energy structure transition comes from further electrification development and much cleaner power structure. In 2070, the electrification rate in industry sector reaches to 62% and clean power generation rate hits more than 90%. Further, 32,200 TWh is expected to be consumed in 2070 which is 3950 TWh more compared to 2060, which will cover the decreasing fossil fuels and increasing industry energy demand needs. Due to negative growth in the world population after 2070, the population starts to drive decreasing emissions, but contributes less than other factors to driving emission reductions.

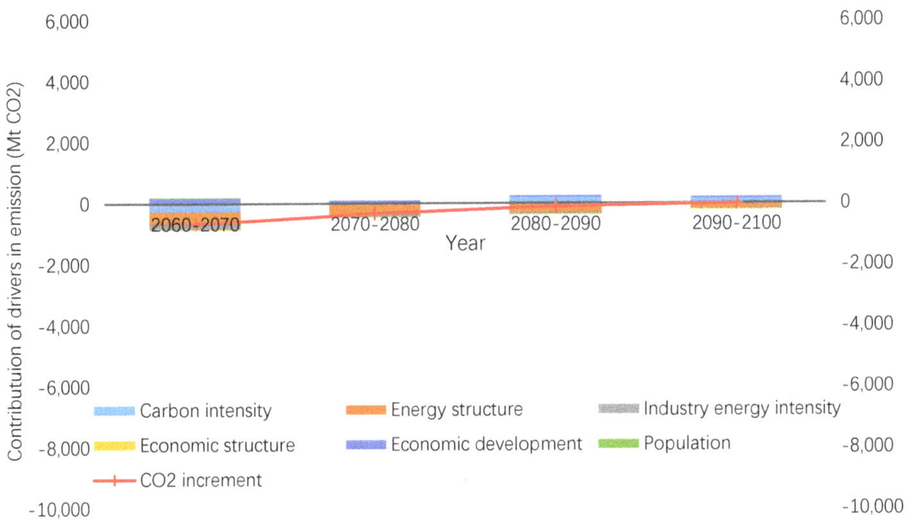

Figure 11. Decomposition of industrial CO_2 emission from 2060 to 2100.

4. Modes for Industrial Emission Mitigation

4.1. Mode 1: Clean Supply-Driven Mode

The clean supply-driven mode refers to the increasing proportion of clean energy power generation and promoting the direct use of clean energy in the industrial sector. The rich clean energy such as water, wind, solar, geothermal energy, etc., could be converted into clean power, which will be used in the industrial sector to realize the support of green power for industrial economic development. Meanwhile, the clean supply-driven mode promotes more modern bioenergy, solar, and hydrogen energy in the industry sector. By replacing fossil fuels with more clean energy to adjust the energy structure, industry sector emissions could be reduced.

Large-scale clean energy development, interconnection and utilization will contribute to the rapid decline of carbon intensity in the power sector, and the direct utilization of clean energy promotes the decline of carbon intensity in the end-use sector [40]. By 2050, clean energy utilization will triple to 6860 Mtoe, three quarters of which will be used for clean energy power generation in the energy supply. CO_2 emissions from the power sector will decrease by more than 95% from 554 g CO_2/kWh in 2015 to 25 g CO_2/kWh in 2050. In terms of regional results, the PAO (Australia, Japan and New Zealand) is expected to have the lowest value of 1.3 g CO_2/kWh owing to 93% clean energy share in power generation and 56% CCS share in fossil fuel power generation. The NAM (North America) is found to have the highest value of 72 g CO_2/kWh, which is attributed to gas power plants in operation with low CCS share as 30%. Globally, the share of CO_2 emissions from the power sector in all energy sectors will fall from 42% in 2015 to 20% by 2050. The direct utilization of clean energy such as solar energy and modern bio-energy can effectively reduce the carbon intensity of energy consumption in end-use sectors. By 2050, the carbon intensity of end-use energy consumption will decrease from 2.21 t CO_2/toe in 2015 to 1.42 t CO_2/toe, a drop of more than 35%. The clean energy consumption flow in 2050 is shown in Figure 12.

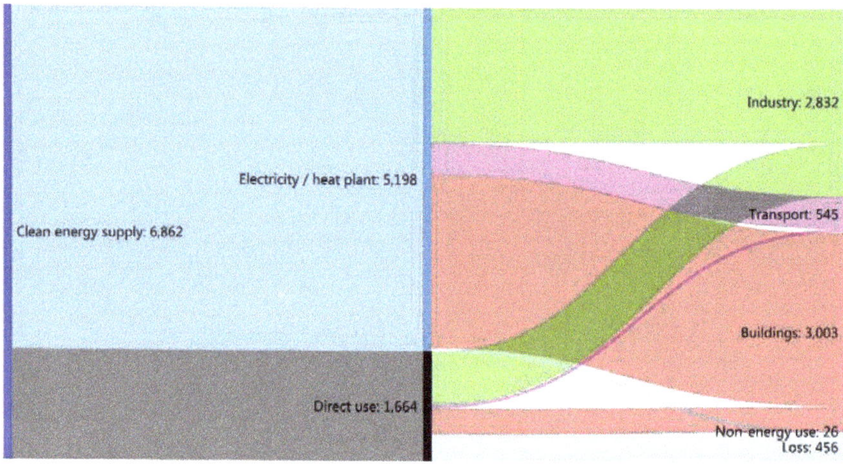

Figure 12. Clean Energy Consumption Flow in 2050 (Unit: 100 Mtoe).

As the cost of clean energy power generation continues to fall, clean electricity will increase significantly, gradually forming a clean dominant power supply landscape. In 2016, clean energy power generation in the world accounted for 35%, with hydropower, wind power, PV, nuclear power, and other clean energy sources accounting for 17%, 4%, 1.3%, 10.5%, and 2.1%, respectively. In the 2 °C scenario, it is estimated that clean energy power generation will account for more than 80% by 2050, broken out as 32% for PV, 23% for wind power, 15% for hydropower, and 6% for nuclear power as indicated from MESSAGE model results.

Traditional biomass combustion will gradually be replaced by other forms of energy utilization due to its adverse effects on air quality and human health. Modern bioenergy would be widely used in industry, transport and building sectors. In 2050, 80% of bioenergy is expected to be used for industrial heating, while the rest will be used in the transport and building sectors for decentralized heating. The direct utilization of solar energy is to provide heating or heating collection for the end-use sectors, which is mainly applied to industrial low temperature heating and hot water and heating in the building sector. In the industrial sector, the utilization of solar energy is expected to be large-scale deployed. With the maturity of solar energy application technology, the cost of solar energy used in the industrial sector has decreased significantly. Solar water heaters, solar air heaters and solar collectors are widely used in low temperature heating and preheating in the industrial process. First, as a mature technology, the solar water heater will be popularized worldwide in the near future. Solar water heaters can increase the water temperature from 25 °C to 80 °C for boilers, thus saving a lot of fossil fuels. Second, solar air heaters provide air in the temperature of 50~80 °C for drying tea leaves or processing fruits, spices, cereals, mushroom, vegetables, seafood etc. Third, solar collector systems can provide steam at up to 300 °C to meet industrial heating needs. Applications include mercerizing, drying and finishing in textile industry; drying, dissolving, thickening, leaching and distillation in chemical industry; cooking, drying and canning in food industry, craft pulping, bleaching and drying in pulp and paper industry; drying and cleaning in leather industry.

4.2. Mode 2: Electricity Consumption-Driven Mode

The electrification in the industrial sector will be further improved, with the electrification rate reaching 48% by 2050. Global industrialization and urbanization further accelerate energy consumption of the industrial sector. The total consumption of fossil energy in the industrial sector will drop to 1120 Mtoe by 2050, down 34% from 2017. The decrease of fossil fuels consumption in the industrial sector is mainly attributed to electricity replacement as discussed in Section 4.3. Compared with 2017, the total electricity consumption in the industrial sector in 2050 will increase by about 1240 Mtoe, providing 90% of energy demand increase (Figure 13).

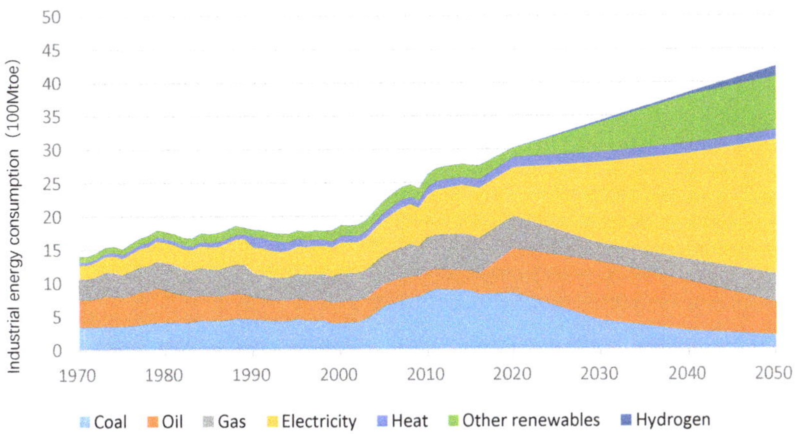

Figure 13. Energy Consumption and Structure in Industry Sector Based on MESSAGE Optimization Analysis.

The industrial sector's carbon dioxide emissions are concentrated in energy-intensive industries, in which electricity replacement has a great potential for mitigation. Traditional energy-intensive industries such as iron and steel, chemical and petrochemical, paper pulp and printing consume the largest amount of energy in the industrial sector, accounting for 69% of end-use energy consumption and 74% of total CO_2 emissions in the industrial sector. Among them, the chemical and petrochemical industry, which mainly relies on oil and natural gas, accounts for 28%, the largest proportion of end-use energy consumption in the industrial sector. The iron and steel industry, which relies heavily on coal, takes the largest share of CO_2 emissions in the industrial sector. At present, the global average electrification rate of traditional energy-intensive industries has much room for improvement. Electrification can significantly reduce the use of fossil fuels such as coal and oil.

Replacing coal fired boilers with electric boilers and replacing coal and oil heating furnaces with electric heating furnace could be the alternative of electricity replacement in traditional energy-intensive industries. In traditional energy-intensive industries, large amounts of coal is burned for heating in the process of cooking, smelting, drying, firing and annealing in the equipment including industrial boilers and kilns. For the iron and steel industry, in the process of steelmaking, the replacement of coal and coke by electricity can be realized by replacing the converter steelmaking with electric steelmaking, which can shorten the steel production process and save energy consumption required by iron smelting. For the non-ferrous metal industry, induction furnaces with high heating efficiency and accurate temperature control are mainly used to replace coke furnaces in the smelting process. For the paper pulp and printing industries, electric boilers and electromagnetic induction heating ovens are used to replace coal fired boilers and steam heating ovens in the pulping and drying of the papermaking process respectively, so as to replace coal, oil, and natural gas with electricity and to save energy.

4.3. Mode 3: Energy Efficiency-Driven Mode

Energy efficiency-driven mode refers to meeting energy demand with less energy consumption through the technical improvement of energy-using equipment in the end-use sectors, technological innovation of energy supply (power generation, oil and gas exploration and refining), and digital development of energy systems, thus reducing CO_2 emissions in industry, transport, building, power generation, and other sectors.

In the industrial sector, in order to reduce energy consumption, it could be very helpful to optimize production processes in energy-intensive industries and to promote the application of digital and intelligent equipment. Three measurements could be taken in energy-intensive industries. First is to optimize the production process of energy-intensive industries such as iron and steel, non-ferrous metals and chemical industry, thus realizing continuous and efficient production process. Enhanced energy efficiency of the production process reduces energy consumption and emissions significantly, and process energy saving is gradually transforming to system energy savings. Second is to explore the characteristics of dynamic balance between supply and demand of energy flow in the production process, thus improving the intelligent level of energy control system. By carrying out fine management of the energy system on the basis of informatization and strengthening energy dynamic prediction and optimal scheduling, the level of energy utilization efficiency will be improved. The third is to promote the application of the recycling economy model, build an ecological chain of industries for product manufacturing, energy conversion, waste disposal and recycling, and promote energy conservation and efficiency in the industrial sector [41].

5. Conclusions

The trends and relationship between industrialization, industrial energy demand and carbon emission are studied in this paper to meet the temperature control goal of 2 °C of the Paris Agreement. The energy system is optimized using the IAM model of MESSAGE and the LMDI approach is adopted to track the historical drivers of emission and to investigate potential drivers of future emission with an expanded Kaya identity. Main conclusions are as follows.

Firstly, historical industrialization of 11 regions is analyzed and a regression method is used to predict the future industrialization for those regions. Sub-Saharan Africa (AFR), South Asia (SAS), and Other Pacific Asia (PAS) are still at the early stage of industrialization, and will keep increasing in terms of industrialization until 2050–2060. There are downward trends for the most developed regions and also the post-industrialization regions such as the North America (NAM), Western Europe (WEU), and Pacific OECD (PAO). With regard to the late stage of industrialization regions, such as East Europe (EEU) and Former Soviet Union (FSU), we find that there are downward trends for EEU and FSU. There is a re-industrialization then downward trend for the Latin American (LAC) region. With regard to Centrally planned Asia and China (CPA) and Middle East and North Africa (MEA), they have passed the peak and goes downward of industrialization.

Secondly, industry energy demand is projected using a hump-shaped function method with consideration of duration of industrialization and urbanization development and energy structure is optimized using MESSAGE model. Industrial energy demands in reference scenario and 2 °C scenario are compared in this paper. In the 2 °C scenario, fossil fuel consumption in the industrial sector fell to 1120 Mtoe in 2050, a decrease of 34% compared with that in 2017 which is mainly due to growing electricity consumption and the direct use of clean energy. When energy consumption for producing electricity, heat, and hydrogen are allocated to the consumer sector, energy consumption in the industrial sector increases from 4230 to 5270 Mtoe in 2050, and adjusted industry energy demand in 2 °C scenario is comparable with that in the reference scenario before 2050, while it is smaller than reference scenario after 2050 due to massive electricity utilization and cleaner energy based power generation in the 2 °C scenario.

Thirdly, an expanded Kaya identity is proposed to decompose carbon emission into six derivers as carbon intensity, energy structure, energy intensity, economic structure, GDP per capita, and population.

As studied from the historical industrial carbon emission from 1995 to 2015, economic development is the biggest driving force for CO_2 emission increase, followed by population. Industrial carbon intensity contributes positive CO_2 emission, but its contribution decreases significantly due to much cleaner gas utilization in end-use sectors, and the use of increasing renewables power generation in the power and heat generation sector. In the 2 °C scenario, carbon intensity and industry energy intensity will contribute most of the decreasing emissions from 2015 to 2060. Contribution of population and economy development in driving emission increase declines due to population saturation. After 2060, it will become increasingly difficult to reduce emissions more, resulting in high marginal costs of emissions reductions, indicating government should seize the opportunity to control emissions in the industrial sector in earlier stages.

Finally, three modes for emission reduction are suggested in this study. Clean supply-driven mode could drive power sector's emission share decreases from 42% in 2015 to 20% by 2050. Clean energy direct use could drive carbon intensity in end use sectors to decrease from 2.21 t CO_2/toe in 2015 to 1.42 t CO_2/toe, with a drop of more than 35%. Electricity consumption-driven mode could drive total consumption of fossil energy in the industrial sector drop to 1120 Mtoe by 2050, down 34% from 2017, and therefore reducing carbon emissions with a clean supply-driven mode will be extensively adopted in power sectors. Energy efficiency-driven mode could drive energy consumption reductions by optimizing production processes in energy-intensive industries, promoting the application of digital and intelligent equipment.

More studies are needed to explore the industry sector, such as to integrate the top-down method with a bottom-up method for the projection of future energy consumption and CO_2 emissions, have a detailed study on the manufacturing industries such as the iron and steel, cement etc., and to investigate global and country-specific industrial emission pathways to meet the 2°C and 1.5 °C goals and their policy implications [42,43].

Author Contributions: C.L. and S.Z. conceived of and designed the proposed scheme. Conceptualization, C.L., S.Z., X.C., F.Y. conceived, designed and performed the experiments, X.T., F.G. and W.J. analyzed the data. F.Y. and Y.Z. supervised the whole project. C.L. and S.Z. wrote the manuscript. F.Y., X.T. and W.J. reviewed the manuscript. All authors have read and agreed to the published version of the manuscript.

Funding: This research was support by the China's National Key Research and Development Project (2016YFA0602602), GEIGC Science and Technology Project (No. 52450018000Q).

Acknowledgments: The authors would like to thank IIASA for providing the scenario data.

Conflicts of Interest: The authors declare no conflict of interest.

References

1. Fischedick, M.; Roy, J.; Acquaye, A.; Allwood, J.; Ceron, J.P.; Geng, Y.; Kheshgi, H.; Lanza, A.; Perczyk, D.; Price, L.; et al. *Industry, in Climate Change 2014: Mitigation of Climate Change. Contribution of Working Group III to the Fifth Assessment Report of the Intergovernmental Panel on Climate Change*; Edenhofer, O., Pichs-Madruga, R., Sokona, Y., Edenhofer, O., Pichs-Madruga, R., Sokona, Y., Eds.; Cambridge University Press: Cambridge, UK; New York, NY, USA, 2014.
2. IEA. IEA Data and Statistics. 2017. Available online: http://data.iea.org (accessed on 1 December 2019).
3. Chen, Z.; Liu, C.; Qu, S. China's Industrialization and the Pathway of Industrial CO_2 Emissions. *Chin. J. Urban Environ. Stud.* **2005**, *3*, 1550019. [CrossRef]
4. Chenery, H.B. Patterns of Industrial Growth. *Am. Econ. Rev.* **1960**, *4*, 624–654.
5. Chen, J.; Huang, Q.; Lv, T.; Li, X. *The Report on Chinese Industrialization (1995–2010)*; Social Sciences Academic Press: Beijing, China, 2012.
6. UNIDO. *Industrial Development Report 2016. The Role of Technology and Innovation in Inclusive and Sustainable Industrial Development*; United Nations Industrial Development Organization: Vienna, Austria, 2015.
7. UNIDO. *Industrial Development Report 2018. Demand for Manufacturing: Driving Inclusive and Sustainable Industrial Development*; United Nations Industrial Development Organization: Vienna, Austria, 2017.

8. Caraiani, C.A.; Lungu, C.I.; Dasc Lu, C. Energy consumption and GDP causality: A Three-Step Analysis for Emerging European Countries. *Renew. Sustain. Energy Rev.* **2015**, *44*, 198–210. [CrossRef]
9. Ntanos, S.; Skordoulis, M.; Kyriakopoulos, G.; Arabatzis, G.; Chalikias, M.; Galatsidas, S.; Batzios, A.; Katsarou, A. Renewable Energy and Economic Growth: Evidence from European Countries. *Sustainability* **2018**, *8*, 2626. [CrossRef]
10. Taeyoung, J.; Jinsoo, K. Coal Consumption and Economic Growth: Panel Cointegration and Causality Evidence from OECD and Non-OECD Countries. *Sustainability* **2018**, *3*, 660. [CrossRef]
11. Wang, H.; Chen, W. Modelling Deep Decarbonization of Industrial Energy Consumption under 2-Degree Target: Comparing China, India and Western Europe. *Appl. Energy* **2019**, 1563–1572. [CrossRef]
12. Van Ruijven, B.J.; Van Vuuren, D.P.; Boskaljon, W.; Neelis, M.L.; Saygin, D.; Patel, M.K. Long-Term Model-Based Projections of Energy Use and CO_2 Emissions from the Global Steel and Cement Industries. *Resour. Conserv. Recycl.* **2016**, *112*, 15–36. [CrossRef]
13. IEA. *Tracking Industrial Energy Efficiency and CO_2 Emissions*; International Energy Agency: Paris, France, 2007.
14. Kaya, Y. *Impact of Carbon Dioxide Emission Control on GNP Growth: Interpretation of Proposed Scenarios*; Intergovernmental Panel on Climate Change (IPCC)/Response Strategies Working Group: Geneva, Switzerland, 1990.
15. Gabriel Blanco, R.G.S.S. Drivers, Trends and Mitigation. In *Climate Change 2014: Mitigation of Climate Change*; Contribution of Working Group III to the Fifth Assessment Report of the Intergovernmental Panel on Climate Change; Cambridge University Press: Cambridge, UK; New York, NY, USA, 2014.
16. IEA. *CO_2 Emissions from Fuel Combustion*; International Energy Agency: Paris, France, 2018.
17. Ang, B.W.; Liu, F.; Chew, E. Perfect Decomposition Techniques in Energy and Environmental Analysis. *Energy Policy* **2003**, *31*, 1561–1566. [CrossRef]
18. Ang, B.W. The LMDI Approach to Decomposition Analysis: A Practical Guide. *Energy Policy* **2005**, *7*, 867–871. [CrossRef]
19. Freitas, L.C.; Kaneko, S. Decomposition of CO_2 Emissions Change from Energy Consumption in Brazil: Challenges and Policy Implications. *Energy Policy* **2011**, *3*, 1495–1504. [CrossRef]
20. Fatima, T.; Xia, E.; Cao, Z.; Khan, D.; Fan, J. Decomposition Analysis of Energy-Related CO_2 Emission in the Industrial Sector of China: Evidence from the LMDI Approach. *Environ. Sci. Pollut. Res.* **2019**, *21*, 21736–21749. [CrossRef]
21. Khan, A.; Jamil, F.; Khan, N.H. Decomposition Analysis of Carbon Dioxide emissions in Pakistan. *SN Appl. Sci.* **2019**, *1*, 1012. [CrossRef]
22. Ouyang, X.; Lin, B. An Analysis of the Driving Forces of Energy-Related Carbon Dioxide Emissions in China's Industrial Sector. *Renew. Sustain. Energy Rev.* **2015**, *45*, 838–849. [CrossRef]
23. Zhang, S.; Wang, J.; Zheng, W. Decomposition Analysis of Energy-Related CO_2 Emissions and Decoupling Status in China's Logistics Industry. *Sustainability* **2018**, *5*, 1340. [CrossRef]
24. Mousavi, B.; Lopez, N.S.A.; Biona, J.B.M.; Chiu, A.S.; Blesl, M. Driving Forces of Iran's CO_2 Emissions from Energy Consumption: An LMDI Decomposition Approach. *Appl. Energy* **2017**, *206*, 804–814. [CrossRef]
25. Sumabat, A.K.; Lopez, N.S.; Yu, K.D.; Hao, H.; Li, R.; Geng, Y.; Chiu, A.S. Decomposition Analysis of Philippine CO_2 Emissions from Fuel Combustion and Electricity Generation. *Appl. Energy* **2016**, *164*, 795–804. [CrossRef]
26. Cansino, J.M.; Sánchez-Braza, A.; Rodríguez-Arévalo, M.L. Driving Forces of Spain's CO_2 Emissions: A LMDI Decomposition Approach. *Renew. Sustain. Energy Rev.* **2015**, *48*, 749–759. [CrossRef]
27. Alves, M.R.; Moutinho, V. *Decomposition analysis for energy-related CO_2 emissions intensity over 1996–2009 in Portuguese Industrial Sectors*; CEF AGE-UE Working Paper; Center for Advanced Studies in Management and Economics: Evora, Portugal, 2013.
28. Hosseini Nasab, E.; Aalami, R.; Foroughi Dahr, S.; Sadeghzadeh, M.A. An Analysis of Energy Consumption in Transportation and Industrial Sectors- a Multiplicative LMDI Approach with Application to Iran. *Iran. Econ. Rev.* **2012**, *16*, 1–17.
29. Akbostancı, E.; Tunç, G.İ.; Türüt-Aşık, S. CO_2 Emissions of Turkish Manufacturing Industry: A Decomposition Analysis. *Appl. Energy* **2011**, *6*, 2273–2278. [CrossRef]
30. Hatzigeorgiou, E.; Polatidis, H.; Haralambopoulos, D. CO_2 Emissions in Greece for 1990–2002: A Decomposition Analysis and Comparison of Results Using the Arithmetic Mean Divisia Index and Logarithmic Mean Divisia Index Techniques. *Energy* **2008**, *3*, 492–499. [CrossRef]

31. Moutinho, V.; Moreira, A.C.; Silva, P.M. The Driving Forces of Change in Energy-Related CO_2 Emissions in Eastern, Western, Northern and Southern Europe: The LMDI Approach to Decomposition Analysis. *Renew. Sustain. Energy Rev.* **2015**, 1485–1499. [CrossRef]
32. IIASA. MESSAGE Model Regions. 2013. Available online: http://www.iiasa.ac.at/web/home/research/researchPrograms/Energy/MESSAGE-model-regions.en.html (accessed on 22 July 2013).
33. World Bank Database. Available online: https://databank.worldbank.org/databases (accessed on 20 December 2019).
34. Fricko, O.; Havlik, P.; Rogelj, J.; Klimont, Z.; Gusti, M.; Johnson, N.; Kolp, P.; Strubegger, M.; Valin, H.; Amann, M.; et al. The Marker Quantification of the Shared Socioeconomic Pathway 2: A Middle-of-the-Road Scenario for the 21st Century. *Glob. Environ. Chang.* **2017**, *42*, 251–267. [CrossRef]
35. Liu, G. *Research on the Model and Application of Sector Final Energy Consumption: Based on the Industry and Transportation Sector*; China University of Geosciences: Beijing, China, 2016; p. 133.
36. Schrattenholzer, L. *The Energy Supply Model MESSAGE*; International Institute for Applied Systems Analysis: Laxenburg, Austria, 1981.
37. Hou, F.; Zhang, S.; Zhao, Z.; Chen, X.; Tan, X.; Huang, H.; Yang, F.; Tan, F. Global Energy Interconnection Scenario Outlook and Analysis in the Context of Achieving the Paris Agreement Goals. *J. Glob. Energy Interconnect.* **2020**, *3*, 34–43.
38. Zhou, Y.; Chen, X.; Tan, X.; Liu, C.; Zhang, S.; Yang, F.; Zhou, W.; Huang, H. Mechanism of CO_2 Emission Reduction by Global Energy Interconnection. *J. Glob. Energy Interconnect.* **2018**, *1*, 409–419.
39. IEA. *Energy Efficiency Indicators Database*; International Energy Agency: Paris, France, 2018.
40. Tan, X.; Liu, C.; Chen, X.; Zhang, S.; Yang, F.; Wei, C. Carbon Flow and Mitigation Benefits Based on Grid Interconnection: A Case Study on Africa Energy Interconnection. *J. Glob. Energy Interconnect.* **2019**, *3*, 291–298.
41. Zhang, Q.; Zhang, W.; Wang, Y.; Xu, J.; Cao, X. Potential of Energy Saving and Emission Reduction and Energy Efficiency Improvement of China's Steel Industry. *Iron Steel.* **2019**, *54*, 7–14.
42. Jiang, K.; He, C.; Xu, X.; Jiang, W.; Xiang, P.; Li, H.; Liu, J. Transition Scenarios of Power Generation in China under Global 2 °C and 1.5 °C Targets. *J. Glob. Energy Interconnect.* **2018**, *4*, 79–88.
43. Wang, M.; Kang, W.; Chen, Z.; Zhang, Y.; Liu, C.; Yang, F. Global Energy Interconnection: An Innovative Solution for Implementing the Paris Agreement—The Significance and Pathway of Integrating GEI into Global Climate Governance. *J. Glob. Energy Interconnect.* **2018**, *4*, 69–78.

energies

MDPI

Article

Influence of Different Biofuels on the Efficiency of Gas Turbine Cycles for Prosumer and Distributed Energy Power Plants

Dariusz Mikielewicz [1], Krzysztof Kosowski [1], Karol Tucki [2,*], Marian Piwowarski [1], Robert Stępień [1], Olga Orynycz [3,*] and Wojciech Włodarski [1]

1 Faculty of Mechanical Engineering, Gdansk University of Technology, Gabriela Narutowicza Street 11/12, 80-233 Gdansk, Poland
2 Department of Organization and Production Engineering, Warsaw University of Life Sciences, Nowoursynowska Street 164, 02-787 Warsaw, Poland
3 Department of Production Management, Bialystok University of Technology, Wiejska Street 45A, 15-351 Bialystok, Poland
* Correspondence: karol_tucki@sggw.pl (K.T.); o.orynycz@pb.edu.pl (O.O.); Tel.: +48-593-45-78 (K.T.); +48-746-98-40 (O.O.)

Received: 27 July 2019; Accepted: 16 August 2019; Published: 19 August 2019

Abstract: The efficiency of a gas turbine can be affected by the use of different biofuels usually with a relatively Lower Heating Value (LHV). The paper evaluates the impact of calorific value of fuel on turbine performance and analyzes the possibilities of optimizing turbine construction from the point of view of maximum efficiency for a particular fuel. The several variants of design of small power microturbines dedicated to various biofuels are analyzed. The calculations were carried out for: gas from biomass gasification (LHV = 4.4 MJ/kg), biogas (LHV = 17.5 MJ/kg) and methane (LHV = 50 MJ/kg). It is demonstrated that analyzed solution enables construction of several kW power microturbines that might be used on a local scale. Careful design of such devices allows for achieving high efficiency with appropriate choice of the turbine construction for specific fuel locally available. Such individually created generation systems might be applied in distributed generation systems assuring environmental profits.

Keywords: thermodynamic cycles; district distributed power plants; effectiveness; sustainability

1. Introduction

The condition of the energy sector determines the state of the national economy and the level of economic growth [1,2]. The power sector is currently experiencing a dynamic transformation, resulting not only from EU conditions, but also from current problems, mainly related to ensuring energy security to consumers [3–5]. In many European Union countries, it was decided to change the model of the electricity market [6–8]. There is a gradual retreat from the energy economy based on the central distribution of oil and other fossil fuels [9–11]. Achieving the synergy effect between the energy sector's potential and its customers is now the overriding goal. As a result, dynamic development on the client side is observed related to photovoltaic panel installations, energy storage, electric vehicles and broadly understood Smart Homes [12–14]. The concept of smart cities (Smart City) is constantly evolving, which is a response to the needs of implementing innovative concepts of city functioning through modern technological solutions and a comprehensive management service, e.g., in the context of sustainable energy. New technologies in the power industry are becoming the driving force of the economy, and an important criterion for their development is the impact on the natural environment [15–17].

The continuous increase in the demand for energy and its carriers as well as the growing normative requirements in the ecological aspects including reduction of CO_2 emissions cause the necessity of systematically increasing the share of energy from renewable sources [18–20]. The level of public awareness along with the increasingly restrictive legal requirements of the European Union in matters of environmental protection affect the attractiveness of activities and technologies that lead to the reduction of adverse human impact on the environment [21–23]. Energy operators, faced with the need to spend large sums on modernization of the power plants and at the same time being aware of the fact that the existing pollutant emission standards will be further tightened, must opt for investments in modern technologies that will produce clean energy without harming human health and the environment [24–26]. The choice of technology is more and more often supported by the Life Cycle Assessment (LCA), which, due to its comprehensive nature, allows for a full assessment of the environmental impact of the entire production process, from obtaining raw materials to final management of waste generated as a result of use of the product [27–29].

Large corporations producing electricity and selling it using international and national distribution networks due to the large installed capacity are not very flexible [30,31]. Currently, the government's actions are aimed at supporting both improving the efficiency of electricity generation and supporting distributed energy based on local and renewable energy sources [32]. As a rule, this eliminates the problem of network losses. Energy is produced in the same place where it is used [33,34]. This also involves the development of smart grids and smart metering [35,36].

Although a constant increase in electricity prices is unavoidable, its freedom of trade and independent production together with storage can be a tool to control its rising costs [37]. Countries where political power is based on the export of raw materials lose their ability to exert pressure, which is a very important element from the point of view of energy security [38]. From the point of view of the challenges facing each national electricity system, such as the energy security of the state, reduction of emission and efficiency, only the coexistence of professional and civic energy is right.

The growing prosumer energetics is a chance for a new shape of the energy system, in which the recipient will be not only a user, but also an active participant. This leads to a visible increase in the number of prosumer polygeneration centers built [39,40]. In these types of units, most often turbines, much attention is paid to durability, reliability, low price and efficiency of components and the entire generator [41,42]. The disadvantage of the domestic power generation sector is the relatively low efficiency of energy production from coal, and in the case of dispersed power engineering, the efficiency of small power plants is even lower [43,44], and thermal power plants based on circuits with organic agents reach efficiency of just a dozen or so percent [45,46]. In the field of prosumer energy, there are no solutions on the market that allow highly efficient energy production around the clock. As part of the development of micro electric generators, it is possible to indicate micro-turbines and bladeless adhesive turbines [47–49].

The search for alternative energy from renewable sources is becoming more and more fashionable and recommended by, among others, the European Union. At present, this group of energy sources includes: tides, sea currents, waves, temperature difference of ocean waters, wind power, solar energy, geothermal energy and energy from biomass. In the case of prosumers, the most popular are solar collectors, wind turbines and biogas plants.

Currently, such popular solar collectors work only during the day [50]. Other solutions, in turn, apart from those using solar energy, have relatively low efficiency, hence their installation and operation is rather unprofitable [51].

An interesting energy source is biogas [52,53]. Its greatest advantage is universality. It can be used both for the production of electricity and heat, and as motor fuel, whereas wind, water or solar power plants only provide electricity.

Biogas is an energy vector formed from the microbiological decomposition of organic raw materials (e.g., of agricultural, industrial or food origin) during the methane fermentation process [54,55]. The composition of biogas depends on its origin (the type of substrate subjected to the fermentation process) [56,57]. A large amount of waste generated by the agri-food industry may contribute to the construction of many biogas installations located at the source of their use (production plant). This solution due to the lack of costs of obtaining the substrate and the possibility of using the produced electricity and heat for technological purposes of the plant is beneficial. The biogas produced can be used to generate thermal or electrical energy or for combined energy production in CHP (Combined Heat and Power) systems. The use of biogas for the production of electricity or heat requires the removal of hydrogen sulphide and water vapor (responsible for the corrosion of equipment) [58,59].

An important place among alternative sources of energy is the wood gas obtained in the gasification process, which, by supplying the CHP system, can be a competitive source of electricity and heat generated in cogeneration. This is of particular importance in the generation of energy in distributed systems, independent of large, centralized energy suppliers [60,61]. Energy is produced directly at the point of demand, and the system is based on clean and environmentally friendly fuel. In addition, the ash produced during the gasification process can be used in the chemical industry or as a natural fertilizer.

The aim of the present work is to estimate the efficiency of several variants of microturbines operating with a number of biofuels in order to evaluate the effect of the fuel calorific value on turbine performance as well as to investigate the possibility of optimization of turbine construction in order to achieve maximum efficiency for a particular fuel. Such optimization should ensure a possibility of fitting the turbine to local needs for distributed energy generation.

Particular biofuels can differ depending on their chemical composition and the heating values which play an important role when thermodynamical cycles are considered. Heating values influence energy balance equations of gas turbine combustion chambers and, as a result, the relations between the temperatures and the mass (and volume) flow rates of the working media (air, gases) are altered. This, in turn, shows some impact on the power plant overall efficiency and the design of turbomachinery flow parts. The paper aims at highlighting this problem, as it has not been discussed thoroughly in the bibliography. It is so because the micro gas turbines (gas turbines of small and very small output) operating on biofuels are only at the beginning of their applications in prosumer and distributed energy power plants. In a typical arrangement of gas turbine engines, the combustion chamber is placed just in front of the turbine, usually high-quality gas or liquid fuels are used and the hot gases flow through the turbine flow part. In the case of various biofuels (especially pellets) so-called "external combustion systems" may be used, which allows the burning of different sorts of fuel (liquid, gas or solid), even of poor quality, because in these units clean air flows through the compressor and the turbine.

2. Materials and Methods

The Computation Algorithm

In the case of small power plants (from several kW to several hundred kW), the maximum temperature 850–900 °C was assumed before the turbine, and the low efficiency of the components was assumed, e.g., turbine efficiency equal to 82%, compressor efficiency—80%, efficiency of the electric generator—90%, efficiency of the combustion chamber—95%. Assumptions adopted for the analysis are presented in Table 1.

Table 1. Assumptions adopted for the design analysis of turbine generator variants [62,63].

Description	Unit	Value
compressor efficiency	(-)	0.800
turbine efficiency	(-)	0.820
mechanical efficiency	(-)	0.980
leakage losses	(-)	0.020
generator efficiency	(-)	0.900
efficiency of the combustion chamber	(-)	0.950
pressure loss in the filter	(-)	0.995
pressure loss in the silencer	(-)	0.995
pressure losses in the combustion chamber	(-)	0.980
pressure loss in the regenerator	(-)	0.980

Currently, it is possible to obtain a stable flame during the combustion of low calorific fuels in a wide range of operating parameters, such as the molar composition of the fuel and the excess air coefficient. However, the biogas must be properly cleaned and dried so that it does not damage the turbine. Depending on the origin, the biogas composition is variable. The calorific value depends primarily on the methane content. Currently, biogas that is combusted in gas turbines has a methane content from 35% to 100%. As a result of continuous combustion with excess air and low pressures in the combustion chamber, turbines as well as microturbines have a significantly lower value of exhaust emissions as compared to the reciprocating engines. The combustion of low calorific gases has a significant impact on the natural environment by reducing the emission of nitrogen oxides [64].

The analyzed variants compare the possibility of using highly efficient exchangers and considered 5 different configurations of gas turbosets (Figure 1):

- Variant 1: turbine set operating according to the simple open cycle,
- Variant 2: turbine set operating according to the open cycle with regenerator,
- Variant 3: turbine set operating according to the open cycle with combustion chamber at turbine exit,
- Variant 4: turbine set operating according to the open cycle with external combustion chamber at turbine exit and high-temperature heat exchanger,
- Variant 5: turbine set operating according to the open cycle with partial bypassing of external combustion chamber at turbine exit and with high-temperature heat exchanger.

Analyses for gases with very different calorific value were carried out in the presented paper. The list of combusted gases analyzed is presented in Table 2. For comparison, the analysis was also carried out for methane (the main component of LNG or natural gas), and hydrogen being an ecological fuel with very high calorific value.

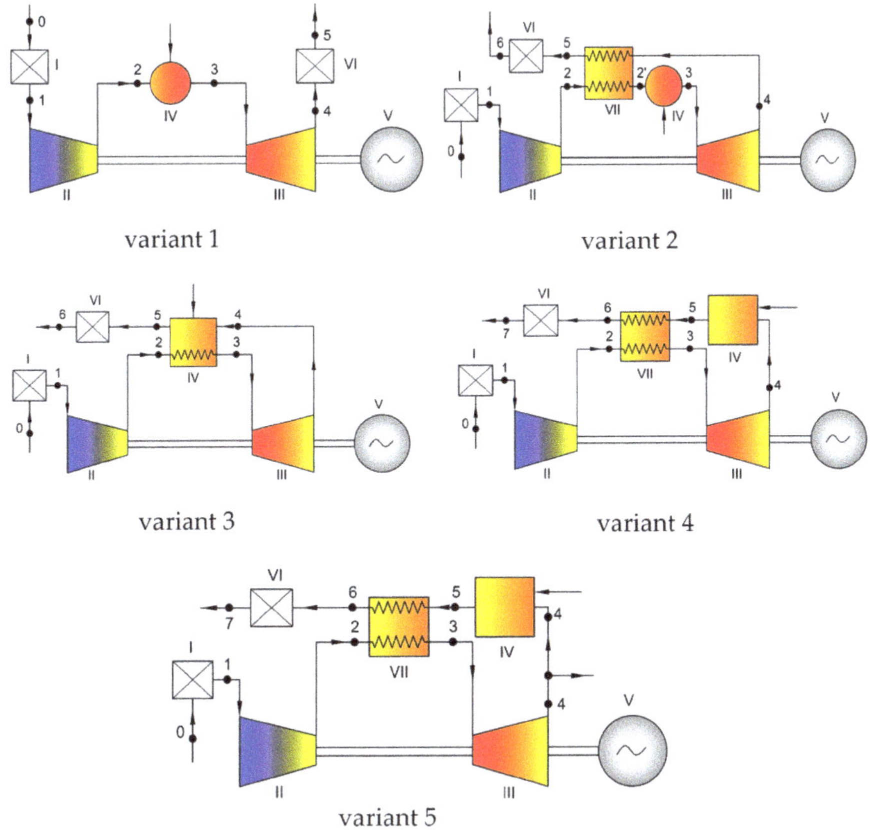

Figure 1. Turbine set arrangements being analyzed. **Variant 1**: turbine set operating according to the simple open cycle; **Variant 2**: turbine set operating according to the open cycle with a regenerator; **Variant 3**: turbine set operating according to the open cycle with a combustion chamber at the turbine exit. The air introduced to the combustion chamber has a temperature equal to the temperature just behind the turbine and it can be compared to the situation when the effectiveness of the regenerator equals 1. Therefore, the efficiency of variant 3 can be higher than the efficiency of other variants. This solution has been well known for years [67,68] but it was not used in practice due to the properties of the materials for regenerators/combustors which did not allow the application of high temperature before the turbines. Nowadays, due to technological progress we can overcome these problems and propose variant 3 as a realistic solution. **Variant 4**: turbine set operating according to the open cycle with an external combustion chamber at the turbine exit and a high-temperature heat exchanger; **Variant 5**: turbine set operating according to the open cycle with partial bypassing of the external combustion chamber at the turbine exit and with a high-temperature heat exchanger.

Table 2. The list of analyzed gases with different heating values [65,66].

Fuel Type	Volumetric Composition (-)	Density (kg/m^3)	Calorific Value (MJ/m^3)	Calorific Value (MJ/kg)
gas from biomass gasification	methane 0.09; carbon dioxide 0.133; carbon monoxide 0.147; hydrogen 0.073; nitrogen 0.42; water 0.137	1.2107	4.8	4.4
wood gas	methane 0.12; carbon dioxide 0.54; carbon monoxide 0.3; hydrogen 0.04	1.4197	12	8.5
biogas	methane 0.4; carbon dioxide 0.23; hydrogen 0.16; carbon monoxide 0.1; nitrogen 0.11	0.9438	16.5	17.5
biogas	methane 0.75; carbon dioxide 0.15; carbon monoxide 0.02; hydrogen sulfide 0.04; nitrogen 0.04	0.9002	22	24.4
city gas	methane 0.25; hydrogen 0.55; carbon monoxide 0.08; nitrogen 0.07; oxygen 0.05	0.4525	17.5	38.7
methane	methane 1.0	0.6660	36	54.1
hydrogen	hydrogen 1.0	0.0835	10.02	120

3. Results

Computations for five variants of the cycles (Figure 1) with seven heating values for each cycle (Table 2) were performed. The thermodynamical calculations were performed following the classical approach well known from the bibliography [62,69–71]. First, the parameters in all the characteristic points of the schemas (turbine cycles) were determined, then the relation between the mass flow rates in particular elements were estimated, and, finally, the overall efficiencies of the cycles were calculated. The calculations were performed in the following order: calculations of the compression process in the compressor, calculations of the expansion line in the turbine, calculations of the regenerator and the combustion chamber energy balance equations. The following main relationships were used in the calculations.

The power and specific work of the gas turbine set:

$$W_{GT} = \eta_m \cdot \dot{m}_T \cdot l_T - \dot{m}_C \cdot l_C \tag{1}$$

$$l_{GT} = \frac{W_{GT}}{\dot{m}_C} = \eta_m \cdot \left(\frac{\dot{m}_T}{\dot{m}_C}\right) \cdot l_T - \left(\frac{\dot{m}_C}{\dot{m}_C}\right) \cdot l_C = \eta_m \cdot \left(\frac{\dot{m}_T}{\dot{m}_C}\right) \cdot l_T - l_C \tag{2}$$

The specific work of the compressor and turbine:

$$l_C = \frac{1}{\eta_C} \cdot c_{pC} \cdot T_1 \cdot \left(\left(\frac{p_2}{p_1}\right)^{\frac{\kappa_C - 1}{\kappa_C}} - 1\right) \tag{3}$$

$$l_T = \eta_T \cdot c_{pT} \cdot T_3 \left(1 - \left(\frac{p_4}{p_3}\right)^{\frac{\kappa_T - 1}{\kappa_T}}\right) \tag{4}$$

The efficiency of the gas turbine cycle:

$$\eta_{GT} = \frac{W_{GT}}{\dot{Q}_1} \tag{5}$$

where the heat flux brought to combustion chamber:

$$\dot{Q}_1 = \left(\dot{m}_T \cdot h_3 - \dot{m}_C \cdot h_{2'}\right) \frac{1}{\eta_{CC}} = \dot{m}_f \cdot LHV \tag{6}$$

The overall efficiency of the gas turbine unit:

$$\eta_{GT} = \frac{\eta_m \cdot \dot{m}_T \cdot l_T - \dot{m}_C \cdot l_C}{\left(\dot{m}_T \cdot i_3 - \dot{m}_C \cdot i_{2'}\right) \cdot \frac{1}{\eta_{CC}}} \tag{7}$$

and:

$$\eta_{TG} = \eta_{CC} \cdot \frac{\eta_m \cdot \eta_C \cdot c_{pT} \cdot \frac{T_3}{T_1}\left[1 - \left(\frac{1}{\Pi_T}\right)^{\frac{\kappa_T - 1}{\kappa_T}}\right] \cdot \frac{\dot{m}_T}{\dot{m}_C} - \frac{1}{\eta_C} \cdot c_{pC} \cdot \left[\left(\Pi_C\right)^{\frac{\kappa_C - 1}{\kappa_C}} - 1\right]}{\frac{\dot{m}_T}{\dot{m}_C} \cdot c_{pCC} \cdot \frac{T_3}{T_1} - \frac{1}{\eta_C} \cdot c_{pC} \cdot \left[\left(\Pi_C\right)^{\frac{\kappa_C - 1}{\kappa_C}} - 1\right]} \tag{8}$$

The heat flux transferred from the exhaust fumes in the regenerator is that:

$$\dot{Q}_{VII,\,T} = \dot{m}_T \cdot (h_4 - h_5) \tag{9}$$

The heat flux received by the air in the regenerator is that:

$$\dot{Q}_{VII,\,C} = \dot{m}_C \cdot (h_{2'} - h_2) \tag{10}$$

The values of specific heat at constant pressure for particular states of working media were determined on the basis of their chemical composition and thermodynamical parameters using REFPROP software.

The most significant results of the calculations are presented in Table 3. The results referred to methane as fuel (row 6 in Table 2), whereas a reference also showed the results of calculations for hydrogen fuel (row 7 in Table 2). Variant 1 (Figure 1) is a basic cycle in the power plants with a gas turbine; therefore, some of the results were related to it. For each of the analyzed variants, the compression was optimized to maximize efficiency. The effect of the compression on the value of the efficiency referring to the value of the maximum efficiency of variant 1 for two exemplary fuels is shown in Figure 2. In this case, the effect of the calorific value of fuel is very clear. For a calorific value equal to 4 MJ/kg, the compression amounts to approximately 17, and for a calorific value of 54 MJ/kg, it amounts to approximately 13, which evidently affects the design of the flow part of the turbine and compressor. As can be easily observed, the calorific value affects the optimum compression value (maximum efficiency) only for variant 1, but in other cases the type of fuel has a small influence on the value of the compression (Figure 3). Therefore, when designing the flow part of a gas turbine, operated in accordance with an open simple cycle (variant 1, often used in low power turbosets), we should pay special attention to the correct selection of compressor pressure depending on the type of fuel expected.

Table 3. Selected calculation results.

Parameter	Variant	W_d (MJ/kg)						
		4.0	8.5	17.5	24.4	38.7	54.1	120.0
η/η_{meth} (-)	V1	1.265	1.109	1.041	1.024	1.008	1	0.989
	V2	0.949	0.972	0.988	0.993	0.998	1	1.003
	V3	0.981	0.992	0.997	0.998	0.999	1	1.001
	V4	0.751	0.893	0.958	0.976	0.992	1	1.011
	V5	0.835	0.929	0.972	0.984	0.995	1	1.007
$m_{fuel}/m_{fuelmeth}$ (-)	V1	16.272	6.793	3.174	2.248	1.402	1	0.448
	V2	20.853	7.496	3.296	2.297	1.409	1	0.441
	V3	13.782	6.416	3.101	2.221	1.399	1	0.450
	V4	18.006	7.124	3.225	2.272	1.409	1	0.446
	V5	17.121	7.020	3.220	2.254	1.405	1	0.448

Table 3. *Cont.*

Parameter	Variant	W_d (MJ/kg)						
		4.0	**8.5**	**17.5**	**24.4**	**38.7**	**54.1**	**120.0**
$m_{spal}/m_{spalmeth}$ (-)	V1	1.181	1.069	1.026	1.015	1.005	1	0.993
	V2	1.118	1.039	1.014	1.008	1.002	1	0.997
	V3	1.039	1.016	1.006	1.004	1.001	1	0.998
	V4	1.107	1.038	1.014	1.008	1.003	1	0.997
	V5	1.094	1.035	1.013	1.007	1.002	1	0.997
$W_{eGT}/W_{eGTmeth}$ (-)	V1	1.506	1.178	1.067	1.037	1.010	1	0.984
	V2	1.460	1.144	1.053	1.029	1.005	1	0.981
	V3	1.000	1.000	1.000	1.000	1.000	1	1.000
	V4	1.000	1.000	1.000	1.000	1.000	1	1.000
	V5	1.058	1.025	1.013	1.000	1.000	1	1.000
η/η_{V1} (-)	V1	1	1	1	1	1	1	1
	V2	0.89	1.04	1.13	1.16	1.18	1.19	1.21
	V3	1.33	1.53	1.64	1.67	1.70	1.71	1.73
	V4	0.73	0.99	1.13	1.17	1.21	1.23	1.25
	V5	0.88	1.12	1.25	1.28	1.32	1.34	1.36
Π_{opt} (-)	V1	17.00	15.00	13.80	13.60	13.40	13.20	13.00
	V2	3.60	3.00	2.85	2.80	2.75	2.75	2.70
	V3	1.75	1.75	1.75	1.75	1.75	1.75	1.75
	V4	2.60	2.60	2.60	2.60	2.60	2.60	2.60
	V5	2.90	2.75	2.70	2.65	2.65	2.65	2.65

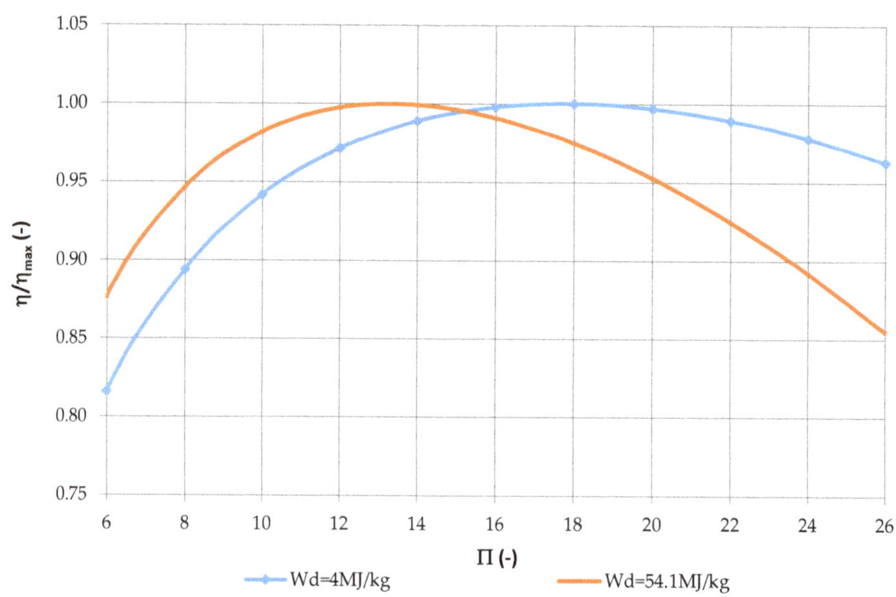

Figure 2. The effect of a compression on the value of the efficiency referred to the maximum value of variant 1 for two exemplary fuels.

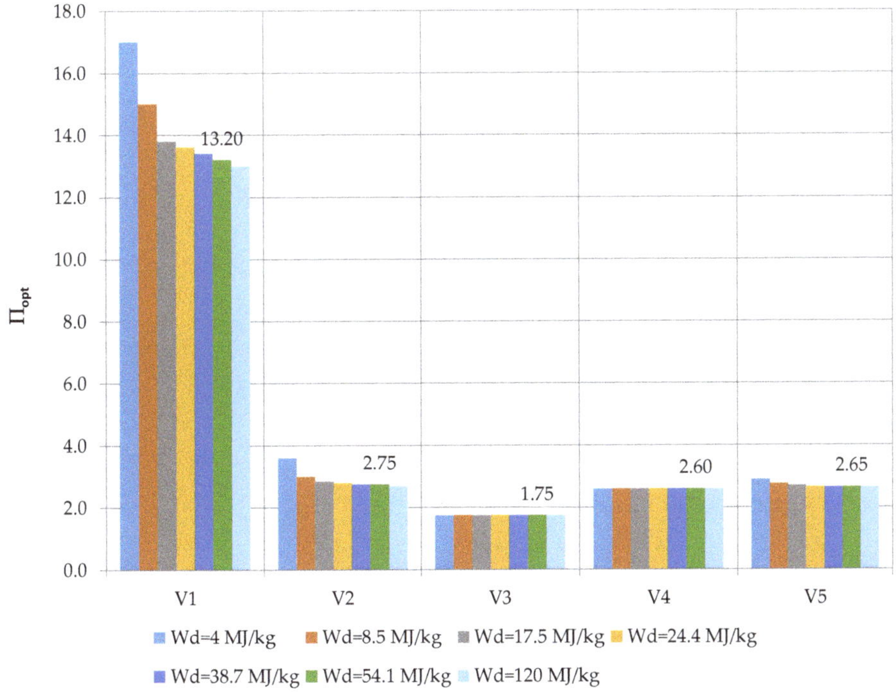

Figure 3. Influence of the structure of the cycle and calorific value of fuel on the optimal compression.

The subsequent drawings (Figures 4–10) show the influence of the cycle's structure and of calorific value on maximum efficiency related to the maximum efficiency of variant 1. Only for the smallest calorific value considered (4 MJ/kg) variant V1 proved itself to be better than variants V2, V4, V5. As far as effectiveness is concerned, the most advantageous is the variant 3. The disproportion of this one as compared to variant 1 increases with an increase of the calorific value of the fuel combusted (collective plot Figure 11). The graph (Figure 12) shows the influence of the structure of the cycle and calorific value of fuel on the efficiency of the system in relation to the variants using methane as a fuel. Only for small heating value of the fuel its effect on the efficiency of turbine sets can be noticed.

The drawings (Figures 13–15) show the influence of the fuel cycle structure and its calorific value on the mass flux of the fuel burned in the combustion chamber as related to the methane mass flux (Figure 13), and on the exhaust gas mass flux again compared to the variants with methane as a fuel (Figure 14), and on the effective power in relation to similar variants with methane as a fuel (Figure 15).

In the case of fuel consumption, the conclusions are obvious: an increase in calorific value is accompanied by a decrease of fuel consumption. This effect is the most visible for variant 2. As a consequence, the exhaust flux also decreases with an increase in fuel's calorific value, but in this case the impact is most pronounced for variant 2. It can also be concluded that in the case of turbine sets with an external combustion chamber (variants 3, 4, and 5), the change in calorific value does not affect the unit power of the turbine set.

Cogeneration systems working with organic media are already for many years [72] available in a wide range of electrical and thermal power. However, only a few examples of ORC (Organic Rankin Cycle) cogeneration installations with an electric power below 5 kW [73] can be found.

Previous research has shown that it is possible to build a set of microturbines with a capacity of about 2 kW with higher efficiency than in existing machines [74]. It is worth noting that the relatively high efficiency of microturbines can be achieved due to a very careful and advanced design process.

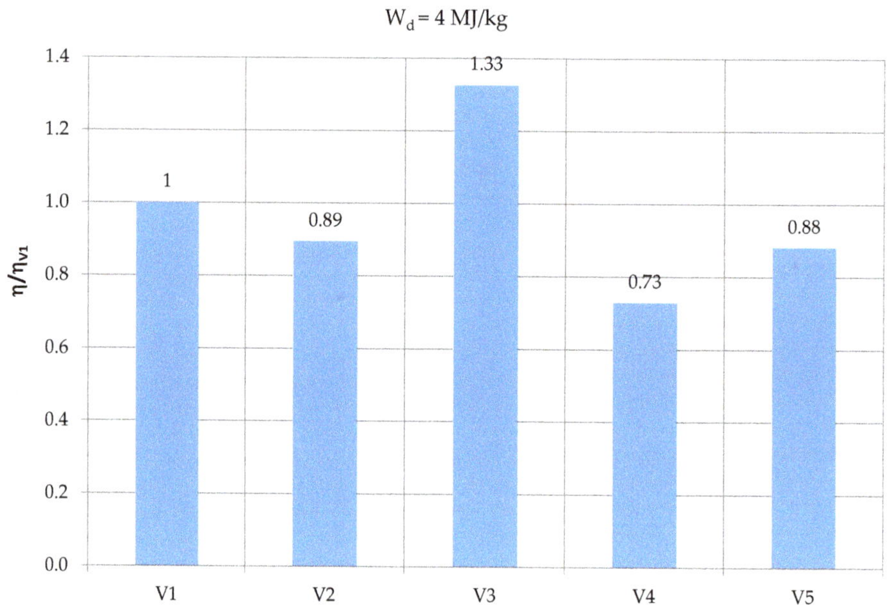

Figure 4. The effect of the cycle structure on its efficiency as compared to variant 1 for heating value W_d = 4 MJ/kg.

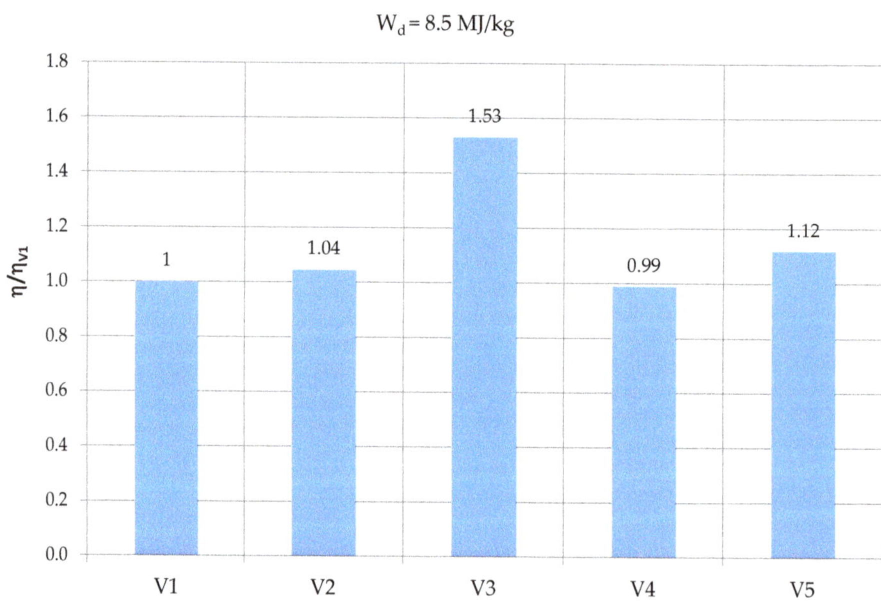

Figure 5. Influence of the cycle structure on its efficiency as compared to variant 1 for heating value W_d = 8.5 MJ/kg.

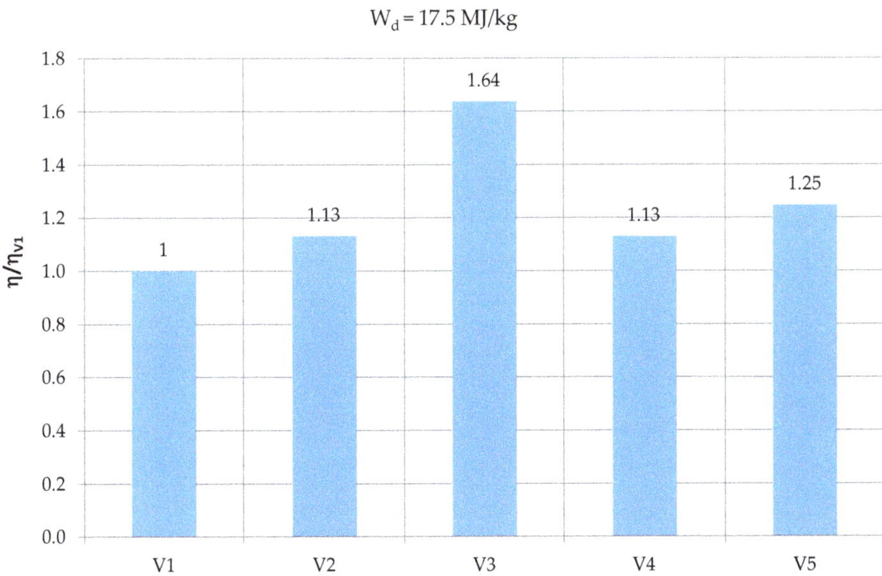

Figure 6. The effect of the cycle structure on its as compared to variant 1 for heating value W_d = 17.5 MJ/kg.

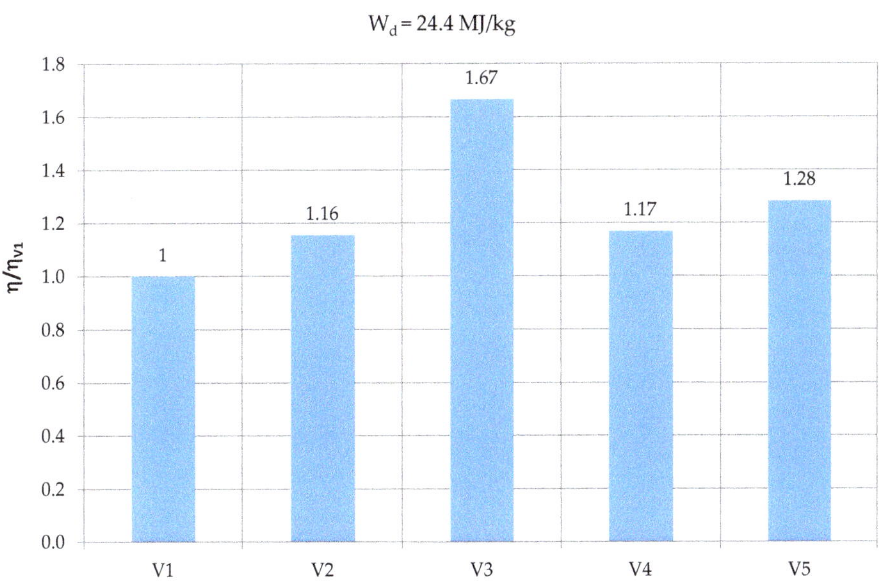

Figure 7. The effect of the cycle structure on its efficiency as compared to variant 1 for heating value W_d = 24.4 MJ/kg.

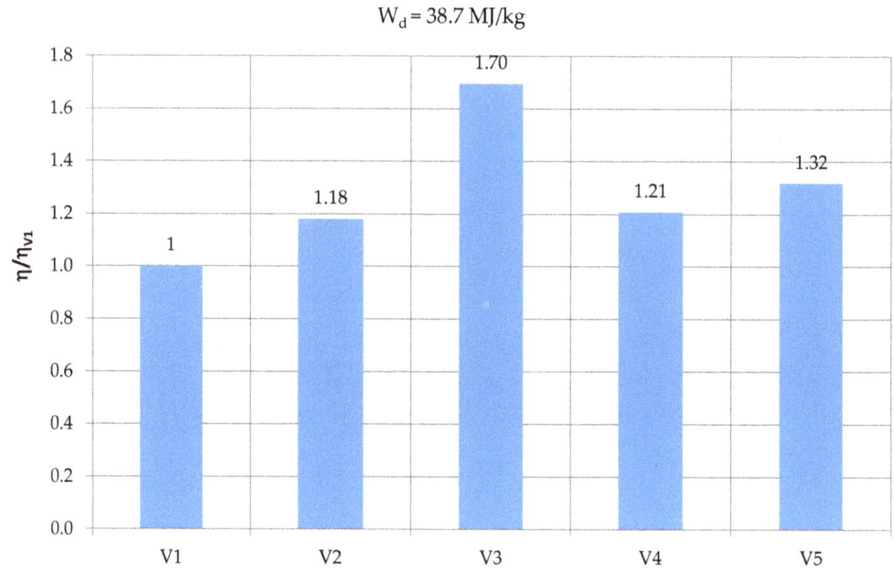

Figure 8. The effect of the cycle structure on its efficiency as compared to variant 1 for heating value W_d = 38.7 MJ/kg.

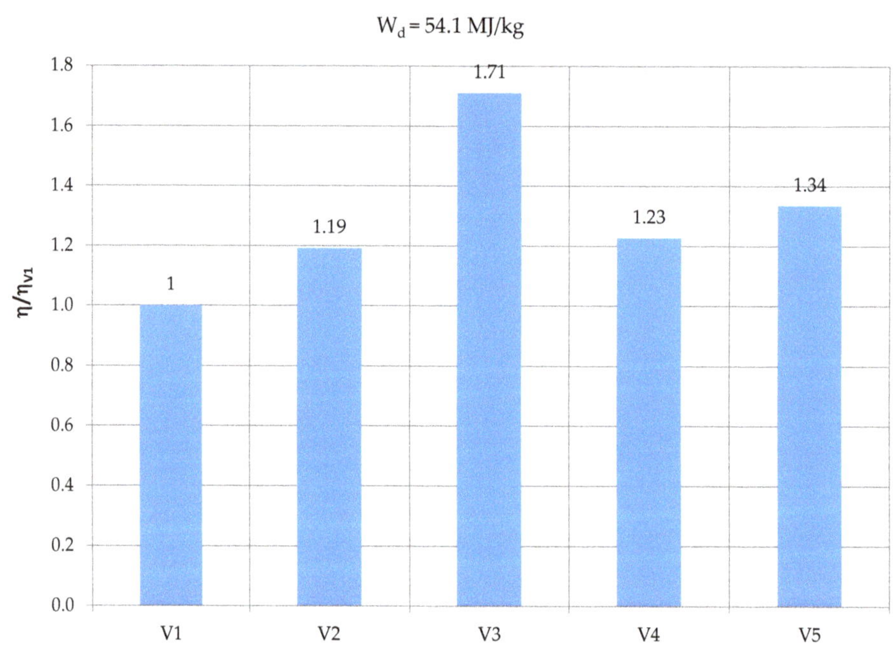

Figure 9. The influence of the cycle structure on its efficiency as compared to variant 1 for methane.

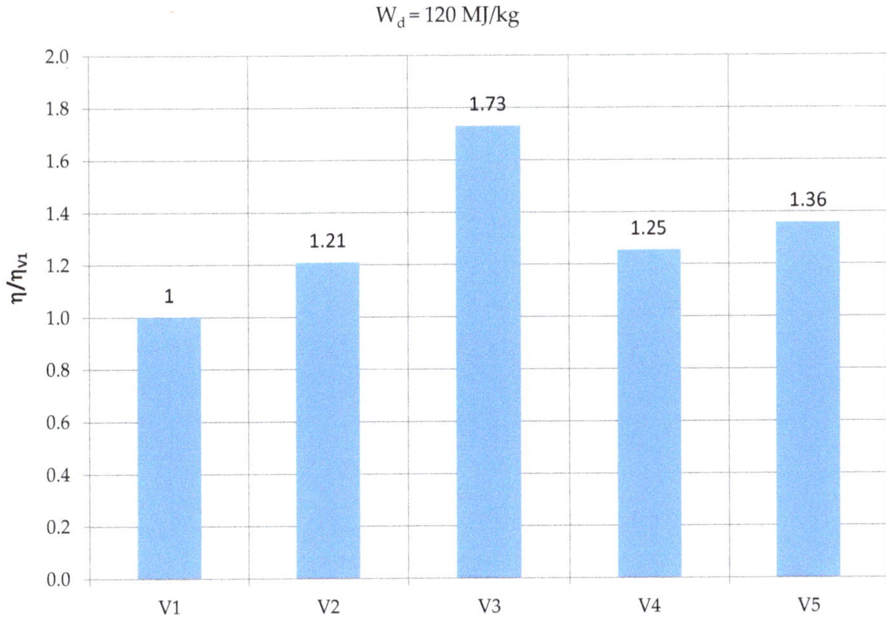

Figure 10. The influence of the cycle structure on its efficiency as compared to variant 1 for hydrogen.

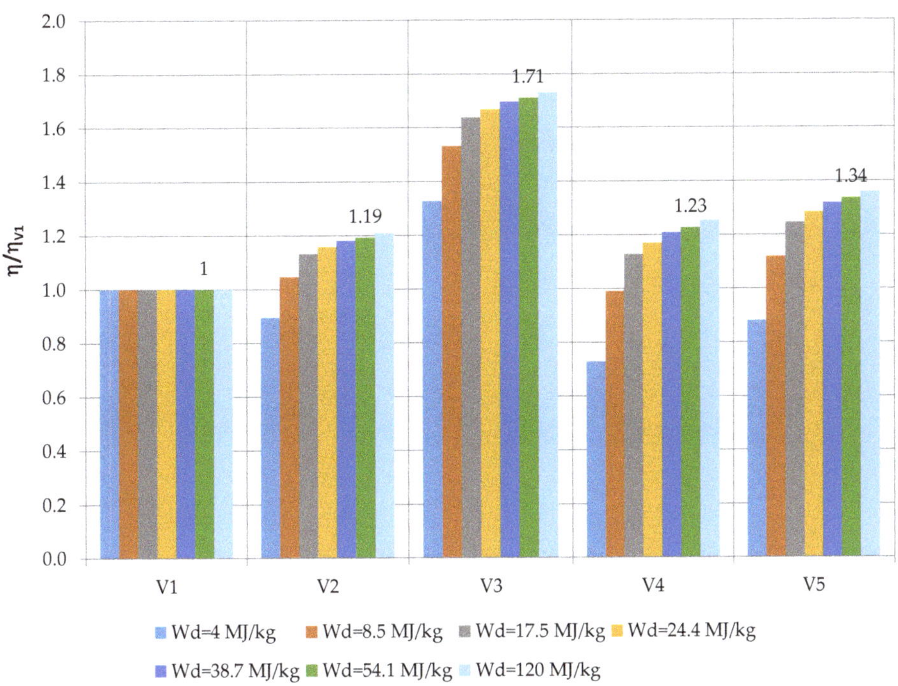

Figure 11. Collective diagram of the influence of the structure of the cycle and calorific value of fuel on the efficiency of the cycle as compared to variant 1.

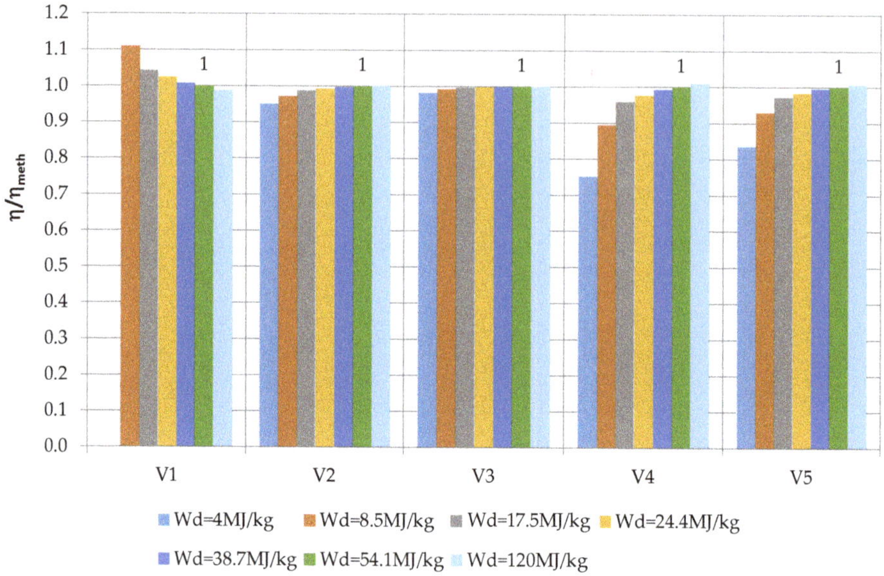

Figure 12. Collective diagram of the influence of the structure of the cycle and calorific value of fuel on its efficiency in relation to the variants of cycles with methane as a fuel.

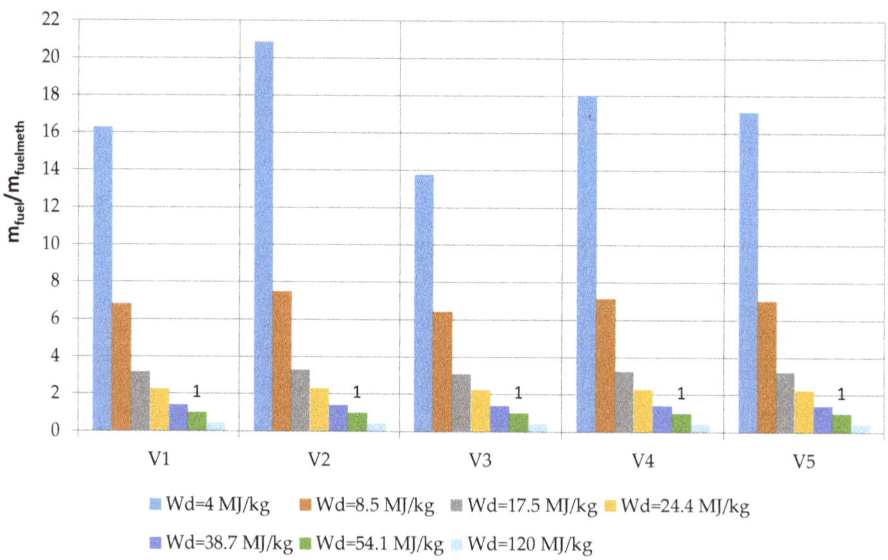

Figure 13. Collective diagram of the effect of the structure of cycle and calorific value of fuel on the mass flux of the fuel recorded in the combustion chamber in relation to the variants of circulation with methane fuel.

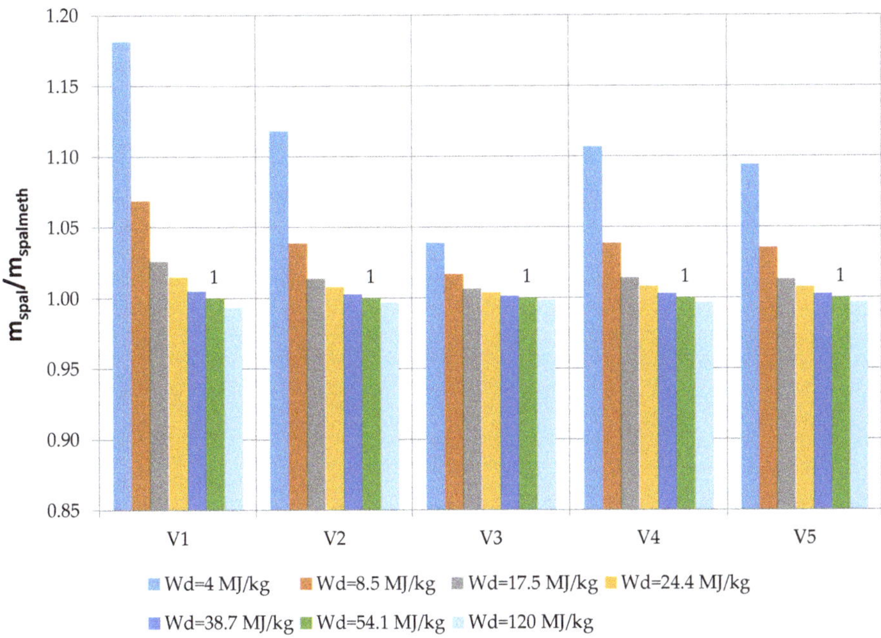

Figure 14. Collective diagram of the influence of the structure of the cycle and the calorific value of fuel on the exhaust gas mass flux as compared to the variants of the cycles with methane.

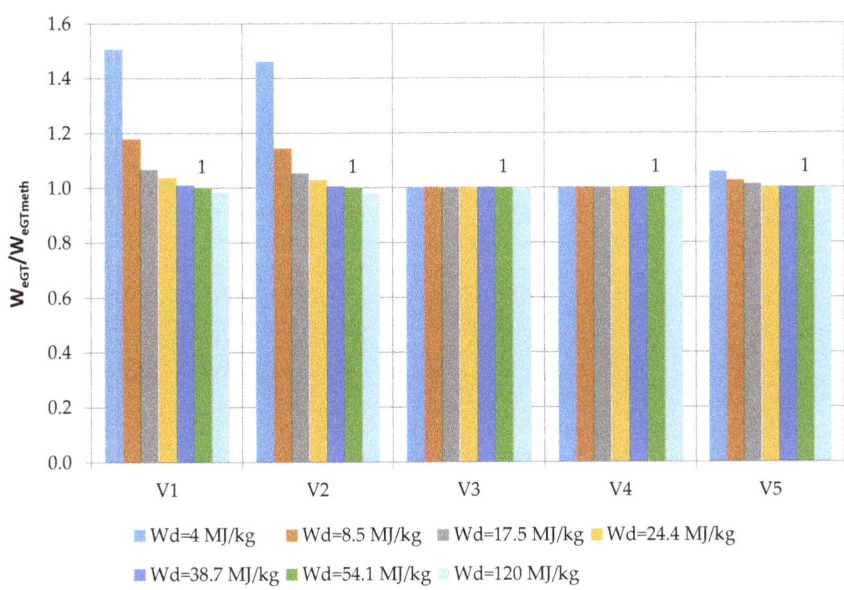

Figure 15. Collective diagram of the effect of the structure of cycle and calorific value of fuel on the effective power in relation to the variants of the cycle with methane as a fuel.

4. Discussion

Variants with external combustion chambers demand particular attention. In fact, this solution has been known for dozens of years and it has, for example, been considered for nuclear power plants with High-Temperature Gas-cooled Reactors (HTGR) but only recently has it appeared in turbomachinery practice, in small output gas turbines for prosumer and distributed energy systems. It is hardly possibly to find any information in the literature about the design problems and the results of technical analysis. However, there are a few companies [75–79] which offer small output power plants for different fuels with external combustion systems. They are characterized by a relatively small compressor ratio which corresponds well to the results shown in the paper. However, the efficiency of these units is rather unimpressive (up to about 24% for sets with regenerators). In spite of that they arouse particular interest due to the fact that they can use various types of fuel, including different biofuels. They can still be treated as pioneer solutions but it is highly likely that they will become quite popular in the near future.

5. Final Conclusions

The analysis carried out allowed for the following conclusions:

- In the design of the flow part of gas turbines that burn biofuel (mainly low power ones, dedicated for distributed energetics), not only the structure of the turbine set, but also the calorific value and type of fuel should be taken into account.
- Variant 3 is the most-advantageous system due to the efficiency achieved; it allows for increasing the efficiency with respect to the reference value of variant 1 by even over 70%. Also, higher values are obtained for fuels with higher heating values (Figure 11).
- The type of fuel affects the cycle efficiency for variant 1 and variant 4 as well as variant 5 (Figure 12). In the case of variant 1, an increase in calorific value reduces efficiency (up to 30%) as related to the efficiency of the cycle with methane as a fuel, while in variant 4 and variant 5 an increase in calorific value increases (up to 20%) the efficiency related to the efficiency of the cycle with methane as a fuel.
- The change in calorific value has a very significant impact on the amount of fuel combusted in the combustion chamber (Figure 13); e.g., for a very low heating value (W_d = 4 MJ/kg), the amount of fuel burned increases by up to 20 times compared to the combusted methane (W_d = 54.1 MJ/kg).
- For fuels with very low calorific values (W_d = 4 MJ/kg and W_d = 8.5 MJ/kg), a clear change in the mass flux of flue gas flowing through the turbine as related to the exhaust mass flux of methane as fuel (Figure 14) can be observed; in other cases this change is minor.
- For variant 1 and variant 2, the effect of calorific value on the effective power referring to effective power with methane used as a fuel (Figure 15) can be seen, while in the remaining variants the heating value of the fuel shows minimal or no effect on the effective power of the turbine set.

The conducted analyses provide knowledge to help to mitigate potential environmental hazards through introduction of biofuels into distributed energy generation and optimization of turbines to such locally available fuels.

The problem still requires further research, but implementation of the findings might contribute to the reduction of environmental burdens.

Author Contributions: Conceptualization, K.K. and M.P.; R.S.; W.W.; methodology, K.T. and O.O.; investigation, R.S. and W.W.; writing—original draft preparation, K.T. and O.O.; funding acquisition, K.K. and D.M.

Funding: The authors wish to express their deep gratitude to Gdansk University of Technology for financial support given to the present publication (Krzysztof Kosowski). The research was carried out under financial support obtained from the research subsidy of the Faculty of Engineering Management (WIZ) of Bialystok University of Technology (Olga Orynycz).

Conflicts of Interest: The authors declare no conflict of interest. The funders had no role in the design of the study; in the collection, analyses, or interpretation of data; in the writing of the manuscript, and in the decision to publish the results.

Nomenclature and Units

The following list contains a collection of the most important quantities used in calculations, together with appropriate symbols and units.Symbols and units used in calculations.

l	work of unit mass	(kJ/kg)
h	enthalpy of unit mass	(kJ/kg)
LHV	lower heating value	(MJ/kg)
Π	compression ratio	(-)
c_p	specific heat at constant pressure	(kJ/kg·K)
i	enthalpy of unit mass	(kJ/kg)
m	mass flow rate	(kg/s)
W	power	(kW)
p	pressure	(Pa)
Q	heat flux	(kW)
R	gas constant	(kJ/kg·K)
s	entropy of unit mass; also: blade pitch	(kJ/kg·K); (mm)
T	temperature	(°C)
v	specific volume	(m^3/kg)
W_d	calorific value	(MJ/m^3) or (MJ/kg);
η	efficiency	(-)
κ	isentropic exponent	(-)

List of used subscripts.

e	an effective
hyd	hydrogen
G	generator
i	internal
j	unit
C	compressor
CC	combustion chamber
m	mechanical
meth	methane
n	leaks
ob	cycle
opt	optimal
spal	exhaust gas
T	turbine
GT	gas turbine set
1, 2, ...	point numbers on diagrams

References

1. Magazzino, C. Electricity Demand, GDP and Employment: Evidence from Italy. *Front. Energy* **2014**, *8*, 31–40. [CrossRef]
2. Magazzino, C. Stationarity of electricity series in MENA countries. *Electr. J.* **2017**, *30*, 16–22. [CrossRef]
3. Herran, D.S.; Tachiiri, K.; Matsumoto, K. Global energy system transformations in mitigation scenarios considering climate uncertainties. *Appl. Energy* **2019**, *243*, 119–131. [CrossRef]
4. Veum, K.; Bauknecht, D. How to reach the EU renewables target by 2030? An analysis of the governance framework. *Energy Policy* **2019**, *127*, 299–307. [CrossRef]
5. Wierzbowski, M.; Filipiak, I.; Lyzwa, W. Polish energy policy 2050—An instrument to develop a diversified and sustainable electricity generation mix in coal-based energy system. *Renew. Sustain. Energy Rev.* **2017**, *74*, 51–70. [CrossRef]

6. Tucki, K.; Orynycz, O.; Wasiak, A.; Świć, A.; Dybaś, W. Capacity market implementation in Poland: Analysis of a survey on consequences for the electricity market and for energy management. *Energies* **2019**, *12*, 839. [CrossRef]
7. European Electricity Market—Diagnosis 2018. Available online: https://www.pse.pl/web/pse-eng/-/the-diagnosis-of-the-european-electricity-market-from-the-point-of-view-of-the-polish-transmission-system-operator (accessed on 24 July 2019).
8. Brown, T.; Schlachtberger, D.; Kies, A.; Schramm, S.; Greiner, M. Synergies of sector coupling and transmission reinforcement in a cost-optimised, highly renewable European energy system. *Energy* **2018**, *160*, 720–739. [CrossRef]
9. Dusonchet, L.; Favuzza, S.; Massaro, F.; Telaretti, E.; Zizzo, G. Technological and legislative status point of stationary energy storages in the EU. *Renew. Sustain. Energy Rev.* **2019**, *101*, 158–167. [CrossRef]
10. The Model of the Optimal Energy Mix for Poland by 2060. Chancellery of the Prime Minister. Available online: https://www.premier.gov.pl/en/news/news/the-model-of-an-optimal-energy-mix-for-poland-by-2060.html (accessed on 24 July 2019).
11. Renn, O.; Marshall, J.P. Coal, nuclear and renewable energy policies in Germany: From the 1950s to the "Energiewende". *Energy Policy* **2016**, *99*, 224–232. [CrossRef]
12. Anifantis, A.S.; Colantoni, A.; Pascuzzi, S.; Santoro, F. Photovoltaic and Hydrogen Plant Integrated with a Gas Heat Pump for Greenhouse Heating: A Mathematical Study. *Sustainability* **2018**, *10*, 378. [CrossRef]
13. Tucki, K.; Orynycz, O.; Świć, A.; Mitoraj-Wojtanek, M. The development of electromobility in Poland and EU States as a tool for management of CO_2 emission. *Energies* **2019**, *12*, 2942. [CrossRef]
14. Caragliu, A.; Del Bo, C.F. Smart innovative cities: The impact of Smart City policies on urban innovation. *Technol. Forecast. Soc. Chang.* **2019**, *142*, 373–383. [CrossRef]
15. Haarstad, H.; Wathne, M.W. Are smart city projects catalyzing urban energy sustainability? *Energy Policy* **2019**, *129*, 918–925. [CrossRef]
16. Tucki, K.; Sikora, M. Technical and logistics analysis of the extension of the energy supply system with the cogeneration unit supplied with biogas from the water treatment plant. *TEKA Comm. Motorization Power Ind. Agric.* **2016**, *16*, 71–75.
17. Raymundo, H.; Dos Reis, J.G.M. Measures for Passenger-Transport Performance Evaluation in Urban Areas. *J. Urban Plan. Dev.* **2018**, *144*, 04018023. [CrossRef]
18. Halkos, G.; Tsilika, K. Understanding transboundary air pollution network: Emissions, depositions and spatio-temporal distribution of pollution in European region. *Resour. Conserv. Recycl.* **2019**, *145*, 113–123. [CrossRef]
19. Gawlik, L.; Szurlej, A.; Wyrwa, A. The impact of the long-term EU target for renewables on the structure of electricity production in Poland. *Energy* **2015**, *92*, 172–178. [CrossRef]
20. Tucki, K.; Mruk, R.; Orynycz, O.; Wasiak, A.; Botwińska, K.; Gola, A. Simulation of the Operation of a Spark Ignition Engine Fueled with Various Biofuels and Its Contribution to Technology Management. *Sustainability* **2019**, *11*, 2799. [CrossRef]
21. Tucki, K.; Orynycz, O.; Wasiak, A.; Swic, A.; Wichlacz, J. The Impact of Fuel Type on the Output Parameters of a New Biofuel Burner. *Energies* **2019**, *12*, 1383. [CrossRef]
22. García-Álvarez, M.T.; Moreno, B. Environmental performance assessment in the EU: A challenge for the sustainability. *J. Clean. Prod.* **2018**, *205*, 266–280. [CrossRef]
23. Capros, C.; Kannavou, M.; Evangelopoulou, S.; Petropoulos, A.; Siskos, P.; Tasios, N.; Zazias, G.; DeVita, A. Outlook of the EU energy system up to 2050: The case of scenarios prepared for European Commission's "clean energy for all Europeans" package using the PRIMES model. *Energy Strategy Rev.* **2018**, *22*, 255–263. [CrossRef]
24. Report on the Polish Power System Vision 2.0 Century Profile. Available online: https://www.agora-energiewende.de/en/ (accessed on 24 July 2019).
25. Kosowski, K.; Tucki, K.; Piwowarski, M.; Stepien, R.; Orynycz, O.; Wlodarski, W. Thermodynamic Cycle Concepts for High-Efficiency Power Plants. Part B: Prosumer and Distributed Power Industry. *Sustainability* **2019**, *11*, 2647. [CrossRef]
26. Bel, G.; Joseph, S. Climate change mitigation and the role of technological change: Impact on selected headline targets of Europe's 2020 climate and energy package. *Renew. Sustain. Energy Rev.* **2018**, *82*, 3798–3807. [CrossRef]

27. Adamczyk, J.; Dzikuć, M. The analysis of suppositions included in the Polish Energetic Policy using the LCA technique—Poland case study. *Renew. Sustain. Energy Rev.* **2014**, *39*, 42–50. [CrossRef]
28. Moslehi, S.; Reddy, T.A. An LCA methodology to assess location-specific environmental externalities of integrated energy systems. *Sustain. Cities Soc.* **2019**, *46*, 1–14. [CrossRef]
29. Cellura, M.; Cusenza, M.A.; Longo, S. Energy-related GHG emissions balances: IPCC versus LCA. *Sci. Total Environ.* **2018**, *628*, 1328–1339. [CrossRef]
30. Cepeda, M. Assessing cross-border integration of capacity mechanisms in coupled electricity markets. *Energy Policy* **2018**, *119*, 28–40. [CrossRef]
31. Manowska, A.; Tobór-Osadnik, K.; Wyganowska, M. Economic and social aspects of restructuring Polish coal mining: Focusing on Poland and the EU. *Resour. Policy* **2017**, *52*, 192–200. [CrossRef]
32. Palm, J.; Eidenskog, M.; Luthander, R. Sufficiency, change, and flexibility: Critically examining the energy consumption profiles of solar PV prosumers in Sweden. *Energy Res. Soc. Sci.* **2018**, *39*, 12–18. [CrossRef]
33. Zafar, R.; Mahmood, A.; Razzaq, S.; Ali, W.; Naeem, U.; Shehzad, K. Prosumer based energy management and sharing in smart grid. *Renew. Sustain. Energy Rev.* **2018**, *82*, 1675–1684. [CrossRef]
34. Picchi, P.; Van Lierop, M.; Geneletti, D.; Stremke, S. Advancing the relationship between renewable energy and ecosystem services for landscape planning and design: A literature review. *Ecosyst. Serv.* **2019**, *35*, 241–259. [CrossRef]
35. Van Aubel, P.; Poll, E. Smart metering in the Netherlands: What, how, and why. *Int. J. Electr. Power Energy Syst.* **2019**, *109*, 719–725. [CrossRef]
36. Wang, Y.; Qiu, H.; Tu, Y.; Liu, Q.; Ding, Y.; Wang, W. A Review of Smart Metering for Future Chinese Grids. *Energy Procedia* **2018**, *152*, 1194–1199. [CrossRef]
37. Jurasz, J.; Dąbek, P.B.; Kaźmierczak, B.; Kies, A.; Wdowikowski, M. Large scale complementary solar and wind energy sources coupled with pumped-storage hydroelectricity for Lower Silesia (Poland). *Energy* **2018**, *161*, 183–192. [CrossRef]
38. Badyda, K.; Niewiński, G.G.; Patrycy, A.; Orzeszek, W. Attempt to Estimate the Costs of Implementing BAT Conclusions for Large Combustion Plants. *Soc. Inequal. Econ. Growth* **2016**, *46*, 315–333.
39. Calise, F.; De Notaristefani di Vastogirardi, G.; D'Accadia, M.D.; Vicidomini, M. Simulation of polygeneration systems. *Energy* **2018**, *163*, 290–337. [CrossRef]
40. Jarnut, M.; Wermiński, S.; Waśkowicz, B. Comparative analysis of selected energy storage technologies for prosumer-owned microgrids. *Renew. Sustain. Energy Rev.* **2017**, *74*, 925–937. [CrossRef]
41. Aiying, R.; Risto, L. Role of polygeneration in sustainable energy system development: Challenges and opportunities from optimization viewpoints. *Renew. Sustain. Energy Rev.* **2016**, *53*, 363–372.
42. Murugan, S.; Horák, B. Tri and polygeneration systems—A review. *Renew. Sustain. Energy Rev.* **2016**, *60*, 1032–1051. [CrossRef]
43. Chmielak, T.; Ziębik, A. *Obiegi Cieplne Nadkrytycznych Bloków Węglowych*; Wydawnictwo Politechniki Śląskiej: Gliwice, Poland, 2010; pp. 19–43. (In Polish)
44. Lampart, P.; Kosowski, K.; Piwowarski, M.; Jędrzejewski, L. Design analysis of tesla micro-turbine operating on a low-boiling medium. *Pol. Marit. Res.* **2009**, *1*, 28–33. [CrossRef]
45. Mikielewicz, J.; Piwowarski, M.; Kosowski, K. Design analysis of turbines for co-generating micro-power plant working in accordance with organic rankine's cycle. *Pol. Marit. Res.* **2009**, *1*, 34–38. [CrossRef]
46. Piwowarski, M.; Kosowski, K. Design analysis of combined gas-vapour micro power plant with 30 kw air turbine. *Pol. J. Environ. Stud.* **2014**, *23*, 1397–1401.
47. Turbines Markets 2016–2024: Steam, Gas Turbines, Wind, Others—Global Strategic Business Report 2018. Available online: https://markets.businessinsider.com/news/stocks/turbines-markets-2016-2024-steam-gas-turbines-wind-others-global-strategic-business-report-2018-1021662041 (accessed on 18 August 2019).
48. Stępniak, D.; Piwowarski, M. Analyzing selection of low-temperature medium for cogeneration micro power plant. *Pol. J. Environ. Stud.* **2014**, *23*, 1417–1421.
49. Weiß, A.P.; Popp, T.; Zinn, G.; Preißinger, M.; Brüggemann, D. A micro-turbine-generator-construction-kit (MTG-c-kit) for small-scale waste heat recovery ORC-Plants. *Energy* **2019**, *181*, 51–55. [CrossRef]
50. Herrando, M.; Pantaleo, A.M.; Wang, K.; Markides, C.N. Solar combined cooling, heating and power systems based on hybrid PVT, PV or solar-thermal collectors for building applications. *Renew. Energy* **2019**, *143*, 637–647. [CrossRef]

51. Hsu, P.C.; Huang, B.J.; Wu, P.H.; Wu, W.H.; Lee, M.J.; Yeh, J.F.; Wang, Y.H.; Tsai, J.H.; Li, K.; Lee, K.Y. Long-term Energy Generation Efficiency of Solar PV System for Self-consumption. *Energy Procedia* **2017**, *141*, 91–95. [CrossRef]
52. Samson, I.; Sikora, M.; Bączyk, A.; Mączyńska, J.; Tucki, K. Technologies used to enhance the biogas and biomethane yield: A review. *Przemysł Chem.* **2017**, *96*, 1605–1611.
53. Saadabadi, S.A.; Thattai, A.T.; Fan, L.; Lindeboom, R.E.F.; Spanjers, H.; Aravind, P.V. Solid Oxide Fuel Cells fuelled with biogas: Potential and constraints. *Renew. Energy* **2019**, *134*, 194–214. [CrossRef]
54. Tonini, D.; Hamelin, L.; Alvarado-Morales, M.; Astrup, T.F. GHG emission factors for bioelectricity, biomethane, and bioethanol quantified for 24 biomass substrates with consequential life-cycle assessment. *Bioresour. Technol.* **2016**, *208*, 123–133. [CrossRef]
55. Shen, X.; Kommalapati, R.R.; Huque, Z. The Comparative Life Cycle Assessment of Power Generation from Lignocellulosic Biomass. *Sustainability* **2015**, *7*, 12974–12987. [CrossRef]
56. Adelt, M.; Wolf, D.; Vogel, A. LCA of biomethane. *J. Nat. Gas Sci. Eng.* **2011**, *3*, 646–650. [CrossRef]
57. Gonzalez-Fernandez, C.; Sialve, B.; Bernet, N.; Steyer, J.P. Impact of microalgae characteristics on their conversion to biofuel. Part ii: Focus on biomethane production. *Biofuels Bioprod. Biorefin.* **2012**, *6*, 205–218. [CrossRef]
58. Cozzolino, R.; Lombardi, L.; Tribioli, L. Use of biogas from biowaste in a solid oxide fuel cell stack: Application to an off-grid power plant. *Renew. Energy* **2017**, *111*, 781–791. [CrossRef]
59. Hijazi, O.; Tappen, S.; Effenberger, M. Environmental impacts concerning flexible power generation in a biogas production. *Carbon Resour. Convers.* **2019**, *2*, 117–125. [CrossRef]
60. Barbuzza, E.; Buceti, G.; Pozio, A.; Santarelli, M.; Tosti, S. Gasification of wood biomass with renewable hydrogen for the production of synthetic natural gas. *Fuel* **2019**, *242*, 520–531. [CrossRef]
61. Coronado, C.R.; Yoshioka, J.T.; Silveira, J.L. Electricity, hot water and cold water production from biomass. Energetic and economical analysis of the compact system of cogeneration run with woodgas from a small downdraft gasifier. *Renew. Energy* **2011**, *36*, 1861–1868. [CrossRef]
62. Kosowski, K.; Domachowski, Z.; Próchnicki, W.; Kosowski, A.; Stępień, R.; Piwowarski, M.; Włodarski, W.; Ghaemi, M.; Tucki, K.; Gardzilewicz, A.; et al. *Steam and Gas Turbines with the Examples of Alstom Technology*, 1st ed.; Alstom: Saint-Quen, France, 2007; ISBN 978-83-925959-3-9.
63. Kosowski, K.; Tucki, K.; Piwowarski, M.; Stępień, R.; Orynycz, O.; Włodarski, W.; Bączyk, A. Thermodynamic Cycle Concepts for High-Efficiency Power Plans. Part A: Public Power Plants 60+. *Sustainability* **2019**, *11*, 2647. [CrossRef]
64. Ślefarski, R.; Jójka, J.; Czyżewski, P.; Grzymisławski, P. Experimental investigation on syngas reburning process in a gaseous fuel firing semi-industrial combustion chamber. *Fuel* **2018**, *217*, 490–498. [CrossRef]
65. Taler, J.; Mruk, A.; Cisek, J.; Majewski, K. Combined heat and power plant with internal combustion engine fuelled by wood gas. *Rynek Energii* **2013**, *4*, 62–67.
66. Kordylewski, W. *Spalanie i paliwa*, 5th ed.; Oficyna Wydawnicza Politechniki Wrocławskiej: Wrocław, Polska, 2008; pp. 10–470. ISBN 978-83-7493-378-0.
67. Masłow, L.A. *Ship Gas Turbines*, 1st ed.; Sudostroene: Leningrad Region, Russia, 1973. (In Russian)
68. Kostiuk, A.G.; Serstiuk, A.N. *Gas Turbines Units*, 1st ed.; Wyschaya Skola: Moscow, Russia, 1979. (In Russian)
69. Traupel, W. *Thermische Turbomachinen*, 2nd ed.; Springer: Berlin/Heidelberg, Germany; New York, NY, USA, 1982; pp. 122–525. Available online: https://link.springer.com/book/10.1007%2F978-3-642-96632-3 (accessed on 12 August 2019).
70. Sawyer, J.W. *Gas Turbine Engineering Handbook*, 3rd ed.; Turbomachinery International Publications: Norwalk, CT, USA, 1985; pp. 98–245. ISBN 0-937506-14-1.
71. Sorensen, H.A. *Gas Turbines (Series in Mechanical Engineering)*, 1st ed.; Ronald Press Company: New York, NY, USA, 1951; pp. 48–440.
72. Gailfuß, M. Private meets Public—Small scale CHP. Technological Developments. In Proceedings of the Workshop BHKW-Infozentrum Rastatt, Berlin, Germany, 9 September 2003.
73. Kosowski, K.; Włodarski, W.; Piwowarski, M.; Stepień, R. Performance characteristics of a micro-turbine. *Adv. Vib. Eng.* **2014**, *2*, 341–350.
74. Kosowski, K.; Piwowarski, M.; Stępień, R.; Włodarski, W. Design and investigations of the ethanol microturbine. *Arch. Thermodyn.* **2018**, *39*, 41–54.

75. AE-T100 Externally Fired Micro Turbine. Available online: https://www.ansaldoenergia.com/business-lines/new-units/microturbines/ae-t100e (accessed on 9 August 2019).
76. High Reliability and Easy Maintenance. Technical Data for Dresser-Rand KG2 Gas Turbines. Available online: https://new.siemens.com/global/en/products/energy/power-generation/gas-turbines/dresser-rand-kg2.html (accessed on 9 August 2019).
77. 1.2 MW Gas Turbine for Biomass Burning. Available online: https://www.braytonenergy.net/our-projects/1-2-mw-gas-turbine/ (accessed on 9 August 2019).
78. Gas Turbine Electric Power Plant of External Combustion 2.5 MW GTEUVS-2.5MS. Available online: http://www.motorsich.com/eng/products/land/vrazrabotke/gteuvc-2.5ms/ (accessed on 9 August 2019).
79. C65. Available online: https://www.capstoneturbine.com/products/c65 (accessed on 9 August 2019).

energies

Article

Inter-Criteria Dependencies-Based Decision Support in the Sustainable wind Energy Management

Paweł Ziemba *

Faculty of Economics and Management, University of Szczecin, Mickiewicza 64, 71-101 Szczecin, Poland

Received: 11 January 2019; Accepted: 20 February 2019; Published: 24 February 2019

Abstract: Decision problems related to the wind energy require considering many, often interrelated and dependent on each other, criteria. To solve such problems, decision systems based on Multi-Criteria Decision Analysis (MCDA) methods are usually used. Unfortunately, most methods assume independence between the criteria, therefore, their application in decision problems related to the wind energy is debatable. This paper presents the use of the Analytic Network Process (ANP) method to solve a decision problem consisting in selecting the location and design of a wind farm. The use of the ANP method allows capturing the complexity of the decision problem by taking into consideration dependencies between criteria. As part of the verification of the solution, the results of the ANP method were compared with those of the Analytic Hierarchy Process (AHP) method, which uses only hierarchical dependencies between criteria. The conducted verification showed that the inter-criteria dependencies may have a significant influence on the obtained solution. On the basis of the conducted sensitivity analysis and the research into robustness of the rankings to the rank reversal phenomenon, it has been found out that the ranking obtained with the use of the ANP is characterized by a higher quality than by means of the AHP.

Keywords: Multi-Criteria Decision Analysis; sustainable wind energy management; sensitivity analysis; rank reversal; Analytic Network Process; Analytic Hierarchy Process

1. Introduction

One of the biggest challenges of today's energy management is adapting it to the demands of low carbon economy characterized by, most of all, the use of renewable energy sources (RES) [1]. The fact can be confirmed in the Polish energy policy, for which major priorities are, among other things: energy efficiency improvement, reduction of pollutions from the energy sector, development of renewable energy and an increase in the use of RES [2,3]. The Polish energy policy in this area is coherent with the policy of the European Union (EU), which assumes that there will be at least a 20% reduction of greenhouse gas emissions by 2020 compared to the 1990 levels and it requires increasing a share of renewable energy in gross final energy consumption to about 20% in 2020 from its member states [4]. In a broader perspective, i.e. by 2030, at least 40% reduction of greenhouse gas emissions is assumed in relation to 1990 and the share of renewable energy in total energy consumption is 32%. The minimum contribution of Member States to the new framework for 2030 should be the achievement of the national targets for 2020 [5]. It should be noted that the objectives set for 2020 have a chance to be met at the level of the European Union (without dividing them into individual Member States). According to the Eurostat data, in 2016 the share of renewable energy in energy consumption in the whole EU amounted to 17% [6]. In turn, the emission of greenhouse gases was reduced in the period 1990-2016 by 22.4% [7]. Nevertheless, the analysis of the quoted sources [6,7] indicates that some Member States have little chance of achieving the targets set for 2020.

The Polish Energy Law Act defines renewable energy as biogas, biomass, geothermal, river fall, sea wave and tidal, solar, and wind energy [8]. Among the above-mentioned RES, the great potential for energy production, both in Poland and in the EU, have wind farms [2,4]. It is because onshore wind farms are characterized by the low capital investment and levelised cost of electricity as well as one of the shortest construction period of all RES power stations [9–11]. The comparison of the most popular RES technologies [12] with regard to costs, construction time, lifetime and installed capacity in Poland and in the EU [9–11,13] are presented in Table 1.

Table 1. Comparison of basic RES technologies with regard to necessary investment and installed capacity.

Energy Source	Capital Investment 2017 (2016 USD/kW)	Levelised Cost of Electricity 2017—World (2016 USD/kWh)	Levelised Cost of Electricity 2017—Europe (2016 USD/kWh)	Construction Time (Year)	Life-Time (Year)	Installed Capacity in Poland—2017 (MW)	Installed Capacity in EU—2017 (MW)
Onshore Wind	1 477	0.06	0.08	1	25	5 798	152 751
Offshore Wind	4 239	0.14	0.15	2	25	-	15 835
Hydropower	1 535	0.05	0.12	3	30	969	130 411
Bioenergy	2 668	0.07	0.07	2	20	1 075	36 341
Geothermal	2 959	0.07	0.08	2	25	-	846
Solar Photovoltaic	1 388	0.1	0.13	0	25	268	106 546
Concentrating Solar Power	5 564	0.22	-	2	25	-	2 308

As for the considerably greater installed power of wind farms with reference to other RES in Poland, apart from lower capital investment, the installed power results from the enormous potential of the wind in Poland. It is assumed that the Polish potential of wind is bigger than in countries, such as Denmark and Sweden, in which an important part of the energy is obtained from the wind. The potential is equal to that of Germany, which is the "world's wind giant" [8]. It is forecasted that by 2020 the installed RES capacity in Poland will amount to circa 10,000 MW, of which about 50% will be wind energy [2]. Therefore, Poland is expected to experience a dynamic increase in wind farm construction investment.

The most important decision problems whose solutions will lead to a successful realization of a wind farm project are selection of a location [14] and selection of a project design [15]. Decision problems concerning wind energy, and similarly other decision issues related to RES management, are multi-criteria decision making problems that require consideration of many contradictory and mutually correlated criteria comprising economic, environmental, social as well as technological and spatial issues [14–20]. Single-criterion decision making methods are not able to deal with such decision problems correctly [21]. In solving such problems, Multi-Criteria Decision Analysis (MCDA) methods can be applied since they are able to deal with multiple and conflicting criteria [22]. However, most MCDA methods assume independence between criteria [23], therefore, it is difficult to apply the methods to complex decision problems in which there are mutual dependencies between criteria [24]. Issues related to RES and sustainable management [24,25], especially problems dealing with wind energy [15], are this kind of decision problems.

The aim of this article is to select a project design and the location of an onshore wind farm in Poland. The selection should be based on a decision model which takes into account dependencies between criteria. The methodological contribution of this article consists in the comparison of the solution obtained using inter-criteria dependencies (based on the ANP approach) with the solution of the decision model which do not take into consideration interdependencies between criteria (based on the AHP approach). This will allow assessing the impact of taking into account the dependencies between criteria on the obtained results. Additionally, an analysis of intrinsic characteristics of the MCDA methods, and thus a formal selection of the MCDA method applied to the decision problem can also be treated as an important contribution. Section 2 contains the analysis of the literature on MCDA methods related to the decision problems in the wind energy field. Section 3 discusses the applicability of MCDA methods in decision problems in the field of wind energy. Additionally in Section 3, the proposed methodology was presented. In Section 4, the approach was applied to the decision problem consisting in selecting the location and design of a wind farm in Poland. Section 5 deals with a summary of research results. Also, further research directions are pointed out.

2. Literature Review

The MCDA methods are often used for solving decision problems related to RES, such as, among other things, the selection of a location for an RES-based power plant, selecting an energy production technology, evaluation of the influence of a renewable energy power station on the environment, an analysis of different scenarios of RES development, optimization of the energy production, etc. [17,26]. The most popular MCDA methods used for solving decision problems in the field of RES are first of all: the AHP and ANP, MAVT, MAUT, TOPSIS, PROMETHEE, ELECTRE [18] as well as fuzzy methods [27]. Also, hybrids were employed, which are a combination of various MCDA methods [26]. Applications of the MCDA methods in the field of RES are presented in the papers [17–19,22,26,27]. In the context of decision support, in the literature, the most often mentioned type of renewable energy is wind energy [18]. Publications concerning decision support in wind energy include, most of all, the construction of decision models and the construction of DSS (Decision Support System) and GIS (Geographical Information System) systems.

An example of constructing a decision model for selecting a wind farm location is [28]. In this paper, evaluation criteria and their weights, presented in a decision model for the sake of selecting a wind farm location, were determined. On the other hand, in [29] a decision model was constructed in order to select a location of an onshore wind farm. The problem of constructing a decision model for selecting an onshore wind farm location is also dealt with in [30], a hybrid farm in [31,32], and an offshore wind farm in [14]. GIS decision systems were proposed, among other things in [33–38]. These systems evaluate the potential of onshore areas with regard to situating wind farms on the areas. Similarly, in [39] a GIS, which allowed evaluating a location of hybrid power stations based on wind and solar energies, were presented. In [40] and [41], a GIS-based DSS systems considering potential onshore wind farm locations were discussed. The problem of constructing a DSS for selecting an offshore wind location was attempted in [42,43]. Decision models dealing with a location selection and the design of a wind farm related to the selection are presented in [44–46]. Similarly, in [47], the location and design of a hybrid power plant were selected. As far as decision problems closely related to the project design of a wind farm are concerned, such an issue was discussed in [15,48]. To the wind farm design are also related technical aspects of wind turbines [16,49,50] as well as, in a broader context, risk assessment [51]. On the other hand, in [52], a decision model used for evaluating the influence of a wind farm on the environment in a given location was presented. Table 2 includes a list of publications dealing with the issue of decision support in the area of wind energy on the basis of MCDA methods.

The analysis of Table 2 indicates that in decision problems concerning wind energy, the AHP method is often used, both in its crisp and fuzzy versions. It is employed to determine the weight of criteria and preference aggregation. A generalization of the AHP, namely the ANP, and different variants of the ELECTRE method are rarely used. Sporadically, other MCDA methods, such as TOPSIS, PROMETHEE, Weighted Overlay, WLC, OWA, DEMATEL, Conjunctive Method, are used. Other MCDA methods are employed in individual cases. It should be noted that in order to solve decision problems related to wind energy, decision models characterized by various complexities are used. The number of criteria ranges from 5 to as many as 35. These criteria are often related to each other and dependent on one another. For example, in the publication [15], the following criteria were used: generating cost, generating profit, and payback period. As it can be easily found out the payback period results from, among other things, the calculation of costs and profits.

Table 2. The use of MCDA in decision support concerning wind energy.

Type of Solution	No. of Criteria	MCDA Approach	Reference
DM	15	Fuzzy DEMATEL (CD); ANP (CW)	[28]
DM	7	Fuzzy AHP (CW); Fuzzy VIKOR (PA)	[29]
DM	5	AHP	[30]
DM	20 *	AHP	[31]
DM	11 *	ELECTRE II	[32]
DM	22	Fuzzy ELECTRE III	[14]
GIS	7	AHP (CW); OWA (PA)	[33]
GIS	6	AHP (CW); WLC (PA)	[34]
GIS	10	LM; ELECTRE TRI	[35]
GIS	10	Fuzzy AHP (CW); Fuzzy TOPSIS (PA)	[36]
GIS	5	AHP (CW); WO (PA)	[37]
GIS	13	CM (EA); AHP (CW); ELECTRE III, ELECTRE TRI, SMAA-TRI (PA)	[38]
GIS	10 *	OWA	[39]
DSS, GIS	14	CM (EA); WO (PA)	[40]
DSS, GIS	8	AHP (CW); WLC (PA)	[41]
DSS	31	Fuzzy DEMATEL (CD); Fuzzy ANP (CW); Fuzzy ELECTRE (PA)	[42]
DSS	10	AHP (CW); PROMETHEE II (PA)	[43]
DM	29	AHP	[44]
DM	9	C-K-Y-L (with indifference threshold)	[45]
DM	10	AHP; PROMETHEE II	[46]
DM	27 *	Fuzzy AHP	[47]
DM	11	NAIADE I	[48]
DM	35	FCI (sub-criteria PA); GIFOGA (criteria PA)	[15]
DM	14	Fuzzy ANP	[16]
DM	9	IFE (CW); Fuzzy TOPSIS (PA)	[49]
DM	11	AHP	[50]
DM	9	Fuzzy ANP	[51]
DM	14	AHP	[52]

Abbreviations: *—wind energy criteria; **Type of solution**: DM—Decision model; DSS—Decision Support System; GIS—Geographical Information System; **MCDA approach (method)**: AHP—Analytic Hierarchy Process; ANP—Analytic Network Process; C-K-Y-L—Condorcet–Kemeny–Young–Levenglick method; CM—Conjunctive Method; DEMATEL—DEcision MAking Trial and Evaluation Laboratory; ELECTRE—ELimination Et Choix Traduisant la REalité; FCI—Fuzzy Choquet Integral; GIFOGA—Generalized Intuitionistic Fuzzy Ordered Geometric Averaging; IFE—Intuitionistic Fuzzy Entropy; LM—Lexicographic Method; NAIADE—Novel Approach to Imprecise Assessment and Decision Environments; OWA—Ordered Weighted Averaging; PROMETHEE—Preference Ranking Organization METHod for Enrichment Evaluation; SMAA—Stochastic Multi-objective Acceptability Analysis; TOPSIS—Technique for Order of Preference by Similarity to Ideal Solution; VIKOR—VIšekriterijumsko KOmpromisno Rangiranje; WLC—Weighted Linear Combination; WO—Weighted Overlay; TRI—Triage; **MCDA approach (application)**: CD—criteria dependencies defining; CW—criteria weighting; EA—elimination of areas; PA—preference aggregation.

Because of inter-criteria dependencies and other elements of specificity of decision problems in the field of RES, an important issue is the selection of a proper MCDA method which can be used in decision problems in the area of wind energy. This is important because solutions to a decision problem may vary depending on the method used [53]. Differences between methods result primarily from: the different way in which weights are taken into account in the decision problem, differences in calculation procedures and the application of additional parameters of the decision problem by the different methods [54].

3. Methodological Background

3.1. Choosing an MCDA Method for Decision Problems in the RES Field

As it was pointed out in [55], appropriateness of an MCDA method to the specificity of a decision problem is essential for its selection. This means that there is no universal method that can be applied to all decision problems [56]. Therefore, determining the specificity of a decision problem is a significant step when selecting an MCDA method, and after this step, to a given problem a method should be selected which complies with specific characteristics.

On the basis of reference sources [24,57–59], in which the issue of an MCDA method selection for decision problems in the fields of RES and sustainability was considered, the following characteristics determining the specificity of decision problems were taken into consideration:

- the issues of a decision which was being considered [58],
- applied preference relations and a way of organizing alternatives [58,59],
- a compensation degree of criteria [24,57–59],
- discrimination power of criteria [57,59],
- a type of applied information [24,57–59],
- applying weights of criteria [24,57–59],
- support for many decision-makers [24,58,59],
- using dependencies between criteria [24,59].

MCDA methods are designed for solving different reference problematics. One can distinguish the following problematics [60]: choice ($P.\alpha$)—aids the decision-maker in choosing a subset that is as small as possible so that a single alternative can eventually be chosen from the subset; sorting ($P.\beta$)—aids the decision-maker by assigning each alternative to a category, where the categories are defined beforehand as a function of certain norms; ranking ($P.\gamma$)—aids the decision-maker by building an order of alternatives, that is obtained by placing alternatives into equivalence classes that are completely or partially ordered according to preferences; description ($P.\delta$)—aids the decision-maker by developing a description of alternatives and their consequences.

In MCDA methods, an order between decision alternatives a and b is expressed by means of relations describing preference situations. Depending on an MCDA method, one can list the following relations: indifference ($a\ I\ b$)—which means that alternatives are equal, strict preference ($a\ P\ b$)—which means that there is a strong advantage of an alternative a over b, weak preference ($a\ Q\ b$)—which means that there is a weak advantage of an alternative a over b, outranking ($a\ S\ b$)—which includes an indifference relation, a weak one and a strict preference relation, incomparability ($a\ R\ b$)—which means that none of the remaining relations takes place [60]. The incomparability relation is related to a way of organizing alternatives in the ranking problematics. That is, when an MCDA method does not take into consideration the incomparability relation, the method usually allows obtaining a total order, i.e., full comparability of all alternatives in the ranking. Otherwise, a partial order is achieved, what means that there may be alternatives which cannot be compared to other ones in the obtained ranking [61].

An important characteristic of multi-criteria methods in the context of sustainability and RES problems is the degree of compensation of the criteria. In the literature [56], there are three basic degrees of compensation: (1) full compensation—meaning that the low values of some criteria can be fully compensated for by the high values of other criteria, (2) no compensation—meaning that the low values of some criteria cannot be compensated for in any way by other criteria, (3) partial compensation—being the intermediate step between full and no compensation. In many methods, absolute compensation is excluded by using a veto threshold (v) as well as an incomparability relation, which is usually related to it. It is important to note that the concept of strong sustainability is reflected by a low degree of compensation and also weak sustainability corresponds to a high degree of criteria compensation [59].

The discriminating power of the criteria refers to how preference relations are established. Absolute discriminating power means that even a minimal advantage of an alternative a over b with regard to a given criterion c results in a strict preference ($a\ P\ b$) relation. On the other hand, applying non-absolute discriminating power results in a situation where indifference ($a\ I\ b$) or weak preference ($a\ Q\ b$) relations take place. Non-absolute discriminating power usually requires using indifference (q) and preference (p) thresholds [60] in an MCDA method.

Both action performance and criterion weights can be expressed on different scales, depending on the nature of the data. Qualitative and quantitative scales are the most common, while Roy [62] indicates that they can be identified with ordinal and cardinal scales respectively. A data nature refers to whether they are certain or uncertain [63]. Certain data (deterministic) have a crisp form, whereas uncertain data (non-deterministic) can be expressed in a fuzzy form [63] or by defining a proper value of indifference or preference thresholds [24,57]. The information type refers to the performance of decision alternatives and weights of criteria. An MCDA method may: not use weights of criteria, accept weights expressed on an ordinal scale or operate on weights presented on a cardinal scale [24]. In addition, some methods offer the possibility of aggregating information from many decision-makers or reflecting different priorities or scenarios.

Most MCDA methods assume independence between criteria what is not a realistic assumption in many real-world problems [23]. These methods cannot be easily applied to more difficult decision problems [24], because omitting existing dependencies between criteria does not allow the problem to be correctly reflected in a decision model what results in obtaining wrong solutions [64].

Table 3 depicts the basic characteristics of the MCDA methods. Table 3 takes into account only method families whose applicability is confirmed by the analysis of reference literature presented in Section 2. Table 3 particularly presents MCDA methods which were used at least twice in decision problems related to wind energy. Consequently, methods which had been incidentally used in this area were eliminated from further analysis. In addition, Table 3 also includes the latest MCDA methods used in sustainability and wind energy issues: BWM (Best Worst Method), COMET (Characteristic Objects Method), NEAT F-PROMETHEE (New Easy Approach To Fuzzy PROMETHEE), PROSA (PROMETHEE for Sustainability Analysis).

In papers dealing with a selection of an MCDA method suitable for applying in RES and sustainability fields, recommendations of characteristics, which such a method should meet, were defined. The characteristics are presented in Table 4.

On the basis of Table 4, it can be stated that the selected method should generate a full ranking of alternatives (total order without incomparability). The MCDA method solving decision problems related to wind energy management should be characterised by non-absolute discriminating power of the criteria, i.e., it should apply q and p thresholds. Additionally, it should apply at most partial compensation of criteria. The method should be able to capture quantitative and qualitative information since in RES problems there are the two types of information [59]. What is more, it is recommended that the method takes into consideration uncertainty of information, makes it possible to support group decisions and allows capturing hierarchical and horizontal dependencies between criteria. When analysing Table 4, one ought to notice that in decision problems dealing with wind energy, no MCDA methods meeting the above-mentioned requirements are usually used. The PROMETHEE II and ANP methods meet the most of the listed characteristics. The PROMETHEE II method does not allow using other than hierarchical dependencies between criteria, whereas the ANP does not make it possible to define indifference and preference thresholds what results in absolute discriminating power of the criteria. Both methods meet seven recommended characteristics. An advantage of the ANP over PROMETHEE is the ability of presenting mutual inter-criteria dependencies in a decision model, thus, such a model could, to a greater extend, present the complexity of a decision problem in the field of RES and sustainability [25].

Table 3. Characteristics of MCDA methods applied in the field of wind energy.

MCDA Method	Reference Problematic	Preference Modelling		Degree of Compensation	Criterion Function		Information Type		Type of Weights	Group Decision Making	Dependencies between Criteria	Reference
		Preference Relations	Order of Alternatives		Discriminating Power of the Criteria	Thresholds	Type of Information	Information Features				
AHP	γ	I, P	TO	PC	AB	n	QL, QN	D, N	C	SG	HD	[65]
ANP	γ	I, P	TO	PC	AB	n	QL, QN	D, N	C	SG	HD, ID	[66]
ELECTRE I	α	S, R	SU	PC	AB	n [1]	QL, QN	D	C	NG	IN	[67]
ELECTRE IS	α	S, R	SU	PC	NA	q, p, v	QL, QN	D, N	C	NG	IN	[67]
ELECTRE II	γ	S, R	PO	PC	AB	v [2]	QL, QN	D	C	NG	IN	[67]
ELECTRE III	γ	S, R	PO	PC	NA	q, p, v	QL, QN	D, N	C	NG	IN	[67]
ELECTRE IV	γ	S, R	PO	PC	NA	q, p, v	QL, QN	D, N	NW	NG	IN	[67]
ELECTRE TRI	β	S, R	PO	PC	NA	q, p, v	QL, QN	D, N	C	NG	IN	[67]
TOPSIS	γ	I, P	TO	FC	AB	n	QN	D	C	SG	IN	[68]
PROMETHEE I	γ	I, P, R	PO	PC	NA	q, p, o [3]	QL, QN	D, N	C	SG [4]	HD [8]	[69]
PROMETHEE II	γ	I, P	TO	PC	NA	q, p, o [3]	QL, QN	D, N	C	SG [4]	HD [8]	[69]
Conjunctive Method	α	I, P	FI	NC	AB	n	QL, QN	D	NW	NG	IN	[68]
WLC [5]	γ	I, P	TO	FC	AB	n	QN	D	C	NG	IN	[70]
Weighted Overlay	γ	I, P	TO	FC	AB	n	QN	D	C	NG	IN	[71]
OWA	γ	I, P	TO	FC, PC, NC [6]	AB	n	QN	D, N	C	SG	IN	[72]
DEMATEL	γ	I, P	PO, TO [7]	FC	AB	n	QL	D, N	NW	SG	ID	[73]
BWM	γ	I, P	TO	PC	AB	n	QL, QN	D, N	C	NG	IN	[74]
COMET	γ	I, P	TO	FC	AB	n	QL, QN	D, N	NW	SG	IN	[75]
NEAT F-PROMETHEE I	γ	I, P, R	PO	PC	NA	q, p, o [3]	QL, QN	D, N	C	NG	IN	[76]
NEAT F-PROMETHEE II	γ	I, P	TO	PC	NA	q, p, o [3]	QL, QN	D, N	C	NG	IN	[76]
PROSA	γ	I, P	TO	PC	NA	q, p, o [3]	QL, QN	D, N	C	NG	HD	[77]

Abbreviations: [1]—ELECTRE Iv applies a veto (v) threshold; [2]—two veto thresholds; [3]—depending on the applied preference function; [4]—supported by applying PROMETHEE GDSS method; [5]—also known as Simple Additive Weighting; [6]—depending on an applied OWA operator; [7]—depending on implementation; [8]—3-level hierarchy is applied in Visual PROMETHEE software; **Reference problematic**: α—choice; β—sorting; γ—ranking; **Preference modelling**: I—indifference; Q—weak preference; P—strict preference; S—outranking; R—incomparability; TO—total order; PO—partial order; SU—subset; FI—filtration; **Degree of compensation**: FC—full compensation; PC—partial compensation; NC—non-compensation; **Criterion function**: AB—absolute; NA—non-absolute; q—indifference; p—preference; v—veto; n—none; o—other; **Information type**: QL—qualitative; QN—quantitative; D—deterministic; N—non-deterministic; **Type of weights**: NW—not applicable; O—ordinal; C—cardinal weights; **Group decision making**: NG—not supported; SG—supported; **Dependencies between criteria**: IN—independence; HD—hierarchical dependencies; ID—interdependencies.

Table 4. Recommendations of characteristics of an MCDA method suitable for applying in RES and sustainability problems.

Reference Problematic (Ch. 1)	Preference Modelling (Ch.2)		Degree of Compensation (Ch. 3)	Criterion Function (Ch. 4)		Type of Information (Ch. 5)		Type of Weights (Ch. 6)	Group Decision Making (Ch. 7)	Dependencies between Criteria (Ch. 8)	References
	Preference Relations	Order of Alternatives		Discriminating Power of the Criteria	Thresholds	Kind of Information	Information Features				
γ	I, P	TO	PC	NA	q, p	QL, QN	-	C	SG	-	[58]
-	I, P	TO	PC	NA	q, p	QL, QN	D, N	NW, O, C	SG	HD	[59]
-	-	-	NC, PC	NA	q, p	QL, QN	D, N	NW, O, C	-	-	[57]
-	-	-	NC	-	-	QL, QN	D, N	C	SG	HD, ID	[24]
γ	I, P	TO	NC ∨ PC	NA	q, p	QL, QN	D, N	NW, O, C	SG	HD, ID	Sum of characteristics

What is more, it should be noted that the ANP is a generalization of the AHP in which a problem of a network structure, as opposed to a hierarchical one, is considered [78]. In other words, the AHP and ANP methods are based on the same calculation apparatus. However, the AHP does not allow modelling other than hierarchical dependencies between criteria [79]. Therefore, solving a decision problem by means of the AHP and ANP methods will allow examining the influence of applying inter-criteria dependencies on an obtained solution, thus the influence of computational algorithms used in individual MCDA methods will be eliminated. Such a comparison is vital since the AHP method is often used in decision problems concerning wind farms. As a consequence, the author of this paper made a decision to use the ANP method as an engine of a proposed solution.

The AHP and ANP methods are based on utility theory. Both methods can be presented in three steps of the calculation procedure:

1. identifying the decision problem and preparing a problem model in the form of a hierarchical structure (AHP) or a network structure (ANP),
2. carrying out pairwise comparisons (alternatives and criteria),
3. achieving a solution with the use of supermatrix [80].

In the hierarchical structure, the decision problem is modelled in the form of objective, criteria, sub-criteria and alternatives. Therefore, only hierarchical relationships are used in it. In the network model, apart from hierarchical dependencies, inner and outer dependencies between criteria/sub-criteria and feedbacks are also modelled [81], i.e., dependencies directed contrary to the classical hierarchy. These differences allow creating more complex decision models using the ANP method. An overview of the network decision model is shown in Figure 1.

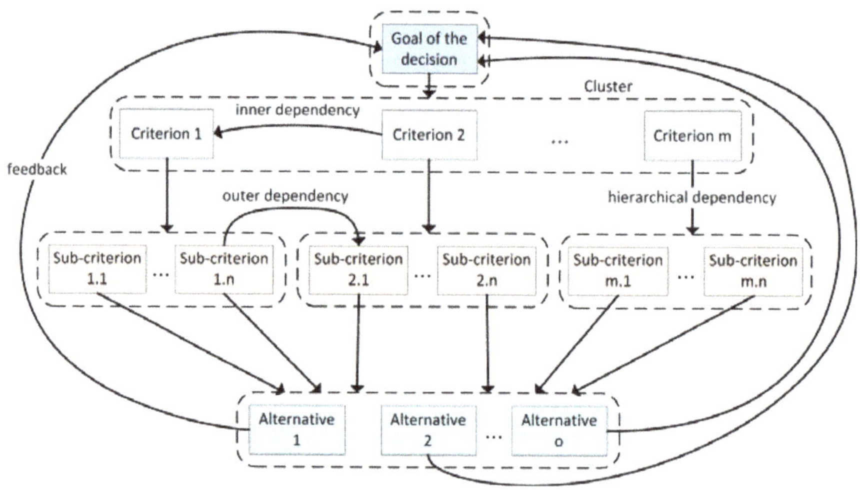

Figure 1. Network structure in the ANP method.

To compare the alternatives or criteria in both methods, pairwise comparison matrices are used. Any such matrix M ought to be positive and reciprocal, according to the formula (1):

$$m_{ji} = \frac{1}{m_{ij}} \quad (1)$$

The main diagonal of the matrix contains unit values. For each matrix M, the preference vector $W = [w_1, \ldots, w_n]$ is calculated, which determines the weights of the criteria considered or the assessment of alternatives. Saaty recommends calculating the vector W using the eigenvector method [82], but other methods of determining the vector W, e.g., column sums, power method, simple geometric mean, are often used in the literature in order to simplify the calculation procedure [80]. The eigenvector method consists in solving Equation (2):

$$MW = \lambda_{max} W \tag{2}$$

where λ_{max} is the highest eigenvalue of the M matrix. It should be emphasized that in the M matrix only minor inconsistencies are allowed, which arise due to incomplete transitory preferences [81]. Moreover, if the data in the matrix are represented on their natural scales, then the M matrix is always consistent.

Both methods allow obtaining a solution to a decision problem with the use of supermatrix [83]. At the beginning, the interfactorial dominance supermatrix [84] is defined, which indicates relations between elements of the decision-making model. In the next step, the unweighted supermatrix is defined, in which eigenvectors W are placed in individual pairwise comparison matrices. A weighted supermatrix is obtained on the basis of the unweighted supermatrix by normalizing column sums to a unit value (stochastic supermatrix). In the last step, the limit supermatrix is reached using the formula (3):

$$LSM = \lim_{k \to \infty} \frac{1}{N} \sum_{k=1}^{N} WSM^k \tag{3}$$

where WSM stands for a weighted supermatrix. A limit supermatrix represents global priorities to solve a decision problem (weighting of criteria, global values of alternatives) [66].

3.2. Proposed Methodology

The used research procedure, whose goal is to select a project design and the location of an onshore wind farm, was based on a decision process model defined by Roy [60]. Therefore, the verification model had the following stages:

1. determining a goal of the decision and alternatives;
2. developing criteria;
3. modelling preferences;
4. investigating and developing the recommendation.

Stages 2–4 were carried out separately for the AHP and ANP methods.

In Stage 2 for the ANP, sub-criteria used for solving a decision problem were defined and dependencies between them were determined. These dependencies were presented in a network ANP decision model, whereas for the AHP, a hierarchical decision model, which was similar to the network model, but did not take into consideration dependencies between criteria, was constructed.

In Stage 3, initial weights of criteria and sub-criteria were attributed. It should be stressed that in the conducted research, each of the criteria was attributed the same weight. Analogically, sub-criteria within one criterion were also considered equally important. After performing the ANP procedure, the influence of the network model on sub-criteria weights, which were finally obtained, was examined and the weights were compared to the initial importance. Additionally, for the network model, the influence of a cluster containing decision alternatives on the obtained sub-criteria weights was examined. It was carried out by determining sub-criteria weights on the basis of a complete model containing the goal of the decision, criteria, sub-criteria and decision alternative clusters, and by determining sub-criteria weights on the basis of a model from which a decision alternative cluster was deleted.

Stage 4 consisted in determining a ranking of alternatives with the use of the ANP and AHP and comparing obtained rankings. The rankings were determined by means of comparing alternatives in pairwise comparison matrices. Usually, in pairwise comparison matrices, Satty's fundamental scale (a qualitative scale [1,2,...,9]) is used. However, the ANP and AHP methods also allow presenting data on their natural scale [85]. In the case of this paper, performances of alternatives for individual criteria are presented on natural scales. The comparison of rankings was carried out with the use of a sensitivity analysis as well as research into the susceptibility of rankings to the rank reversal phenomenon. In this way, the influence of dependencies between sub-criteria on the obtained solution was examined. It should be noted that comparing the results of the operation of the ANP with any other method, different from the AHP, would not enable such research. It results from the fact that the ANP and AHP methods are based on the same computational apparatus, but they differ in that the ANP takes into consideration dependencies between criteria. However, other methods employ different computational algorithms and, therefore, differences in obtained results could be caused by the very differences and not by inter-criteria dependencies. The research procedure is depicted in Figure 2.

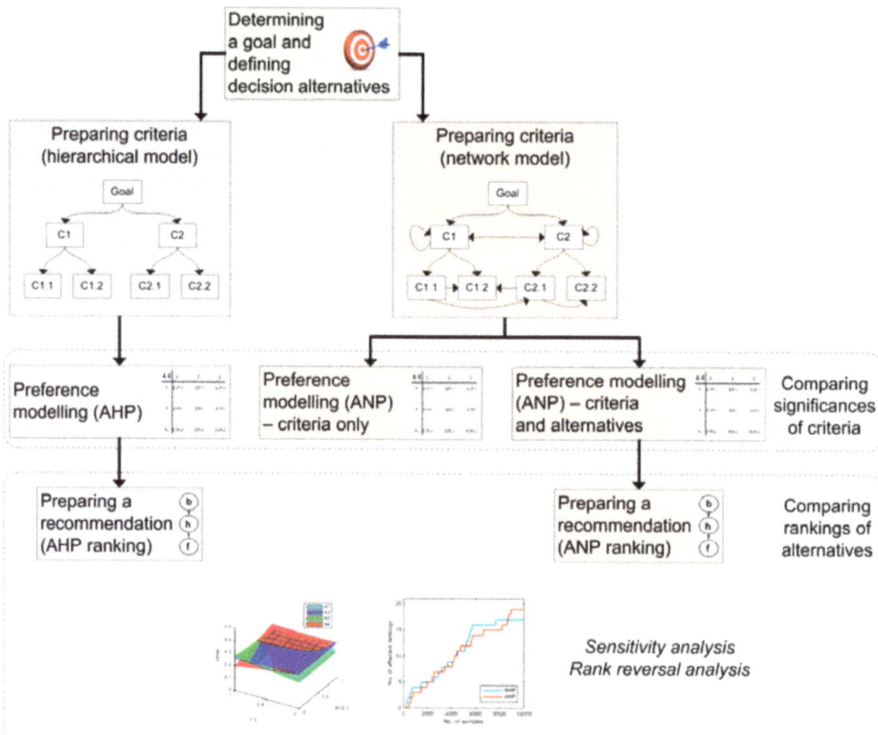

Figure 2. Research procedure

4. Results

4.1. Determining a Goal of the Decision and Developing Criteria

The goal of the decision was the selection of a location for an onshore wind farm in Poland related to its specific design. Four decision alternatives were considered, which had different values of individual sub-criteria.

Another step was to select criteria and sub-criteria against which decision alternatives would be evaluated. On the basis of the literature analysis presented in Section 2, a set of criteria and sub-criteria for evaluating locations and designs of onshore wind farms, presented in Table 5, was prepared. Technical, economic, spatial, social and environmental criteria were singled out. For each criterion, detailed sub-criteria taken from the literature were attributed. Here, a certain level of sub-criterion generalization was accepted in order to avoid many sub-criteria having the same or similar meaning which could occur in the set. For instance, in the paper [48], an "energy production capacity" criterion was used, whereas in [43], "energy production", in [51], "power production", and in [15] and [31], "annual on-grid energy" were applied. All the criteria in this work were generalized as "annual energy production", since, in fact, they refer to it.

Table 5. Criteria and sub-criteria for evaluating locations and designs of onshore wind farms.

Criteria	Sub-Criteria		Reference
C1—Technical	C1.1	Annual mean wind speed (at the height of 100 m)	[14,15,28,31,33–37,40,42,44,46,47,52]
	C1.2	Output power of wind turbine	[15,16,29,44,47,50]
	C1.3	Power transmission grid voltage	[46,52]
C2—Economic	C2.1	Annual energy production	[15,31,38,43,44,46–48,50,51]
	C2.2	Investment cost	[14–16,29–32,43,46–48,50]
	C2.3	Annual operation and maintenance costs	[14–16,29,30,32,46–48,50]
	C2.4	Annual profit	[14,15,31,41,42,46,49]
	C2.5	Payback period	[14,15,31,43]
C3—Social	C3.1	Social acceptability	[29,31,32,42,43,46,48]
	C3.2	Employment	[14,15,49]
C4—Spatial	C4.1	Distance to main roads	[28,33–40,46]
	C4.2	Distance to power transmission grid	[14,31,32,35,36,38–40,42,43,46]
C5—Environmental	C5.1	Distance to protected areas (i.e., Nature 2000)	[34,38–41,43,46]

Table 6 contains a full characteristic of alternatives, with regard to the criteria.

Table 6. Criteria and sub-criteria for evaluating locations and designs of onshore wind farms

Criteria	Sub-Criteria		Alternatives			
			A1	A2	A3	A4
C1—Technical	C1.1	Annual mean wind speed (at the height of 100 m) (m/s)	6.75	7.12	6.95	6.04
	C1.2	Output power of wind turbine (MW)	0.53	0.58	0.57	0.38
	C1.3	Power transmission grid voltage (kV)	220	400	220	220
C2—Economic	C2.1	Annual energy production (MWh)	106 784	86 374	104 857	46 603
	C2.2	Investment cost (mln PLN)	455.40	336.60	415.80	277.20
	C2.3	Annual operation and maintenance costs (mln PLN)	8.86	7.17	8.70	3.87
	C2.4	Annual profit (mln PLN)	27.98	22.63	27.47	12.21
	C2.5	Payback period (years)	16.3	14.9	15.1	22.7
C3—Social	C3.1	Social acceptability (%)	59	24	61	26
	C3.2	Employment (number)	1062	606	831	533
C4—Spatial	C4.1	Distance to main roads (km)	6	10	7	3
	C4.2	Distance to power transmission grid (km)	2	3	60	2
C5—Environmental	C5.1	Distance to protected areas (binary)	1	9	1	9

What needs explaining are the values of alternatives for sub-criterion C5.4—a distance to protected areas. Because there is no possibility of attributing the "0" value to a criterion in the ANP and AHP, binary values reflected on Saaty's fundamental scale [66] was used here. When a location was situated outside, protected areas it got a "9", otherwise it was given a "1".

Another step was to determine dependencies between sub-criteria. The dependencies were formulated on the basis of the literature analysis. The dependencies between sub-criteria are presented in Table 7. In cells of Table 7, literature sources, from which information was taken about individual dependencies, were marked. The direction of dependency is defined from rows to columns, e.g., the sub-criterion C1.1 has influence on the C1.2.

Table 7. Dependencies between sub-criteria based on the literature sources

	C1.2	C2.1	C2.2	C2.3	C2.4	C2.5	C3.1
C1.1	[38,41]						
C1.2		[38,41]					
C2.1				[86]	[86]		
C2.2				[48]		[87]	
C2.3					[52]		
C2.4						[87]	
C3.2							[14,28]
C4.1			[38]	[38]			
C4.2					[38]		

For the considered decision problem were constructed two network models considering dependencies between sub-criteria and a hierarchical model which does not consider dependencies of this type. One of the network models, which can be determined as complete, contained the considered decision alternatives. The second network model included only the aim of the decision, criteria and sub-criteria. Therefore, the evaluation of decision alternatives was not possible in the second model and it was constructed only for the sake of obtaining the weight of criteria and sub-criteria for comparison. The complete decision model containing decision alternatives and considering dependencies between sub-criteria is presented in Figure 3.

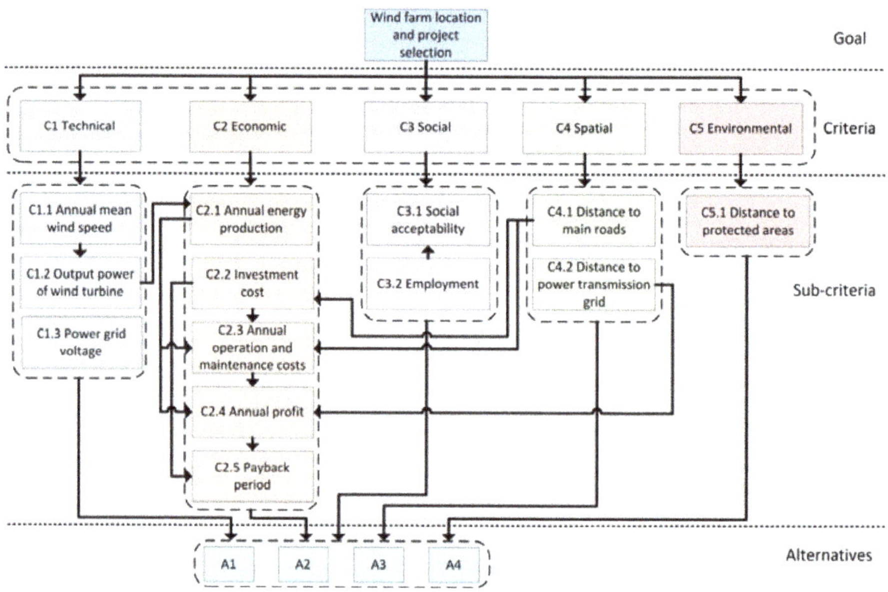

Figure 3. Decision model with alternatives and dependencies between sub-criteria

4.2. Modelling Preferences

The criteria and sub-criteria were attributed initial weights and preference directions presented in Table 8. Also, Table 8 contains sub-criteria weights obtained as a result of conducting the ANP computational procedures for network models with and without presented decision alternatives. It should be stressed that the AHP method, unlike the ANP, in the whole computational procedure uses criteria and sub-criteria weights defined at the beginning and it does not set their new values.

The analysis of Table 8 points out that the final weight of criteria (sub-criteria) changes significantly with reference to predefined values, as a result of applying the ANP calculation procedure. In general, the weight of a sub-criterion which other sub-criteria influence is increasing. Consequently, the weight of another sub-criteria in a given cluster is decreasing. Moreover, taking into consideration the cluster of alternatives in the decision model influences the result of sub-criteria weights. This effect can be seen in the case of an interdependent pair of sub-criteria C3.1 and C.3.2. The effect of weight changes takes also place in the case of sub-criteria which are not mutually dependent on each other, what can be illustrated by a criterion C1.3.

Table 8. Weights of criteria

Criteria	Predefined Weight	Sub-Criteria	Preference Direction	Weight—Predefined and AHP	Weight—ANP without Alternatives	Weight—ANP with Alternatives
C1	0.2	C1.1	max	0.333	0.25	0.286
		C1.2	max	0.333	0.5	0.428
		C1.3	max	0.333	0.25	0.286
C2	0.2	C2.1	max	0.2	0.118	0.16
		C2.2	min	0.2	0.061	0.116
		C2.3	min	0.2	0.151	0.185
		C2.4	max	0.2	0.306	0.293
		C2.5	min	0.2	0.364	0.246
C3	0.2	C3.1	max	0.5	0.667	0.6
		C3.2	max	0.5	0.333	0.4
C4	0.2	C4.1	min	0.5	0.5	0.5
		C4.2	min	0.5	0.5	0.5
C5	0.2	C5.1	max	1	1	1

4.3. Investigating and Developing the Recommendation

4.3.1. Preference Aggregation

The last stage was to determine the utilities and ranking of alternatives as well as the analysis of the utilities and ranking. The rankings obtained for the AHP and ANP methods are shown in Table 9.

Table 9. Utility and rank of alternatives.

Alternative		A1	A2	A3	A4
Utility	AHP	0.237	0.275	0.195	0.293
	ANP	0.235	0.279	0.218	0.268
Rank	AHP	3	2	4	1
	ANP	3	1	4	2

When analysing Table 9, one can notice that the AHP and ANP rankings are different from one another with regard to the obtained values of alternative utilities. The differences result from, most of all, different weights of sub-criteria obtained by means of the AHP and ANP methods. The presentation of dependencies between sub-criteria also influences the obtained differences. The dependencies influence the form of a weighted supermatrix used in the AHP and ANP methods. Unweighted, weighted and limit supermatrices obtained for the AHP and ANP methods are shown in Appendix A. The most significant differences with regard to utilities can be noticed in the case of alternatives A3 and A4, since they amount to 0.023 and 0.025 respectively. The differences seem to be insignificant, however, they can influence their positions in the ranking. It is shown in the case of alternatives A2 and A4, for which a slight change in the utilities obtained by means of the ANP method with reference to the utilities determined by the AHP influenced the exchange of their positions in the rankings.

As for the comparison of qualities of rankings obtained with the AHP and ANP methods, it is not possible in a mathematical sense [88]. It results from the fact that MCDA methods, such as the AHP and ANP, are used in problems in which individual decision alternatives are Pareto-optimal solutions [89]. Also, in the decision problem which is being examined, all considered alternatives belong to Pareto-front solutions (they are Pareto-optimal solutions). However, Pareto-optimal solutions are not comparable in a mathematical programming sense. This means that one cannot formally decide which alternative is better than another one [90]. That's why, in the present paper, the comparison of the results obtained with the use of the AHP and ANP methods was conducted on the basis of a sensitivity analysis [91] and the examination of occurrence of the rank reversal phenomenon [92,93].

4.3.2. Sensitivity Analysis

The sensitivity analysis carried out by the author of the present paper is more precise than the analysis proposed by the author of the AHP and ANP methods, i.e. Saaty. Saaty finds out that because of the complexity of a network or hierarchical structure, the sensitivity analysis of precise numerical values is difficult to carry out, that's why he suggests using an abstract value P (perturbation) based on linear optimization and describing a trend of changes [66] (Chapter 15). However, in the sensitivity analysis presented in this paper, the precise numerical value of a weight of criteria and sub-criteria were obtained. A dedicated solution to a decision problem is obtained for these weights. Hence, such stability intervals based on the numerical value of weights were also determined.

In the presented model of a decision problem, there are two levels of a hierarchy of criteria. This fact has been used in the present author's sensitivity analysis. As a result, in this case, the sensitivity analysis is an issue consisting of three dimensions which are: possible weights of a criterion (e.g., C1) and a sub-criterion (e.g., C1.1) belonging to it as well as utilities of alternatives. It results from the fact that the real weight of a sub-criterion is the product of the weight of a criterion and a sub-criterion. For instance, for weights depicted in Table 8, the real predefined weight of sub-criteria C1.1, C1.2 and C1.3 amount to 0.0666. However, for a weight C1 = 0.1 and a weight C1.1 = 0.666, the real weight amounts to 0.0666, whereas for C1.2 and C1.3 the real weight C1.1 amounts to 0.01665. In consequence, for the two cases, where the real weight is C1.1 = 0.0666, the result will be different utility values of individual alternatives of individual values.

The sensitivity analysis was carried out on the basis of implementation of the AHP and ANP methods in the MATLAB software. Plane graphs presenting utility values of individual alternatives in the function of the weight of the criterion C1 and sub-criterion C1.1 are shown in Figure 4a (for the AHP method) and 4b (for the ANP method). Other graphs are depicted in Supplementary Materials.

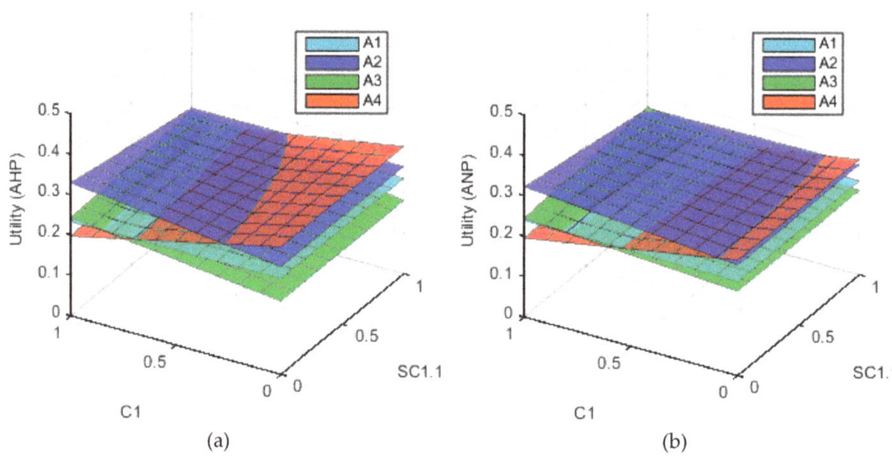

Figure 4. Utility of alternatives determined by means of: (**a**) the AHP method; (**b**) the ANP method; depending on the weight of the criterion C1 and sub-criterion C1.1.

Table 10 presents selected intervals of stability of the AHP and ANP solutions. Stability intervals for criteria were determined on the assumption that all sub-criteria contained in a given criterion have equal predefined weights. On the other hand, stability intervals for sub-criteria were determined on the assumption that superior criteria weights are equal. As a result of the sensitivity analysis it was found out that, assuming an equal weights of criteria, only changes to weights of sub-criteria presented in Table 10 can influence the ranking of alternatives.

Table 10. Stability intervals of rankings

AHP					ANP				
Criterion/ Sub-Criterion	Stability Interval (Weight)			Nominal Weight	Criterion/ Sub-Criterion	Stability Interval (Weight)			Nominal Weight
	Min	Max	Range			Min	Max	Range	
C1	0	0.32	0.32	0.2	C1	0.11	0.52	0.41	0.2
C2	0	0.63	0.63	0.2	C2	0	0.57	0.57	0.2
C3	0	0.33	0.33	0.2	C3	0	0.31	0.31	0.2
C4	0.13	0.4	0.27	0.2	C4	0.02	0.29	0.27	0.2
C5	0.11	1	0.89	0.2	C5	0.13	0.99	0.86	0.2
C2.2	0	0.95	0.95	0.2	C2.3	0	0.97	0.97	0.2
C2.6	0	0.95	0.95	0.2	C2.4	0	0.72	0.72	0.2
C4.1	0.01	1	0.99	0.5	C4.1	0	0.85	0.85	0.5
C4.2	0	0.99	0.99	0.5	C4.2	0.15	1	0.85	0.5

When analysing Table 10, it can be noticed that the ranking obtained with the AHP method is almost as stable as the solution generated by means of the ANP method. In both cases, the ranges of stability intervals, with relation to a predefined weight of criteria and sub-criteria (see Table 8), is wide enough to acknowledge the rankings obtained by means of both AHP and ANP as stable rankings.

4.3.3. Rank Reversal Phenomenon Analysis

The analysis of ranking robustness to the rank reversal phenomenon was based on the implementation of the AHP and ANP methods in the MATLAB software. The analysis was conducted by constructing an alternative A5 which was examined along with alternatives A1–A4. A ranking, which was obtained in this way, of five alternatives, was compared with the reference ranking presented in Table 9. If the sequence of alternatives A1–A4 was changed with relations to the reference ranking, the rank reversal phenomenon was considered to have taken place. The alternative A5 was constructed on the basis of random data in two ways, that is both (1) taking into consideration dependencies between sub-criteria and (2) without taking into account the above-mentioned dependencies.

In the case of constructing the alternative A5, in which inter-criteria dependencies were taken into account, only the values of independent sub-criteria were random, i.e. C1.1, C1.3, C3.2, C4.1, C4.2, C5.1. On the basis of the values of independent criteria, the values of the remaining sub-criteria were determined. The value interval of the sub-criterion C1.1 was determined on the basis of the annual average wind speed in Poland [94]. The value of C1.2 as a dependent sub-criterion was determined on the basis of the formula for the wind power energy for a generic turbine: $P = (1/2)dACv^3$, where d is the air density (equal to 1.225 kg/m3), A is the rotor's blades swept area (equal to 6362m2), Cp is the power coefficient (equal to 0.45), and v representing the wind speed [95]. The values of individual parameters of the formula were taken from the specification of one wind turbine [96]. Three permissible values of C1.3 result from the fact that the voltages of the national power grid of higher voltages used in Poland amount to 110kV, 220kV, and 400kV [97]. The value of C2.1 (annual energy production) was calculated as the product of the number of wind turbines (N), energy generated by a single turbine (C1.2) and the number of hours in a year (8760) [41]. Determining the value of C2.2 (investment cost) was based on the fact that the value of capital investment for an onshore wind farm in Poland amounts to 6.6 million PLN/MW [98]. Approximate costs of constructing a service road to the wind farm were added to the amount (1 million PLN/km). The calculations of C2.3 (operation and maintenance costs) take into consideration the fact that the operation costs of a wind farm in Poland amounts to 83 PLN per one MWh of the energy generated by the wind farm [98]. To the operation costs, the costs, estimated at 10.000 PLN/km, related to the transport and delivery of service elements were also added. To put it simply, a yearly profit (C2.4) from selling the energy can be determined as the difference between incomes from the energy sales and operational costs incurred to generate the energy sold. Incomes are above all influenced by a current energy price. However, new law defining so-called RES auctions has been recently introduced in Poland. The reference price, for an RES auction, of the onshore wind energy generated in a wind farm of combined power greater than 1MW in 2016 amounts to 385PLN/MWh [99].

The reference price in the bidding can be lowered to some extent, therefore, in the prepared simulation, the price of 345 PLN/MWh was taken. Moreover, it was assumed that the profit is reduced by the maintenance cost of the connection to the grid, estimated at 10.000 PLN/km. The payback period (C2.5) was calculated as a ratio of the investment costs to annual profits. Social acceptability (C3.1) is a random value increased by an employment factor amounting to 1% per 100 potential work places related to the construction and maintenance of a power plant. The number of potential work places (C3.2) was determined with the use of a conversion factor, according to which 1MW generates ca. 15.4 work places [98]. The distance to main roads (C4.1) and the distance to the power transmission grid (C4.2) were determined as random values within the range of 1–15 km, whereas for a criterion C5.1 (distance to protected areas) the value of yes or no was drawn. In addition, for some independent sub-criteria, deviation in the range of up to 15% of the calculated value was admitted.

The construction of the alternative A5 without taking into consideration dependencies between sub-criteria consisted in drawing the value of every sub-criterion C1.1–C5.1 from determined ranges of values. The ranges for the sub-criteria were calculated on the basis of minimal and maximal values which could be obtained with the use of the dependencies described above. The way of determining sub-criterion values for the alternative A5 is presented in detail in Table 11.

Table 11. The way of generating the alternative A5 which does not take into consideration and considers dependencies between sub-criteria.

Sub-Criterion	No Dependencies between Sub-Criteria	Implementation of Dependencies between Sub-Criteria
C1.1	*Random <5, 7.5> (m/s)*	*Random <5, 7.5> (m/s)*
C1.2	*Random <0.22, 0.74> (MW)*	$\left(\frac{1}{2} * 1.225 * 6362 * 0.45 * C1.1^3\right)/1000000$ *(MW)*
C1.3	*Random {110; 220; 400} (kV)*	*Random {110; 220; 400} (kV)*
C2.1	*Random <16321, 186311> (MWh)*	$(N * C1.2 * 8760)$ *+/- 15% (MWh)*
C2.2	*Random <169.2, 586.5> (mln PLN)*	$(P * 6.6 + C4.1)$ *+/- 15% (mln PLN)*
C2.3	*Random <1.16, 17.96> (mln PLN)*	$\left(\frac{C2.1*83}{1000000} + 0.01 * C4.1\right)$ *+/- 15% (mln PLN0*
C2.4	*Random <4.66, 73.91> (mln PLN)*	$\left(\frac{C2.1*345-C2.3}{1000000} - 0.01 * C4.2\right)$ *+/- 15% (mln PLN)*
C2.5	*Random <7.9, 36.3> (years)*	C2.2/C2.4 *(years)*
C3.1	*Random <19, 93> (%)*	*Random <15, 80>* +0.01 ∗ C3.2 *(%)*
C3.2	*Random <393, 1328> (number)*	P ∗ 15.4 *+/- 15% (number)*
C4.1	*Random <1, 15> (km)*	*Random <1, 15> (km)*
C4.2	*Random <1, 15> (km)*	*Random <1, 15> (km)*
C5.1	*Random {1; 9} (binary)*	*Random {1; 9} (binary)*
N	-	*Random <10, 25> (number)*
P	-	N ∗ 3 *(MW)*

Abbreviations: N—number of turbines; P—installed power.

The results of examining the robustness of the rankings to the rank reversal phenomenon are shown in Table 12. It should be noted that the generator of random numbers before each test was reset, what guarantees the repeatability of results.

Table 12. The occurrence of the rank reversal phenomenon in examined rankings depending on the way of generating the alternative A5.

Dependencies between Sub-Criteria of the Alternative A5	The Number of Samples of the Alternative A5	AHP Ranking		ANP Ranking	
		The Number of Changed Rankings	The Number of Changes in Rankings	The Number of Changed Rankings	The Number of Changes in Rankings
Independent	1000 samples	4	7	0	0
	10,000 samples	17	32	1	2
Dependent	1000 samples	10	19	0	0
	10,000 samples	66	127	1	2

Figure 5a,b present the number of changed rankings depending on the number of samples of the alternative A5.

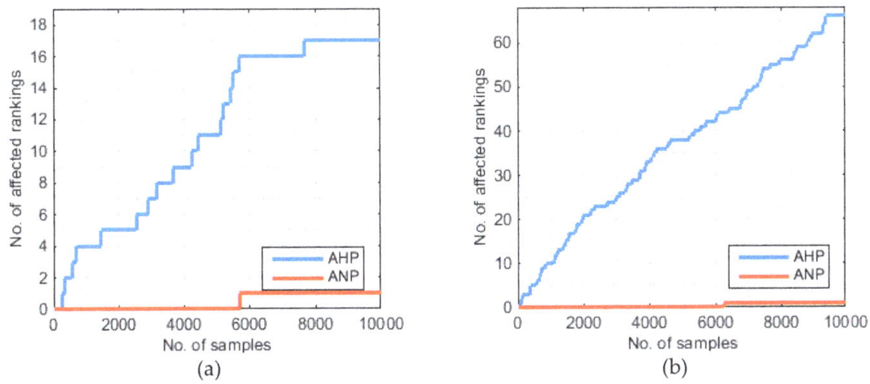

Figure 5. The number of changed rankings for the alternative A5 generated: (**a**) without considering dependencies between sub-criteria; (**b**) when taking into account dependencies between sub-criteria

The analysis of research results points out that when the alternative A5 was constructed without or with the implementation of dependencies between criteria, the ANP ranking had much higher robustness to the rank reversal. It might be assumed that the higher robustness resulted from the fact that the ANP method, unlike the AHP, took real dependencies between sub-criteria into consideration. The robustness of the rankings to the rank reversal phenomenon is especially important in decision problems in which new alternatives may appear for consideration. Decision problems related to the selection of wind farm location are this kind of problems.

It should be pointed out that conclusions drawn from the sensitivity analysis as well as the research into the rank reversal phenomenon occurrence refer only to the comparison of ranking shown in Table 9 and obtained by means of the AHP and ANP methods. Without further research, the results cannot be generalized and related to the quality of other solutions obtained by means of the indicated MCDA methods. However, on the basis of the conducted research in should be noted that the application of the ANP method in the decision problems concerning wind farm location and design makes it possible to obtain a solution with a higher value than by means of the AHP. Furthermore, after analysing the literature [23,25,64,100,101], one can find out that taking into consideration real dependencies between criteria in the decision model makes the model precisely reflect the real decision problem and allows obtaining a more reliable solution.

5. Discussion

The application analysis, presented in Section 2, of the MCDA methods in decision problems in the field of wind energy points out that in order to solve these problems the MCDA methods are used, although usually they do not take into consideration dependencies between criteria or they allow considering only hierarchical dependencies (criteria – sub-criteria). It takes place even though many authors notice that in such decision problems, methods which allow modelling dependencies between criteria ought to be used [15,24,25]. The ANP is a method of this kind. Due to this fact, the decision model, prepared for the problem of the selection of the location and design of a wind farm was based on the ANP method. Therefore, it takes into consideration complex dependencies between decision-making criteria. However, the ANP method does not meet all expectations with regard to the MCDA methods used in the RES issues [24,57–59], since it does not allow applying indifference and preference thresholds, therefore, the ANP is characterized by absolute discriminating power of the criteria. It indicates the need for further development of the MCDA methods and requires working out a method combining the features of the ANP (dependencies between criteria) and outranking methods (indifference and preference thresholds) such as PROSA (PROMETHEE for Sustainability Analysis) [77] and NEAT F-PROMETHEE (New Easy Approach To Fuzzy PROMETHEE) [76].

As part of the verification of the obtained solution, it has been demonstrated that taking into consideration dependencies between criteria in the decision model can influence the obtained recommendation of the decision. It was obtained by comparing solutions gained for the network ANP model and the hierarchical AHP model. A significant observation refers to the differences between sub-criteria weights (see Section 4.2). The weights were obtained by means of the ANP method with the considered cluster of alternatives or without it. The differences, in the author's opinion, question the publications in which the ANP method is only used to obtain the weight of criteria and sub-criteria, e.g., [28].

On the basis of the conducted sensitivity analysis and the research into the robustness of the rankings to the rank reversal phenomenon, it has been found out that the ranking obtained with the use of the ANP method is characterized by higher quality. Moreover, on the basis of the literature one can state that the solution obtained with the ANP method is more reliable owing to the fact that it considers inter-criteria dependencies [102]. However, it is obvious that the network model (ANP) allows reflecting the decision problem more precisely than the hierarchical model (AHP).

As regards the policy recommendations that can be defined on the basis of the study carried out, a number of issues need to be identified here, mainly related to the analysis of potential decision-making alternatives (see Table 6). First of all, it should be noted that in many areas of Poland a power grid with relatively low voltages is available (220kV). This grid should be systematically developed to operate at 400 kV and 750 kV voltages, which will improve the robustness of transmission lines against voltage and power fluctuations generated in the grid by high-capacity wind turbines. As regards the economics of this type of investment, it should be noted that the availability of the road network makes it possible to reduce investment costs and operating costs to a certain extent. Therefore, onshore wind farms should be located close to the main roads. However, in Poland, areas with good wind conditions are usually poorly covered with the road network. Therefore, in addition to the development of the power grid, the development of the road network is also important. Another issue is the relatively low public acceptance of wind energy investments. This is due to the fact that a few years ago in Poland wind energy was developed without taking into account social costs. Wind farms were built very close to human settlements, which resulted in a continuous decrease in public acceptance of this type of investment. Meanwhile, the development of wind energy should be carried out with respect for the inhabitants of the areas in the vicinity of which wind power plants are being built. This development should take into account not only economic and environmental but also social issues. Therefore, this development should be sustainable in economic, environmental and social terms.

Among the limitations of the study presented, it should be pointed out that the comparison of the AHP and ANP methods was carried out in a single case study. Therefore, conclusions drawn from the sensitivity analysis as well as the research into the rank reversal phenomenon occurrence only refer the decision problem presented in this study. Without further research, the results cannot be generalized and related to the quality of other solutions obtained by means of the indicated MCDA methods. However, on the basis of the conducted research it should be noted that the application of the ANP method in the decision problems concerning wind farm location and design makes it possible to obtain a solution with a higher value than by means of the AHP. Furthermore, after analysing the literature [23,25,64,100,101], one can find out that taking into consideration real dependencies between criteria in the decision model makes the model precisely reflect the real decision problem and allows obtaining a more reliable solution. Another constraint linked to the decision problem itself is that the decision-maker, when considering a similar decision problem, may formulate other criteria, the inclusion of which may, of course, be given a different ranking of alternatives. Finally, it should be noted that this is an ex-ante study so that the values of the alternatives in terms of evaluation criteria are uncertain and may change to some extent [103].

6. Conclusions

When summing up the studies carried out, it is important to note their methodological and practical contribution to management science, and in particular to the decision analysis. One should mention here:

- formal selection of the MCDA method for decision problems in the field of RES and sustainability, based on the analysis of intrinsic characteristics of individual methods,
- consideration of different locations and projects for the construction of onshore wind farms,
- comparison of rankings obtained using the AHP (without dependencies between criteria) and ANP (with inter-criteria dependencies) methods in order to assess the impact of such dependencies on the solution obtained,
- study of the quality of the solutions obtained through a sensitivity analysis and rank reversal phenomenon analysis.

Obviously, the prepared solution should be continuously developed. A natural direction of further work is to extend the decision model with other criteria and sub-criteria of evaluating onshore wind farms as well as to include other RES decision problems in the metamodel. Furthermore, an interesting issue would be presenting the decision model in the form of an ontology, what would make it possible to infer new knowledge from the model [104]. It would also be a natural development of the functionality of the model in the direction of an ontological knowledge base.

Supplementary Materials: The following are available online at http://www.mdpi.com/1996-1073/12/4/749/s1, Figure S1–12: Utility of alternatives determined by means of: (a) the AHP method; (b) the ANP method; depending on the weight of the criterion Cx and sub-criterion Cx.y.

Funding: This research received no external funding.

Conflicts of Interest: The author declare no conflict of interest.

Appendix A Unweighted, Weighted and Limit Supermatrices

Table A1. AHP unweighted and weighted supermatrix

	A1	A2	A3	A4	C1.1	C1.2	C1.3	C2.1	C2.2	C2.3	C2.4	C2.5	C3.1	C3.2	C4.1	C4.2	C5.1	C1	C2	C3	C4	C5	Goal
A1	0	0	0	0	0.251	0.257	0.208	0.310	0.196	0.180	0.310	0.257	0.347	0.350	0.224	0.370	0.05	0	0	0	0	0	0
A2	0	0	0	0	0.265	0.282	0.377	0.251	0.266	0.223	0.251	0.281	0.141	0.200	0.135	0.247	0.45	0	0	0	0	0	0
A3	0	0	0	0	0.259	0.277	0.208	0.304	0.215	0.184	0.304	0.277	0.359	0.274	0.192	0.012	0.05	0	0	0	0	0	0
A4	0	0	0	0	0.225	0.184	0.208	0.135	0.323	0.413	0.135	0.185	0.153	0.176	0.449	0.370	0.45	0	0	0	0	0	0
C1.1	0	0	0	0	0	0	0	0	0	0	0	0	0	0	0	0	0	0.333	0	0	0	0	0
C1.2	0	0	0	0	0	0	0	0	0	0	0	0	0	0	0	0	0	0.333	0	0	0	0	0
C1.3	0	0	0	0	0	0	0	0	0	0	0	0	0	0	0	0	0	0.333	0	0	0	0	0
C2.1	0	0	0	0	0	0	0	0	0	0	0	0	0	0	0	0	0	0	0.2	0	0	0	0
C2.2	0	0	0	0	0	0	0	0	0	0	0	0	0	0	0	0	0	0	0.2	0	0	0	0
C2.3	0	0	0	0	0	0	0	0	0	0	0	0	0	0	0	0	0	0	0.2	0	0	0	0
C2.4	0	0	0	0	0	0	0	0	0	0	0	0	0	0	0	0	0	0	0.2	0	0	0	0
C2.5	0	0	0	0	0	0	0	0	0	0	0	0	0	0	0	0	0	0	0.2	0	0	0	0
C3.1	0	0	0	0	0	0	0	0	0	0	0	0	0	0	0	0	0	0	0	0.5	0	0	0
C3.2	0	0	0	0	0	0	0	0	0	0	0	0	0	0	0	0	0	0	0	0.5	0	0	0
C4.1	0	0	0	0	0	0	0	0	0	0	0	0	0	0	0	0	0	0	0	0	0.5	0	0
C4.2	0	0	0	0	0	0	0	0	0	0	0	0	0	0	0	0	0	0	0	0	0.5	0	0
C5.1	0	0	0	0	0	0	0	0	0	0	0	0	0	0	0	0	0	0	0	0	0	1	0
C1	0	0	0	0	0	0	0	0	0	0	0	0	0	0	0	0	0	0	0	0	0	0	0.2
C2	0	0	0	0	0	0	0	0	0	0	0	0	0	0	0	0	0	0	0	0	0	0	0.2
C3	0	0	0	0	0	0	0	0	0	0	0	0	0	0	0	0	0	0	0	0	0	0	0.2
C4	0	0	0	0	0	0	0	0	0	0	0	0	0	0	0	0	0	0	0	0	0	0	0.2
C5	0	0	0	0	0	0	0	0	0	0	0	0	0	0	0	0	0	0	0	0	0	0	0.2
Goa	0	0	0	0	0	0	0	0	0	0	0	0	0	0	0	0	0	0	0	0	0	0	0

Table A2. ANP unweighted supermatrix

	A1	A2	A3	A4	C1.1	C1.2	C1.3	C2.1	C2.2	C2.3	C2.4	C2.5	C3.1	C3.2	C4.1	C4.2	C5.1	C1	C2	C3	C4	C5	Goal
A1	0	0	0	0	0.251	0.257	0.208	0.310	0.196	0.180	0.310	0.257	0.347	0.350	0.224	0.370	0.05	0	0	0	0	0	0
A2	0	0	0	0	0.265	0.282	0.377	0.251	0.266	0.223	0.251	0.281	0.141	0.200	0.135	0.247	0.45	0	0	0	0	0	0
A3	0	0	0	0	0.259	0.277	0.208	0.304	0.215	0.184	0.304	0.277	0.359	0.274	0.192	0.012	0.05	0	0	0	0	0	0
A4	0	0	0	0	0.225	0.184	0.208	0.135	0.323	0.413	0.135	0.185	0.153	0.176	0.449	0.370	0.45	0	0	0	0	0	0
C1.1	0	0	0	0	0	0	0	0	0	0	0	0	0	0	0	0	0	0.333	0	0	0	0	0
C1.2	0	0	0	0	1	0	0	0	0	0	0	0	0	0	0	0	0	0.333	0	0	0	0	0
C1.3	0	0	0	0	0	0	0	0	0	0	0	0	0	0	0	0	0	0.333	0	0	0	0	0
C2.1	0	0	0	0	0	1	0	0	0	0	0	0	0	0	0	0	0	0	0.2	0	0	0	0
C2.2	0	0	0	0	0	0	0	0	0	0	0	0	0	0	0.5	0	0	0	0.2	0	0	0	0
C2.3	0	0	0	0	0	0	0	0.5	0.5	0	0	0	0	0	0.5	0	0	0	0.2	0	0	0	0
C2.4	0	0	0	0	0	0	0	0.5	0	1	0	0	0	0	0	1	0	0	0.2	0	0	0	0
C2.5	0	0	0	0	0	0	0	0	0.5	0	1	0	0	0	0	0	0	0	0.2	0	0	0	0
C3.1	0	0	0	0	0	0	0	0	0	0	0	0	0	1	0	0	0	0	0	0.5	0	0	0
C3.2	0	0	0	0	0	0	0	0	0	0	0	0	0	0	0	0	0	0	0	0.5	0	0	0
C4.1	0	0	0	0	0	0	0	0	0	0	0	0	0	0	0	0	0	0	0	0	0.5	0	0
C4.2	0	0	0	0	0	0	0	0	0	0	0	0	0	0	0	0	0	0	0	0	0.5	0	0
C5.1	0	0	0	0	0	0	0	0	0	0	0	0	0	0	0	0	0	0	0	0	0	1	0
C1	0	0	0	0	0	0	0	0	0	0	0	0	0	0	0	0	0	0	0	0	0	0	0.2
C2	0	0	0	0	0	0	0	0	0	0	0	0	0	0	0	0	0	0	0	0	0	0	0.2
C3	0	0	0	0	0	0	0	0	0	0	0	0	0	0	0	0	0	0	0	0	0	0	0.2
C4	0	0	0	0	0	0	0	0	0	0	0	0	0	0	0	0	0	0	0	0	0	0	0.2
C5	0	0	0	0	0	0	0	0	0	0	0	0	0	0	0	0	0	0	0	0	0	0	0.2
Goa	0	0	0	0	0	0	0	0	0	0	0	0	0	0	0	0	0	0	0	0	0	0	0

Table A3. ANP weighted supermatrix

	A1	A2	A3	A4	C1.1	C1.2	C1.3	C2.1	C2.2	C2.3	C2.4	C2.5	C3.1	C3.2	C4.1	C4.2	C5.1	C1	C2	C3	C4	C5	Goal
A1	0	0	0	0	0.126	0.129	0.208	0.155	0.098	0.090	0.155	0.257	0.347	0.175	0.112	0.185	0.05	0	0	0	0	0	0
A2	0	0	0	0	0.133	0.141	0.377	0.125	0.133	0.111	0.125	0.281	0.141	0.100	0.067	0.123	0.45	0	0	0	0	0	0
A3	0	0	0	0	0.129	0.138	0.208	0.152	0.108	0.092	0.152	0.277	0.359	0.137	0.096	0.006	0.05	0	0	0	0	0	0
A4	0	0	0	0	0.112	0.092	0.208	0.068	0.161	0.206	0.068	0.185	0.153	0.088	0.224	0.185	0.45	0	0	0	0	0	0
C1.1	0	0	0	0	0	0	0	0	0	0	0	0	0	0	0	0	0	0.333	0	0	0	0	0
C1.2	0	0	0	0	0.5	0	0	0	0	0	0	0	0	0	0	0	0	0.333	0	0	0	0	0
C1.3	0	0	0	0	0	0	0	0	0	0	0	0	0	0	0	0	0	0.333	0	0	0	0	0
C2.1	0	0	0	0	0	0.5	0	0	0	0	0	0	0	0	0	0	0	0	0.2	0	0	0	0
C2.2	0	0	0	0	0	0	0	0	0	0	0	0	0	0	0.25	0	0	0	0.2	0	0	0	0
C2.3	0	0	0	0	0	0	0	0.25	0.25	0	0	0	0	0	0.25	0	0	0	0.2	0	0	0	0
C2.4	0	0	0	0	0	0	0	0.25	0	0.5	0	0	0	0	0	0.5	0	0	0.2	0	0	0	0
C2.5	0	0	0	0	0	0	0	0	0.25	0	0.5	0	0	0	0	0	0	0	0.2	0	0	0	0
C3.1	0	0	0	0	0	0	0	0	0	0	0	0	0	0.5	0	0	0	0	0	0.5	0	0	0
C3.2	0	0	0	0	0	0	0	0	0	0	0	0	0	0	0	0	0	0	0	0.5	0	0	0
C4.1	0	0	0	0	0	0	0	0	0	0	0	0	0	0	0	0	0	0	0	0	0.5	0	0
C4.2	0	0	0	0	0	0	0	0	0	0	0	0	0	0	0	0	0	0	0	0	0.5	0	0
C5.1	0	0	0	0	0	0	0	0	0	0	0	0	0	0	0	0	0	0	0	0	0	1	0
C1	0	0	0	0	0	0	0	0	0	0	0	0	0	0	0	0	0	0	0	0	0	0	0.2
C2	0	0	0	0	0	0	0	0	0	0	0	0	0	0	0	0	0	0	0	0	0	0	0.2
C3	0	0	0	0	0	0	0	0	0	0	0	0	0	0	0	0	0	0	0	0	0	0	0.2
C4	0	0	0	0	0	0	0	0	0	0	0	0	0	0	0	0	0	0	0	0	0	0	0.2
C5	0	0	0	0	0	0	0	0	0	0	0	0	0	0	0	0	0	0	0	0	0	0	0.2
Goa	0	0	0	0	0	0	0	0	0	0	0	0	0	0	0	0	0	0	0	0	0	0	0

Table A4. AHP limit supermatrix

	A1	A2	A3	A4	C1.1	C1.2	C1.3	C2.1	C2.2	C2.3	C2.4	C2.5	C3.1	C3.2	C4.1	C4.2	C5.1	C1	C2	C3	C4	C5	Goal
A1	0	0	0	0	0.251	0.257	0.208	0.310	0.196	0.180	0.310	0.257	0.347	0.350	0.224	0.370	0.05	0.119	0.125	0.174	0.149	0.025	0.079
A2	0	0	0	0	0.265	0.282	0.377	0.251	0.266	0.223	0.251	0.281	0.141	0.200	0.135	0.247	0.45	0.154	0.127	0.085	0.095	0.225	0.092
A3	0	0	0	0	0.259	0.277	0.208	0.304	0.215	0.184	0.304	0.277	0.359	0.274	0.192	0.012	0.05	0.124	0.128	0.158	0.051	0.025	0.065
A4	0	0	0	0	0.225	0.184	0.208	0.135	0.323	0.413	0.135	0.185	0.153	0.176	0.449	0.370	0.45	0.103	0.119	0.082	0.205	0.225	0.098
C1.1	0	0	0	0	0	0	0	0	0	0	0	0	0	0	0	0	0	0.167	0	0	0	0	0.022
C1.2	0	0	0	0	0	0	0	0	0	0	0	0	0	0	0	0	0	0.167	0	0	0	0	0.022
C1.3	0	0	0	0	0	0	0	0	0	0	0	0	0	0	0	0	0	0.167	0	0	0	0	0.022
C2.1	0	0	0	0	0	0	0	0	0	0	0	0	0	0	0	0	0	0	0.1	0	0	0	0.013
C2.2	0	0	0	0	0	0	0	0	0	0	0	0	0	0	0	0	0	0	0.1	0	0	0	0.013
C2.3	0	0	0	0	0	0	0	0	0	0	0	0	0	0	0	0	0	0	0.1	0	0	0	0.013
C2.4	0	0	0	0	0	0	0	0	0	0	0	0	0	0	0	0	0	0	0.1	0	0	0	0.013
C2.5	0	0	0	0	0	0	0	0	0	0	0	0	0	0	0	0	0	0	0.1	0	0	0	0.013
C3.1	0	0	0	0	0	0	0	0	0	0	0	0	0	0	0	0	0	0	0	0.25	0	0	0.033
C3.2	0	0	0	0	0	0	0	0	0	0	0	0	0	0	0	0	0	0	0	0.25	0	0	0.033
C4.1	0	0	0	0	0	0	0	0	0	0	0	0	0	0	0	0	0	0	0	0	0.25	0	0.033
C4.2	0	0	0	0	0	0	0	0	0	0	0	0	0	0	0	0	0	0	0	0	0.25	0	0.033
C5.1	0	0	0	0	0	0	0	0	0	0	0	0	0	0	0	0	0	0	0	0	0	0.5	0.067
C1	0	0	0	0	0	0	0	0	0	0	0	0	0	0	0	0	0	0	0	0	0	0	0.067
C2	0	0	0	0	0	0	0	0	0	0	0	0	0	0	0	0	0	0	0	0	0	0	0.067
C3	0	0	0	0	0	0	0	0	0	0	0	0	0	0	0	0	0	0	0	0	0	0	0.067
C4	0	0	0	0	0	0	0	0	0	0	0	0	0	0	0	0	0	0	0	0	0	0	0.067
C5	0	0	0	0	0	0	0	0	0	0	0	0	0	0	0	0	0	0	0	0	0	0	0.067
Goa	0	0	0	0	0	0	0	0	0	0	0	0	0	0	0	0	0	0	0	0	0	0	0

Table A5. ANP limit supermatrix

	A1	A2	A3	A4	C1.1	C1.2	C1.3	C2.1	C2.2	C2.3	C2.4	C2.5	C3.1	C3.2	C4.1	C4.2	C5.1	C1	C2	C3	C4	C5	Goal
A1	0	0	0	0	0.134	0.142	0.208	0.157	0.131	0.133	0.189	0.257	0.347	0.232	0.121	0.187	0.05	0.095	0.101	0.155	0.098	0.025	0.068
A2	0	0	0	0	0.136	0.140	0.377	0.139	0.157	0.140	0.177	0.281	0.141	0.114	0.105	0.147	0.45	0.112	0.099	0.069	0.080	0.225	0.081
A3	0	0	0	0	0.138	0.147	0.208	0.157	0.140	0.136	0.194	0.277	0.359	0.211	0.115	0.087	0.05	0.098	0.105	0.150	0.065	0.025	0.063
A4	0	0	0	0	0.104	0.095	0.208	0.099	0.165	0.164	0.107	0.185	0.153	0.110	0.197	0.152	0.45	0.074	0.084	0.071	0.112	0.225	0.078
C1.1	0	0	0	0	0	0	0	0	0	0	0	0	0	0	0	0	0	0.126	0	0	0	0	0.019
C1.2	0	0	0	0	0.256	0	0	0	0	0	0	0	0	0	0	0	0	0.190	0	0	0	0	0.029
C1.3	0	0	0	0	0	0	0	0	0	0	0	0	0	0	0	0	0	0.126	0	0	0	0	0.019
C2.1	0	0	0	0	0.128	0.262	0	0	0	0	0	0	0	0	0	0	0	0.095	0.078	0	0	0	0.026
C2.2	0	0	0	0	0	0	0	0	0	0	0	0	0	0	0.134	0	0	0	0.078	0	0.045	0	0.019
C2.3	0	0	0	0	0.032	0.066	0	0.138	0.148	0	0	0	0	0	0.168	0	0	0.016	0.111	0	0.056	0	0.030
C2.4	0	0	0	0	0.048	0.098	0	0.207	0.074	0.286	0	0	0	0	0.084	0.286	0	0.024	0.146	0	0.117	0	0.048
C2.5	0	0	0	0	0.024	0.049	0	0.103	0.185	0.143	0.333	0	0	0	0.076	0.143	0	0.012	0.171	0	0.070	0	0.040
C3.1	0	0	0	0	0	0	0	0	0	0	0	0	0	0.333	0	0	0	0.032	0.026	0.333	0	0	0.043
C3.2	0	0	0	0	0	0	0	0	0	0	0	0	0	0	0	0	0	0	0	0.222	0	0	0.029
C4.1	0	0	0	0	0	0	0	0	0	0	0	0	0	0	0	0	0	0	0	0	0.178	0	0.029
C4.2	0	0	0	0	0	0	0	0	0	0	0	0	0	0	0	0	0	0	0	0	0.178	0	0.029
C5.1	0	0	0	0	0	0	0	0	0	0	0	0	0	0	0	0	0	0	0	0	0	0.5	0.058
C1	0	0	0	0	0	0	0	0	0	0	0	0	0	0	0	0	0	0	0	0	0	0	0.058
C2	0	0	0	0	0	0	0	0	0	0	0	0	0	0	0	0	0	0	0	0	0	0	0.058
C3	0	0	0	0	0	0	0	0	0	0	0	0	0	0	0	0	0	0	0	0	0	0	0.058
C4	0	0	0	0	0	0	0	0	0	0	0	0	0	0	0	0	0	0	0	0	0	0	0.058
C5	0	0	0	0	0	0	0	0	0	0	0	0	0	0	0	0	0	0	0	0	0	0	0.058
Goa	0	0	0	0	0	0	0	0	0	0	0	0	0	0	0	0	0	0	0	0	0	0	0

References

1. Halicka, K. Designing routes of development of renewable energy technologies. *Procedia Soc. Behav. Sci.* **2014**, *156*, 58–62. [CrossRef]
2. Paska, J.; Surma, T. Electricity generation from renewable energy sources in Poland. *Renew. Energy* **2014**, *71*, 286–294. [CrossRef]
3. Mesjasz-Lech, A. Planning of production resources use and environmental effects on the example of a thermal power plant. *Procedia Soc. Behav. Sci.* **2015**, *213*, 539–545. [CrossRef]
4. Scarlat, N.; Dallemand, J.F.; Monforti-Ferrario, F.; Banja, M.; Motola, V. Renewable energy policy framework and bioenergy contribution in the European Union—An overview from National Renewable Energy Action Plans and Progress Reports. *Renew. Sustain. Energy Rev.* **2015**, *51*, 969–985. [CrossRef]
5. Directive (EU) 2018/2001 of the European Parliament and of the Council of 11 December 2018 on the Promotion of the Use of Energy From Renewable Sources. PE/48/2018/REV/1. Official Journal of the European Union. 21.12.2018. Available online: https://eur-lex.europa.eu/eli/dir/2018/2001/oj (accessed on 3 February 2019).
6. IEA Bioenergy. European Union—2018 Update. Bioenergy Policies and Status of Implementation. Country Reports. 09.2018. Available online: https://www.ieabioenergy.com/wp-content/uploads/2018/10/CountryReport2018_EU_final.pdf (accessed on 3 February 2019).
7. European Environment Agency. Environmental Indicator Report 2018. Greenhouse Gas Emissions. 07.12.2018. Available online: https://www.eea.europa.eu/airs/2018/resource-efficiency-and-low-carbon-economy/greenhouse-gas-emission (accessed on 3 February 2019).
8. Paska, J.; Sałek, M.; Surma, T. Current status and perspectives of renewable energy sources in Poland. *Renew. Sustain. Energy Rev.* **2009**, *13*, 142–154. [CrossRef]
9. International Renewable Energy Agency. Global Trends. Available online: https://www.irena.org/ourwork/Knowledge-Data-Statistics/Data-Statistics/Costs/Global-Trends (accessed on 3 February 2019).
10. International Renewable Energy Agency. Query Tool. Available online: https://www.irena.org/ourwork/Knowledge-Data-Statistics/Data-Statistics/Capacity-and-Generation/Query-Tool (accessed on 3 February 2019).
11. EU Commission. Energy sources, production costs and performance of technologies for power generation, heating and transport, Commission staff working document accompanying the communication from the comission to the european parliament, the council, the european economic and social committee and the committee of the regions. Second strategic energy review. SEC (2008) 2892 final, 13.11.2008. Available online: http://aei.pitt.edu/39570/ (accessed on 3 February 2019).
12. Ioannou, K.; Tsantopoulos, G.; Arabatzis, G.; Andreopoulou, Z.; Zafeiriou, E. A Spatial Decision Support System Framework for the Evaluation of Biomass Energy Production Locations: Case Study in the Regional Unit of Drama, Greece. *Sustainability* **2018**, *10*, 531. [CrossRef]
13. International Renewable Energy Agency. Renewable Power Generation Costs in 2017. Available online: https://www.irena.org/publications/2018/Jan/Renewable-power-generation-costs-in-2017 (accessed on 3 February 2019).
14. Wu, Y.; Zhang, J.; Yuan, J.; Geng, S.; Zhang, H. Study of decision framework of offshore wind power station site selection based on ELECTRE-III under intuitionistic fuzzy environment: A case of China. *Energy Convers. Manag.* **2016**, *113*, 66–81. [CrossRef]
15. Wu, Y.; Geng, S.; Xu, H.; Zhang, H. Study of decision framework of wind farm project plan selection under intuitionistic fuzzy set and fuzzy measure environment. *Energy Convers. Manag.* **2014**, *87*, 274–284. [CrossRef]
16. Lee, A.H.I.; Hung, M.C.; Kang, H.Y.; Pearn, W.L. A wind turbine evaluation model under a multi-criteria decision making environment. *Energy Convers. Manag.* **2012**, *64*, 289–300. [CrossRef]
17. Taha, R.A.; Daim, T. Multi-Criteria Applications in Renewable Energy Analysis, a Literature Review. In *Research and Technology Management in the Electricity Industry*; Daim, T., Oliver, T., Kim, J., Eds.; Springer: London, UK, 2013; pp. 281–304.
18. Strantzali, E.; Aravossis, K. Decision making in renewable energy investments: A review. *Renew. Sustain. Energy Rev.* **2016**, *55*, 885–898.
19. Wimmler, C.; Hejazi, G.; de Oliveira Fernades, E.; Moreira, C.; Connors, S. Multi-Criteria Decision Support Methods for Renewable Energy Systems on Islands. *J. Clean Energy Technol.* **2015**, *3*, 185–195. [CrossRef]

20. Wang, J.J.; Jing, Y.Y.; Zhang, C.F.; Zhao, J.H. Review on multi-criteria decision analysis aid in sustainable energy decision-making. *Renew. Sustain. Energy Rev.* **2009**, *13*, 2263–2278. [CrossRef]
21. San Cristobal, J.R. Multi-criteria decision making in the selection of a renewable energy project in Spain: The Vikor method. Renew. *Energy* **2011**, *36*, 498–502. [CrossRef]
22. Henggeler Antunes, C.; Oliveira Henriques, C. Multi-Objective Optimization and Multi-Criteria Analysis Models and Methods for Problems in the Energy Sector. In *Multiple Criteria Decision Analysis. State of the Art Surveys*, 2nd ed.; Greco, S., Ehrgott, M., Figueira, J.R., Eds.; Springer: New York, NY, USA, 2016; pp. 1067–1165.
23. Golcuk, I.; Baykasoglu, A. An analysis of DEMATEL approaches for criteria interaction handling within ANP. *Expert Syst. Appl.* **2016**, *46*, 346–366. [CrossRef]
24. de Montis, A.; De Toro, P.; Droste-Franke, B.; Omann, I.; Stagl, S. Assessing the quality of different MCDA method. In *Alternatives for Environmental Valuation*; Getzner, M., Spash, C.L., Stagl, S., Eds.; Taylor & Francis: New York, NY, USA, 2005; pp. 99–133.
25. Chen, C.R.; Huang, C.C.; Tsuei, H.J. A Hybrid MCDM Model for Improving GIS-Based Solar Farms Site Selection. *Int. J. Photoenergy* **2014**, *2014*, 925370. [CrossRef]
26. Mardani, A.; Jusoh, A.; Zavadskas, E.K.; Cavallaro, F.; Khalifah, Z. Sustainable and Renewable Energy: An Overview of the Application of Multiple Criteria Decision Making Techniques and Approaches. *Sustainability* **2015**, *7*, 13947–13984. [CrossRef]
27. Suganthi, L.; Iniyan, S.; Samuel, A.A. Applications of fuzzy logic in renewable energy systems—A review. *Renew. Sustain. Energy Rev.* **2015**, *48*, 585–607. [CrossRef]
28. Yeh, T.M.; Huang, Y.L. Factors in determining wind farm location: Integrating GQM, fuzzy DEMATEL and ANP. *Renew. Energy* **2014**, *66*, 159–169. [CrossRef]
29. Kaya, T.; Kahraman, C. Multicriteria renewable energy planning using an integrated fuzzy VIKOR & AHP methodology: The case of Istanbul. *Energy* **2010**, *35*, 2517–2527.
30. San Cristobal, J.R. *Multi-Criteria Analysis in the Renewable Energy Industry*; Springer: London, UK, 2012; pp. 11–17.
31. Wu, Y.; Geng, S. Multi-criteria decision making on selection of solar-wind hybrid power station location: A case of China. *Energy Convers. Manag.* **2014**, *81*, 527–533.
32. Jun, D.; Tian-tian, F.; Yi-sheng, Y.; Yu, M. Macro-site selection of wind/solar hybrid power station based on ELECTRE-II. *Renew. Sustain. Energy Rev.* **2014**, *35*, 194–204. [CrossRef]
33. Al-Yahyai, S.; Charabi, Y.; Gastli, A.; Al-Badi, A. Wind farm land suitability indexing using multi-criteria analysis. *Renew. Energy* **2012**, *44*, 80–87. [CrossRef]
34. Latinopoulos, D.; Kechagia, K. A GIS-based multi-criteria evaluation for wind farm site selection. A regional scale application in Greece. *Renew. Energy* **2015**, *78*, 550–560. [CrossRef]
35. Sanchez-Lozano, J.M.; Garcia-Cascales, M.S.; Lamata, M.T. Identification and selection of potential sites for onshore wind farms development in Region of Murcia, Spain. *Energy* **2014**, *73*, 311–324. [CrossRef]
36. Sanchez-Lozano, J.M.; Garcia-Cascales, M.S.; Lamata, M.T. GIS-based onshore wind farm site selection using Fuzzy Multi-Criteria Decision Making methods. Evaluating the case of Southeastern Spain. *Appl. Energy* **2016**, *171*, 86–102. [CrossRef]
37. Jangid, J.; Bera, A.K.; Joseph, M.; Singh, V.; Singh, T.P.; Pradhan, B.K.; Das, S. Potential zones identification for harvesting wind energy resources in desert region of India—A multi criteria evaluation approach using remote sensing and GIS. *Renew. Sustain. Energy Rev.* **2016**, *65*, 1–10. [CrossRef]
38. Atici, K.B.; Simsek, A.B.; Ulucan, A.; Tosun, M.U. A GIS-based Multiple Criteria Decision Analysis approach for wind power plant site selection. *Util. Policy* **2015**, *37*, 86–96. [CrossRef]
39. Aydin, N.Y.; Kentel, E.; Duzgun, H.S. GIS-based site selection methodology for hybrid renewable energy systems: A case study from western Turkey. *Energy Convers. Manag.* **2013**, *70*, 90–106. [CrossRef]
40. Noorollahi, Y.; Yousefi, H.; Mohammadi, M. Multi-criteria decision support system for wind farm site selection using GIS. Sustain. *Energy Technol. Assess.* **2016**, *13*, 38–50.
41. Ramirez-Rosado, I.J.; Garcia-Garrido, E.; Fernandez-Jimenez, L.A.; Zorzano-Santamaria, P.J.; Monteiro, C.; Miranda, V. Promotion of new wind farms based on a decision support system. *Renew. Energy* **2008**, *33*, 558–566. [CrossRef]
42. Fetanat, A.; Khorasaninejad, E. A novel hybrid MCDM approach for offshore wind farm site selection: A case study of Iran. *Ocean Coast. Manag.* **2015**, *109*, 17–28. [CrossRef]

43. Wątróbski, J.; Ziemba, P.; Wolski, W. Methodological Aspects of Decision Support System for the Location of Renewable Energy Sources. *Ann. Comput. Sci. Inf. Syst.* **2015**, *5*, 1451–1459.
44. Lee, A.H.I.; Chen, H.H.; Kang, H.Y. Multi-criteria decision making on strategic selection of wind farms. *Renew. Energy* **2009**, *34*, 120–126. [CrossRef]
45. Gamboa, G.; Munda, G. The problem of windfarm location: A social multi-criteria evaluation framework. *Energy Policy* **2007**, *35*, 1564–1583. [CrossRef]
46. Wątróbski, J.; Ziemba, P.; Jankowski, J.; Zioło, M. Green Energy for a Green City—A Multi-Perspective Model Approach. *Sustainability* **2016**, *8*, 702. [CrossRef]
47. Chen, H.H.; Kang, H.Y.; Lee, A.H.I. Strategic selection of suitable projects for hybrid solar-wind power generation systems. *Renew. Sustain. Energy Rev.* **2010**, *14*, 413–421. [CrossRef]
48. Cavallaro, F.; Ciraolo, L. A multicriteria approach to evaluate wind energy plants on an Italian island. *Energy Policy* **2005**, *33*, 235–244. [CrossRef]
49. Gumus, S.; Kucukvar, M.; Tatari, O. Intuitionistic fuzzy multi-criteria decision making framework based on life cycle environmental, economic and social impacts: The case of U.S. wind energy. *Sustain. Prod. Consum.* **2016**, *8*, 78–92. [CrossRef]
50. Shirgholami, Z.; Zangeneh, S.N.; Bortolini, M. Decision system to support the practitioners in the wind farm design: A case study for Iran mainland. *Sustain. Energy Technol. Assess.* **2016**, *16*, 1–10. [CrossRef]
51. Shafiee, M. A fuzzy analytic network process model to mitigate the risks associated with offshore wind farms. *Expert Syst. Appl.* **2015**, *42*, 2143–2152. [CrossRef]
52. Tian, W.; Bai, J.; Sun, H.; Zhao, Y. Application of the analytic hierarchy process to a sustainability assessment of coastal beach axploitation: A case study of the wind power projects on the coastal beaches of Yancheng, China. *J. Environ. Manag.* **2013**, *115*, 251–256. [CrossRef] [PubMed]
53. Hajkowicz, S.; Higgins, A. A comparison of multiple criteria analysis techniques for water resource management. *Eur. J. Oper. Res.* **2008**, *184*, 255–265. [CrossRef]
54. Wątróbski, J.; Jankowski, J.; Ziemba, P.; Karczmarczyk, A.; Zioło, M. Generalised framework for multi-criteria method selection. *Omega*. In Press. [CrossRef]
55. Hanne, T. Meta Decision Problems in Multiple Criteria Decision Making. In *Multicriteria Decision Making: Advances in MCDM Models, Algorithms, Theory, and Applications*; Gal, T., Stewart, T.J., Hanne, T., Eds.; Springer Science: New York, NY, USA, 1999; pp. 147–171.
56. Guitouni, A.; Martel, J.M. Tentative guidelines to help choosing an appropriate MCDA method. *Eur. J. Oper. Res.* **1998**, *109*, 501–521. [CrossRef]
57. Cinelli, M.; Coles, S.R.; Kirwan, K. Analysis of the potentials of multi criteria decision analysis methods to conduct sustainability assessment. *Ecol. Indic.* **2014**, *46*, 138–148. [CrossRef]
58. Bagheri Moghaddam, N.; Nasiri, M.; Mousavi, S.M. An appropriate multiple criteria decision making method for solving electricity planning problems, addressing sustainability issue. *Int. J. Environ. Sci. Technol.* **2011**, *8*, 605–620. [CrossRef]
59. Polatidis, H.; Haralambopoulos, D.A.; Munda, G.; Vreeker, R. Selecting an Appropriate Multi-Criteria Decision Analysis Technique for Renewable Energy Planning. *Energy Sources Part B Econ. Plan. Policy* **2006**, *1*, 181–193. [CrossRef]
60. Roy, B. *Multicriteria Methodology for Decision Aiding*; Springer Science: Dordrecht, Germany, 1996.
61. Bouyssou, D.; Vincke, P. Binary Relations and Preference Modeling. In *Decision-making Process: Concepts and Methods*; Bouyssou, D., Dubois, D., Pirlot, M., Prade, H., Eds.; ISTE Ltd.: London, UK, 2009; pp. 49–84.
62. Roy, B. Paradigms and Challenges. In *Multiple Criteria Decision Analysis. State of the Art Surveys*, 2nd ed.; Greco, S., Ehrgott, M., Figueira, J.R., Eds.; Springer: New York, NY, USA, 2016; pp. 19–39.
63. Moretti, S.; Ozturk, M.; Tsoukias, A. Preference Modeling. In *Multiple Criteria Decision Analysis. State of the Art Surveys*, 2nd ed.; Greco, S., Ehrgott, M., Figueira, J.R., Eds.; Springer-Verlag: New York, NY, USA, 2016; pp. 43–95.
64. Mandic, K.; Bobar, V.; Delibasić, B. Modeling Interactions Among Criteria in MCDM Methods: A Review. *Lect. Notes Bus. Inf. Process.* **2015**, *216*, 98–109.
65. Saaty, T.L. *The Analytic Hierarchy Process*; McGraw-Hill: New York, NY, USA, 1980.
66. Saaty, T.L.; Vargas, L.G. *Decision Making with the Analytic Network Process. Economic, Political, Social and Technological Applications with Benefits, Opportunities, Costs and Risks*, 2nd ed.; Springer Science: New York, USA, NY, 2013.

67. Figueira, J.R.; Mousseau, V.; Roy, B. ELECTRE Methods. In *Multiple Criteria Decision Analysis. State of the Art Surveys*, 2nd ed.; Greco, S., Ehrgott, M., Figueira, J.R., Eds.; Springer-Verlag: New York, NY, USA, 2016; pp. 155–185.
68. Hwang, C.L.; Yoon, K. *Multiple Attribute Decision Making: Methods and Applications*; Springer: Berlin/Heidelberg, Germany, 1981.
69. Brans, J.P.; Mareschal, B.; Vincke, P. Promethee: A new family of outranking methods in multicriteria analysis. In Proceedings of the International conference on Operational Research OR'84, Washington, CA, USA, 6–10 August 1984; pp. 408–421.
70. MacCrimmon, K.R. *Decision making among multiple-attribute alternatives: a survey and consolidated approach*; The Rand Corporation: Santa Monica, CA, USA, 1968; pp. 17–44.
71. Shit, P.K.; Bhunia, G.S.; Maiti, R. Potential landslide susceptibility mapping using weighted overlay model (WOM). *Model. Earth Syst. Environ.* **2016**, *2*, 21. [CrossRef]
72. Yager, R.R.; Kacprzyk, J. *The Ordered Weighted Averaging Operators. Theory and Applications*; Springer Science: New York, NY, USA, 1997.
73. Wu, W.W. Choosing knowledge management strategies by using a combined ANP and DEMATEL approach. *Expert Syst. Appl.* **2008**, *35*, 828–835. [CrossRef]
74. Rezaei, J. Best-worst multi-criteria decision-making method: Some properties and a linear model. *Omega* **2016**, *64*, 126–130. [CrossRef]
75. Sałabun, W. The Characteristic Objects Method: A New Distance-based Approach to Multicriteria Decision-making Problems. *J. Mult. Criteria Decis. Anal.* **2015**, *22*, 37–50. [CrossRef]
76. Ziemba, P. NEAT F-PROMETHEE—A New Fuzzy Multiple Criteria Decision Making Method Based on the Adjustment of Mapping Trapezoidal Fuzzy Numbers. *Expert Systems with Applications* **2018**, *110*, 363–380. [CrossRef]
77. Ziemba, P.; Wątróbski, J.; Zioło, M.; Karczmarczyk, A. Using the PROSA Method in Offshore Wind Farm Location Problems. *Energies* **2017**, *10*, 1755. [CrossRef]
78. Ziemba, P.; Wątróbski, J. Selected Issues of Rank Reversal Problem in ANP Method. In *Selected Issues in Experimental Economics. Proceedings of the 2015 Computational Methods in Experimental Economics (CMEE) Conference, Międzyzdroje, Poland, 17-19 September 2015*; Nermend, K., Łatuszyńska, M., Eds.; Springer: Cham, Switzerland, 2016; pp. 203–225.
79. Yang, C.-L.; Yuan, B.J.C.; Huang, C.-Y. Key Determinant Derivations for Information Technology Disaster Recovery Site Selection by the Multi-Criterion Decision Making Method. *Sustainability* **2015**, *7*, 6149–6188. [CrossRef]
80. Ziemba, P.; Wątróbski, J.; Jankowski, J.; Piwowarski, M. Research on the Properties of the AHP in the Environment of Inaccurate Expert Evaluations. In *Selected Issues in Experimental Economics. Proceedings of the 2015 Computational Methods in Experimental Economics (CMEE) Conference, Międzyzdroje, Poland, 17–19 September 2015*; Nermend, K., Łatuszyńska, M., Eds.; Springer: Cham, Switzerland, 2016; pp. 227–243.
81. Saaty, T.L. The Analytic Hierarchy and Analytic Network Process for the Measurement of Intangible Criteria and for Decision-Making. In *Multiple Criteria Decision Analysis. State of the Art Surveys*, 2nd ed.; Greco, S., Ehrgott, M., Figueira, J.R., Eds.; Springer: New York, NY, USA, 2016; pp. 363–419.
82. Saaty, T.L. Decision-making with the AHP: Why is the principal eigenvector necessary. *Eur. J. Oper. Res.* **2003**, *145*, 85–91. [CrossRef]
83. Saaty, T.L. Fundamentals of the analytic network process—Dependence and feedback in decision-making with a single network. *J. Syst. Sci. Syst. Eng.* **2004**, *13*, 129–157. [CrossRef]
84. Cortes-Aldana, F.A.; Garcia-Melon, M.; Fernandez-de-Lucio, I.; Aragones-Beltran, P.; Poveda-Bautista, R. University objectives and socioeconomic results: A multicriteria measuring of alignment. *Eur. J. Oper. Res.* **2009**, *199*, 811–822. [CrossRef]
85. Whitaker, R. Criticisms of the Analytic Hierarchy Process: Why they often make no sense. *Math. Comput. Model.* **2007**, *46*, 948–961. [CrossRef]
86. Giberson, M. *Assessing Wind Power Cost Estimates*; Institute for Energy Research: Washington, DC, USA, 2013; Available online: http://instituteforenergyresearch.org/wp-content/uploads/2013/10/Giberson-study-Final.pdf (accessed on 8 February 2019).
87. Yang, J.; Chen, W.; Chen, B.; Jia, Y. Economic feasibility analysis of a renewable energy project in the rural China. *Procedia Environ. Sci.* **2012**, *13*, 2280–2283. [CrossRef]

88. Zitzler, E.; Knowles, J.; Thiele, L. Quality Assessment of Pareto Set Approximations. *Lect. Notes Comput. Sci.* **2008**, *5252*, 373–404.
89. Brans, J.P.; Mareschal, B. The PROMCALC & GAIA decision support system for multicriteria decision aid. *Decis. Support Syst.* **1994**, *12*, 297–310.
90. Makowski, M.; Wierzbicki, A.P. Modeling Knowledge: Model-based Decision Support and Soft Computations. In *Applied Decision Support with Soft Computing*; Yu, X., Kacprzyk, J., Eds.; Springer: Berlin/Heidelberg, Germany, 2003; pp. 3–60.
91. Tanino, T. Sensitivity Analysis in MCDM. In *Multicriteria Decision Making: Advances in MCDM Models, Algorithms, Theory, and Applications*; Gal, T., Stewart, T.J., Hanne, T., Eds.; Springer Science: New York, NY, USA, 1999; pp. 173–201.
92. Wang, Y.M.; Luo, Y. On rank reversal in decision analysis. *Math. Comput. Model.* **2009**, *49*, 1221–1229. [CrossRef]
93. Maleki, H.; Zahir, S. A Comprehensive Literature Review of the Rank Reversal Phenomenon in the Analytic Hierarchy Process. *J. Multi-Criteria Decis. Anal.* **2013**, *20*, 141–155. [CrossRef]
94. Global Atlas for renewable energy. Available online: http://irena.masdar.ac.ae (accessed on 8 February 2019).
95. Shokrzadeh, S.; Jozani, J.; Bibeau, E.; Molinski, T. A statistical algorithm for predicting the energy storage capacity for baseload wind power generation in the future electric grids. *Energy* **2015**, *89*, 793–802. [CrossRef]
96. Vestas. V90 3.0 MW. Available online: https://www.ceoe.udel.edu/File%20Library/Research/Wind%20Power/ProductbrochureV90_3_0_UK.pdf (accessed on 8 February 2019).
97. PSE. Plan sieci elektroenergetycznej najwyższych napięć. Available online: https://www.pse.pl/documents/20182/32630243/plan_sieci_elektroenergetycznej_najwyzszych_napiec.jpg (accessed on 8 February 2019).
98. Ernst & Young. Wpływ energetyki wiatrowej na wzrost gospodarczy w Polsce, 2012. Available online: http://www.domrel.pl/_publikacje/raport_psew_2012.pdf (accessed on 8 February 2019).
99. Dziennik Ustaw Rzeczypospolitej Polskiej. Rozporządzenie Ministra Gospodarki w sprawie ceny referencyjnej energii elektrycznej z odnawialnych źródeł energii w 2016 roku. 13.11.2015. Available online: http://dziennikustaw.gov.pl/du/2015/2063/1 (accessed on 8 February 2019).
100. Lu, M.T.; Lin, S.W.; Tzeng, G.H. Improving RFID adoption in Taiwan's healthcare industry based on a DEMATEL technique with a hybrid MCDM model. *Decis. Support Syst.* **2013**, *56*, 259–269. [CrossRef]
101. Hung, Y.H.; Chou, S.C.; Tzeng, G.H. Knowledge management adoption and assessment for SMEs by a novel MCDM approach. *Decis. Support Syst.* **2011**, *51*, 270–291. [CrossRef]
102. Carlsson, C.; Fuller, R. Multiple criteria decision making: The case for interdependence. *Comput. Oper. Res.* **1995**, *22*, 251–260. [CrossRef]
103. Ziemba, P.; Becker, J. Analysis of the Digital Divide Using Fuzzy Forecasting. *Symmetry* **2019**, *11*, 166. [CrossRef]
104. Ziemba, P.; Jankowski, J.; Wątróbski, J.; Wolski, W.; Becker, J. Integration of domain ontologies in the repository of website evaluation methods. In Proceedings of the 2015 Federated Conference on Computer Science and Information Systems (FedCSIS), Lodz, Poland, 13–16 September 2015; pp. 1585–1595.

energies

Article

Analyzing Carbon Emissions Embodied in Construction Services: A Dynamic Hybrid Input–Output Model with Structural Decomposition Analysis

Xi Zhang [1,2], Zheng Li [1,2], Linwei Ma [1,2,*], Chinhao Chong [1] and Weidou Ni [1]

1 State Key Laboratory of Power Systems, Department of Energy and Power Engineering, Tsinghua-BP Clean Energy Research & Education Centre, Tsinghua University, Beijing 100084, China; zhangxi14@mails.tsinghua.edu.cn (X.Z.); lz-dte@tsinghua.edu.cn (Z.L.); zjh08@tsinghua.org.cn (C.C.); niwd@tsinghua.edu.cn (W.N.)

2 Tsinghua-Rio Tinto Joint Research Centre for Resources, Energy and Sustainable Development, Laboratory for Low Carbon Energy, Tsinghua University, Beijing 100084, China

* Correspondence: malinwei@tsinghua.edu.cn; Tel.: +86-10-6279-5734 (ext. 302); Fax: +86-10-6279-5736

Received: 8 March 2019; Accepted: 13 April 2019; Published: 17 April 2019

Abstract: The energy embodied in construction services consumed by industrial sectors used to increase capacities has led to massive energy-related carbon emissions (ERCE). From the perspective of consumer responsibility, ERCE embodied in construction services is driven by technological changes and the increases in final demand of various sectors, including final consumption, fixed assets investment, and net export. However, little attention has been paid to decomposing sectoral responsibilities from this perspective. To fill this research gap, we propose a dynamic hybrid input–output model combined with structural decomposition analysis (DHI/O-SDA model). We introduce DHI/O modeling into the estimation of ERCE embodied in construction services from the perspective of consumer responsibility and introduce SDA into DHI/O models to improve the resolution of the estimate. Taking China as a case study, we verified the DHI/O-SDA model and present the bilateral relationships among sectoral responsibilities for ERCE embodied in construction services. A major finding is that the "Other Tertiary Industry" sector is most responsible for ERCE embodied in construction services and strongly influences other sectors. Therefore, controlling the final demand increase of the service industry will be the most effective policy to reduce the ERCE embodied in construction services.

Keywords: energy-related carbon emissions; embodied energy; fixed assets investment; dynamic hybrid input–output model; structural decomposition analysis

1. Introduction

The Intergovernmental Panel on Climate Change (IPCC) has pointed out that reducing greenhouse gas emissions can help mitigate global greenhouse effects [1]. According to the Fifth Assessment Report of the IPCC [2], the energy-related carbon emissions (ERCE) of industrial processes accounted for approximately 65% of anthropogenic greenhouse gas emissions worldwide in 2010. To reduce ERCE, energy conservation remains an important task in addition to the development of low-carbon energy sources.

Energy conservation not only includes improving technological efficiency (such as reducing energy consumption per unit of energy services provided), but also reducing demand for energy services. Energy services can be categorized as "operation services" and "construction services" [3]. The former indicates the services required by an economy to maintain current living standards with existing infrastructure and energy devices, and the latter indicates the services required for expanding the capacities of various industrial sectors. Construction services normally involve investments in infrastructure and energy devices, which cause direct and indirect energy consumption and ERCE. Therefore, the conservation of construction services (for example, by avoiding the insufficient use of infrastructure and production capacities caused by overcapacity or frequent construction and demolition) must not be neglected in energy conservation analysis and planning.

The conservation of construction services is especially important in developing countries that are experiencing rapid industrialization, and in those whose demands for energy services are far from saturated, as exemplified by China after 2000. Using fixed assets investment as an indicator of construction services, two previous studies [4,5] verified that energy consumed for construction services accounted for a large proportion in China, and found that a considerable amount of potential energy savings can be captured by conserving construction services. In existing work related to estimating energy consumption and ERCE embodied in construction services [4–8], the responsibility for energy consumption and ERCE embodied in construction services was mainly attributed to the sectors that directly provide construction services. Therefore, the policy implication was mainly to control the investments in the construction and manufacturing sectors to reduce energy consumption and ERCE embodied in construction services. However, much of the construction services in these two sectors are not due to their own needs, but to the increase in other sectors' final demand, including final consumption, fixed assets investment, and net export. The increase in the final demand of various sectors is the deeper reason for the need for construction services, and for energy consumption and ERCE embodied in construction services. Therefore, restricting the final demand increase of the sectors that mainly consume construction services may be more effective. Accordingly, we needed to clarify the drivers of construction services from the perspective of each sector's final demand increase, meaning from the perspective of consumer responsibility. This approach involves the redistributing the investment responsibility of each sector based on the final demand increases. To this end, we must introduce the dynamic hybrid input–output (DHI/O) method, because there are limitations in the static hybrid input–output method applied in previous studies.

This paper introduces the DHI/O model into the decomposition of sectoral responsibility for energy consumption and the ERCE embodied in construction services from the perspective of consumer responsibility. Moreover, structural decomposition analysis (SDA) is combined with the DHI/O model to further improve the resolution of the model, and shapes a dynamic hybrid input–output model combined with structural decomposition analysis (DHI/O-SDA model). Being the largest ERCE emitter in the world, China is used as a case study to demonstrate the model.

This paper provides two key contributions. First, this is an early attempt to introduce the DHI/O model into the analysis of energy consumption and the ERCE embodied in construction services. We also firstly introduce SDA into the DHI/O model to differentiate the energy consumption and the ERCE embodied in construction services caused by technology changes from those caused by final demand increases. Such a model is essential for formulating energy conservation policies in developing countries that are experiencing rapid industrialization to reduce energy consumption and the ERCE embodied in construction services. Second, using the DHI/O-SDA model, we present the sectoral responsibilities for the ERCE embodied in the construction services consumed by China from 2007 to 2012. Based on these outcomes, we discussed the policy implications of reducing ERCE embodied in construction services in China. These policies are presented from a new perspective that emphasizes more control of the final demand increase of the sectors that mainly consume construction services.

The rest of this paper is organized as follows: Section 2 reviews the related literature, Section 3 introduces the methodology and data, Section 4 presents the results, uncertainties, and discussions of our study, and Section 5 provides our conclusions and policy implications.

2. Literature Review

Construction services directly consumed by various sectors can be quantified as a proportion of fixed assets investment. Generally, fixed assets are invested in sectors for two reasons: primarily to increase the production capacities for meeting the increasing demand on production, and to compensate for the loss in the existing production capacities of the sectors [9]. The former reason can be regarded as the demand for construction services, or construction services-related fixed assets investment (CSFAI), whereas the latter as the demand for operation services. Therefore, as an assumption, estimating the energy consumption and ERCE embodied in construction services consumed by various sectors is equivalent to calculating the energy consumption and the ERCE embodied in the CSFAI of various sectors.

The total quantities of direct and indirect energy consumed by various sectors to produce final demands (including CSFAI) are usually estimated using embodied energy analysis. The concept of embodied energy analysis, which was initially called vertical analysis, was introduced by Bullard and Herendeen [10] to describe both direct and indirect primary energy invested in production. Some studies applied embodied energy analysis to estimate the energy consumption or ERCE embodied in the fixed assets investment of various economies. For example, Fu et al. [5] estimated the embodied energy and the ERCE of various types of final demand products in China. Acquaye et al. [6] calculated the ERCE of the Irish construction sector in 2005. Liu et al. [7] calculated the energy consumption embodied in the final demand products of China's economic sectors and found that considerable amounts of energy were consumed in the "Construction" and "Other Service Activities" sectors. Huang et al. [8] estimated the ERCE embodied in the construction activities of 40 countries in 2009, which accounted for 23% of the total ERCE embodied in the global economic activities. Most of these studies applied static hybrid input–output models, which assumed fixed assets investment as a given exogenous variable rather than a result of output increases. Therefore, the responsibilities for energy consumption and the ERCE embodied in fixed assets investment were often taken by the sectors mainly providing fixed assets investment, such as the construction and manufacturing sectors.

To treat fixed assets investment as an endogenous variable and establish the relationship between fixed assets investment and the final demand increase of each sector, we decided to adopt a DHI/O method for this study. Because in the dynamic input–output model, which was developed by Leontief [11], the fixed assets investment in various sectors is represented as a set of coefficients related to the increase in each sector's output between time series [12]. Therefore, the responsibilities of fixed assets investment are allocated to various sectors based on the total output increases. By establishing the relationship between the final demand increase and the total output changes, the fixed assets investment can be estimated from the final demand increase. To this end, Pauliuk et al. [13] built a dynamic input–output model with a comprehensive set of life cycle inventories to analyze the influence of final demand increase on fixed assets investment, and differentiated the fixed assets investment caused by construction and operation services. This is the only study we found that tried to analyze the connection between CSFAI and final demand increase. However, that study did not calculate the energy consumption and the ERCE embodied in the CSFAI.

In the field of energy analysis, the DHI/O model was introduced by Rhoten [14]. Researchers have applied DHI/O models in predicting the total energy consumption or energy consumption of different energy sources (such as coal, oil, gas, and non-fossil electricity) under multiple optimization objectives [15–18]. The DHI/O models have also been adopted for analyzing the development of renewable energy [19] and bioenergy [20]. However, decomposing the responsibilities for the energy and the ERCE embodied in the fixed assets investment (especially the CSFAI) of each sector according to the final demand increase of each sector has received little attention. We speculate that this is mainly because scholars in the field of dynamic input–output modeling have not paid enough attention to the energy consumption and the ERCE embodied in the construction services sectors in developing countries. Therefore, the application of the DHI/O method in the decomposition of sectoral responsibility for energy consumption and ERCE embodied in construction services from the perspective of each sector's final demand increase is still an open field for research.

In addition to the increase in the final demand of each sector, technological changes may also lead to increases in the total output of various sectors, which may lead to CSFAI and energy consumption and ERCE embodied in construction services. Several previous studies have shown that technological changes can result in changes in the total output as the final demand increases of various sectors [21–23]. Therefore, by distinguishing the total output changes caused by technological changes and that caused by final demand increase, we can further improve the accuracy of the decomposition of CSFAI responsibility and improve the resolution of the calculation. This improvement can be achieved by using SDA, which was first proposed by Chenery et al. [24] to identify the influence of each factor (including the final demand increase of each sector and technological changes) on the total output change. Subsequently, many studies applied SDA to identify the contributions of the final demand increase of each sector and technological changes on the output change for various economies [25–31]. By introducing SDA into the dynamic input–output model, we can further separate the CSFAI responsibilities of each sector into one part caused by the final demand increase and the other part caused by technological changes. To the best of our knowledge, this type of study has not been conducted because previous SDA studies have overlooked energy consumption and the ERCE embodied in construction services.

Our literature review indicates that the energy consumption and the ERCE embodied in construction services can be estimated by calculating the energy consumption and the ERCE embodied in the CSFAI of various sectors. To redistribute the responsibilities for energy consumption and the ERCE embodied in CSFAI to sectors that consume the CSFAI for increasing their final demand, the DHI/O method must be introduced. Furthermore, SDA can be introduced to differentiate the output increase caused by the final demand increase from that caused by technological changes. The combination of SDA and the DHI/O model marks a new direction in dynamic input–output analysis that can improve the resolution of estimates of energy consumption and the ERCE embodied in construction services from the perspective of consumer responsibility.

3. Methodology and Data Preparation

3.1. DHI/O-SDA Model

This section describes the development of a DHI/O-SDA model for estimating the energy consumption and the ERCE embodied in CSFAI from the perspective of consumer responsibility. Given the strong correlation between ERCE and energy consumption, the results are displayed in terms of ERCE only. The three steps for establishing a DHI/O-SDA model are illustrated in Figure 1.

The first step (Step 1; the dark blocks with slanting lines on the left side of Figure 1) involves establishing the hybrid input-output tables and preparing the data for further decomposition and estimation. Using the energy balance tables and input–output tables for different years, a set of energy-economy hybrid input–output tables are built.

The second step (Step 2; the blue blocks with vertical lines in the middle of Figure 1) involves using SDA to distinguish the output changes caused by the final demand increase of various sectors from those caused by technological changes. The changes in total output of various sectors are calculated by comparing the hybrid input–output tables from different years. The results of Step 2 are the changes in the output of various sectors caused by final demand increases of various sectors and by technological changes.

The last step (Step 3; the red blocks on the right side of Figure 1) involves estimating the energy consumption and ERCE embodied in CSFAI. Step 3 includes three parts: (1) estimating the CSFAI quantities by the dynamic input–output method based on the output changes caused by the final demand increases of various sectors and the technological changes; (2) estimating the energy consumption embodied in the CSFAI using embodied energy analysis; and (3) calculating the ERCE caused by energy consumption using the IPCC-recommended carbon accounting method [32].

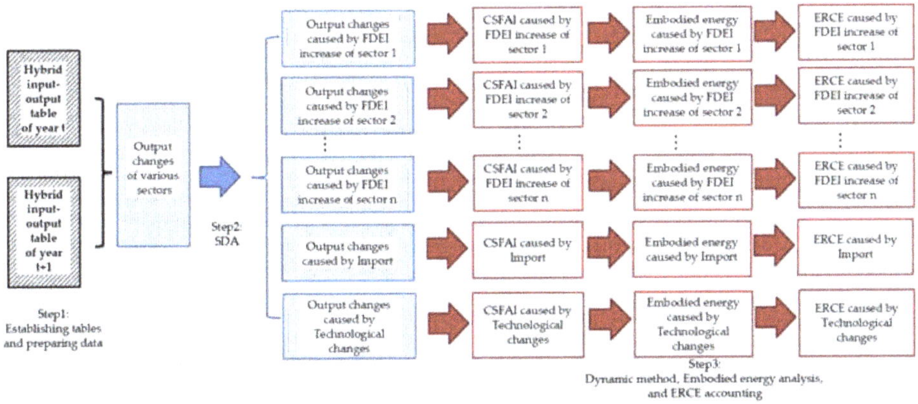

Figure 1. Three steps for establishing a dynamic hybrid input–output combined with structural decomposition analysis (DHI/O-SDA) model.

3.1.1. Establishment of Energy-Economic Hybrid Tables

The energy-economy hybrid input–output table is obtained by replacing the economic value of the energy sector in the conventional economic input–output table with the physical quantity of various types of energy. The final demands of each sector can be classified into final demand excluding import (FDEI) and import. The reason why import is separated from final demand is that import values are negative relative to other final demands. Therefore, in sectors with large amounts of imports, the total final demand including import may be negative, which prevents the analysis of how many CSFAI are needed in those sectors.

The direct input coefficient of sector i to sector j, $a_{ij}{}^{t}$, can be defined as the ratio of the economic value produced by sector i and consumed by sector j to the total output of sector j in year t. If sector i or sector j is an energy sector, the economic value should be replaced by the quantity of energy product. The Leontief inverse matrix of year t can be calculated by:

$$L^t = (I - A^t)^{-1} \quad (1)$$

where L^t represents the Leontief inverse matrix of year t, I represents the identity matrix, A^t is the direct input matrix composed of $a_{ij}{}^{t}$, and $(I - A^t)$ must be a non-singular matrix. At least two hybrid input–output tables for two consecutive years must be established for further analysis.

3.1.2. SDA

Comparing the hybrid input–output tables of years t and $t + 1$, the changes in some parameter matrices from t to $t + 1$ can be estimated by Equations (2)–(4):

$$\Delta L^t = L^{t+1} - L^t \quad (2)$$

$$\Delta x^t = x^{t+1} - x^t \quad (3)$$

$$\Delta f^t = f^{t+1} - f^t \quad (4)$$

where ΔL^t is the change in the Leontief inverse matrix between the two years, $x^t = (x_1{}^t, x_2{}^t, \ldots, x_n{}^t)^T$ stands for the total output of year t, Δx^t is the change in the outputs between the two years, $f^t = (f_1{}^t, f_2{}^t, \ldots, f_n{}^t)^T$ represents the final demands of year t including FDEI and import, and Δf^t is the change in the final demands between the two years. The physical meaning of x^t can be considered as the outputs of all sectors embodied in the final demand. Carbon emissions embodied in the outputs of all primary energy sectors in x^t is the ERCE.

In SDA, the change in the outputs between the two years consists of two parts: changes due to the final demand increase and changes due to technological changes. We assumed that the weight of the two parts are equal. Thus, the output changes from year t to $t + 1$ year can be described as:

$$\Delta x^t = 1/2\Delta L - (f^t + f^{t+1}) + 1/2(L^t + L^{t+1}) - \Delta f \tag{5}$$

Equation (5) produces a decomposition of the total change in outputs into two parts. The first part is attributable to technological changes and can be defined as $\Delta x^t{}_{byL}$, as shown in Equation (6). The second part represents the contribution of the final demand increase to the output changes and can be defined as $\Delta x^t{}_{byf}$, as shown in Equation (7):

$$\Delta x^t_{byL} = 1/2\Delta L - (f^t + f^{t+1}) \tag{6}$$

$$\Delta x^t_{byf} = 1/2(L^t + L^{t+1}) - \Delta f^t \tag{7}$$

$\Delta x^t{}_{byf}$ can be further decomposed into the changes in the outputs caused by the FDEI increase of each sector and the change in the outputs caused by the change of import, as shown in Equation (8):

$$\Delta x^t_{byf} = \sum_i \Delta x^t_{byFDEI,i} + \Delta x^t_{byimports} = \sum_i 1/2(L^t + L^{t+1}) - \Delta f^t_{byFDEI,i} + 1/2(L^t + L^{t+1}) - \Delta f^t_{byimports} \tag{8}$$

The change in the outputs between the two years is decomposed into three categories: the change due to technological changes, the changes due to the FDEI increase of each sector, and the change due to the change of import. A more detailed derivation process can be found in Miller et al. [12].

3.1.3. CSFAI Estimation, Embodied Energy Analysis and ERCE Calculation

In the traditional DHI/O model, fixed assets investment is taken as the endogenous variable of the output changes between two consecutive time instances. As discussed by Holz et al. [9], fixed assets investment is produced for both CSFAI and operation services-related fixed assets investment. We assumed that the ratio of the CSFAI to the total fixed assets investment in various sectors is $pcsfai^t$.

In the conventional dynamic input–output model, the capital coefficient $b^t{}_{ij}$ is defined as the ratio of "the value of the output of sector i that is held by sector j as stock" [12] to the total output increase of sector j in year t. However, only the CSFAI rather than the total fixed assets investment can be taken as the result of changes in output. This capital coefficient is meaningful only when the total output increases of sector i are positive. To solve these two problems, a similar coefficient, called "positive capital coefficient" $c^t{}_{ij}$, is proposed in this paper. This coefficient is defined as the ratio of the CSFAI in sector i that is held by sector j to the sum of the positive output increases.

The definition of $c^t{}_{ij}$ is based on the assumption that the CSFAI is only caused by positive output increase, and negative output changes have no effect on the CSFAI in various sectors. As energy consumption and the ERCE embodied in the CSFAI occur during the installation of fixed assets and cannot be recycled even if the fixed assets are either idle or retired, only the CSFAI caused by a positive demand increase leads to new energy consumption and ERCE. Otherwise, energy consumption and the ERCE embodied in the CSFAI in a sector with a decreasing demand is zero, but not negative. Therefore, the aforementioned assumption is reasonable when analyzing historical data because the fixed assets have been installed. $c^t{}_{ij}$ can be calculated with historical CSFAI data and energy-economy hybrid input–output tables.

Because CSFAI is correlated with positive changes in outputs corresponding to the decomposition of the change in the outputs, CSFAI can also be decomposed into three categories: CSFAI due to technological changes, CSFAI due to the FDEI increase of each sector, and CSFAI due to the change in imports. The CSFAI caused by various factors can be estimated, as shown in Equations (9)–(11):

$$CSFAI^t_{byL,ij} = c^t_{ij} - \Delta x^t_{byL,j} \tag{9}$$

$$CSFAI^t_{byFDEI,ij} = c^t_{ij} - \Delta x^t_{byFDEI,j} \tag{10}$$

$$CSFAI^t_{byimports,ij} = c^t_{ij} - \Delta x^t_{byimports} \tag{11}$$

The output embodied in the CSFAI can be calculated by:

$$x^t_{CSFAI} = L^t - CSFAI^t \tag{12}$$

where $x^t{}_{CSFAI}$ is the output embodied in CSFAI. The output of the primary energy sector represents the embodied energy of CSFAI.

The method we adopted for estimating ERCE from energy consumption is recommended by the IPCC [32]. As carbon dioxide (CO_2) is the most important greenhouse gas, the ERCE in this paper only focuses on CO_2. The ERCE resulting from the embodied energy can be estimated by Equation (13) as follows:

$$ERCE = \sum_{EE} Q_{EE} \times LHV_{EE} \times f_{CO_2,EE} \tag{13}$$

where Q_{EE} is the quantity of one type of embodied energy, LHV_{EE} represents the low heat value of the embodied energy, and $f_{CO2,EE}$ is the CO_2 emission factor of the embodied energy. ERCE is the sum of the CO_2 emissions of all types of embodied energy, and $f_{CO2,EE}$ is determined by the carbon content of the embodied energy and the carbon oxidation rate.

The basic assumptions of DHI/O models still apply, such as energy supply equals energy consumption, and there is only one type of product in most sectors. There are three other assumptions. First, the installation process of fixed assets begins and ends within the same time series. Second, the fixed assets invested in the last time series play a role in the industrial production of the next time series. Third, each sector operates on a certain activity in each time series.

3.2. Data Preparation

As a rapidly developing country, China is the largest carbon emitter in the world. From 2008 to 2010, the Chinese government spent a lot of money to promote economic development, a large part of which were used for construction services, including infrastructure construction and expansion of industrial capacity [33]. This period is special and important for the analysis of China's construction services. The energy embodied in construction services constitutes a large part of the total energy consumption in China [4,5], which has led to a massive ERCE. Estimating the ERCE embodied in construction services and identifying which sector is key to reducing the ERCE embodied in construction services consumed by China are necessary. Therefore, taking China from 2007 to 2012 as an example can demonstrate this model, and can also provide some data support for China to create and change relevant policies.

In this study, we adopted the input–output tables for China for 2007 [34], 2010 [35], and 2012 [36] published by the National Bureau of Statistics (NBS) of China. To be consistent with the sectors settings in the energy balance tables published by NBS, the input–output tables for each year were merged into 21 sectors for the DHI/O-SDA model, as shown in Appendix A. By replacing the economic value of the original input–output tables with the energy quantities in the energy balance tables for China and the final energy consumption by industrial sector tables for China for 2007, 2010, and 2012 [37], the hybrid input–output tables for China for 2007, 2010, and 2012 were established, respectively. Because the statistical data of energy balance tables published by NBS are not high enough in resolution, it is difficult to separate the specific energy utilization data of some sectors from their merged sectors, such as other tertiary industry. Due to lack of input–output tables for China for 2008, 2009, and 2011, we linearly expanded the HI/O tables for these three years according to the ratio of gross domestic product. The tables of investment in fixed assets by industry for China from 2007 to 2012 [38] were adopted to set the proportion of fixed assets invested by each sector. The *pcsfai* of sector 04 in 2007, 2010, and 2012 and related assumptions were obtained from the study of Zhang et al. [39]. Due to lack of *pcsfai* data, we set other sectors' *pcsfai* to the same value as that of Sector 04 (Electric Power). This assumption is reasonable as discussed by Zhang et al. [39].

For simplicity, we assumed the carbon oxidation rate to be 100%. We also adopted a low heat value and CO_2 emission factor of each primary fossil energy source from IPCC [32], as shown in Table 1.

Table 1. Low heat value and CO_2 emission factor.

Primary Fossil Fuel	Low Heat Value (TJ/t)	CO_2 Emission Factor (kg/TJ)
Coal	0.0293	98,300
Crude oil	0.0423	73,300
Natural gas	0.048	56,100

4. Results, Uncertainties and Discussions

4.1. ERCE Embodied in CSFAI Caused by Each Sector

The ERCE embodied in the CSFAI caused by the final demand increase of each sector (including FDEI increases and import) and technological changes is shown in Figure 2. From 2007 to 2011, the FDEI increase of sector 21 (Other Tertiary Industry) always caused the most ERCE embodied in CSFAI in China. From 2007 to 2009, sector 18 (Construction) caused the second most ERCE embodied in CSFAI, followed by sector 16 (Other Manufacture). From 2010 to 2011, Tech (technological changes) overtook sector 18 (Construction) as the second largest cause of ERCE embodied in CSFAI, while sector 18 (Construction) was third. Sectors 21 (Other Tertiary Industry) and 18 (Construction) made large sectoral contributions to the ERCE embodied in CSFAI from 2007 to 2011 because the FDEI increases in these two sectors were the largest among all the sectors. Therefore, the policy implication is that these sectors should receive priority for reducing the ERCE embodied in construction services.

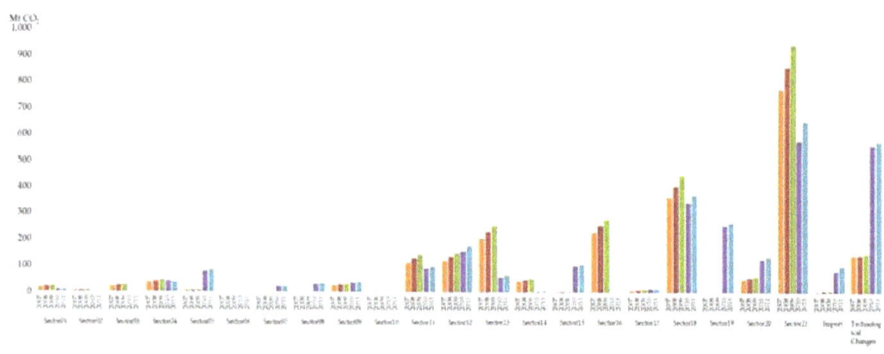

Figure 2. Energy-related carbon emission (ERCE) embodied in the constructive services-related fixed assets investment (CSFAI) caused by the final demand increase of each sector and technological changes.

Import and Tech (technological changes) have been causing the ERCE embodied in CSFAI during each time period. Classically, the overall effect of technological changes should be to reduce the ERCE of an economy. However, based on the assumptions in this paper, Import and Tech (technological changes) may cause the ERCE emitted. Because technological changes also increase the demand of some sectors, which causes an increase in construction services; in turn, energy consumption and ERCE embodied in CSFAI increase. In this study, technological changes mainly caused output increases in sector 21 (Other Tertiary Industry), sector 04 (Electric Power), and five other sectors. The increase in the output demand of sector 21 (Other Tertiary Industry) is because many tertiary industries related to technology development, such as technical services and scientific research, are included in sector 21 (Other Tertiary Industry). More high-tech products were powered by electricity, which resulted in the output demand increase of sector 04 (Electric Power). The ERCE embodied in the CSFAI caused by technological changes was greater after 2010, which may be related to China's technological innovation policy. The Chinese government paid more attention to scientific and technological innovation after 2010. At the 18th National Congress of the Communist Party of China in 2012, innovation was considered an important development strategy for China [40].

Based on the distribution of sectoral responsibility from the perspective of consumer responsibility, the responsibilities for energy consumption and the ERCE embodied in construction services undertaken by some sectors were different than those reported in previous studies. For example, sectors 06 (Ferrous Metals) and 08 (Non-metallic) were considered to be responsible for the huge energy consumption and the ERCE embodied in the construction services, as argued in this paper. These two sectors consumed huge amounts of energy to provide fixed assets investment (11% and 5% of the total intermediate energy consumption by all the sectors in 2012, respectively). However, the ERCE embodied in the CSFAI caused by the FDEI increase in sectors 06 (Ferrous Metals) and 08 (Nonmetallic) only accounted for less than 1% of all the ERCE embodied in CSFAI in all sectors. This occurred because the huge quantities of energy consumption and ERCE in these two sectors for producing CSFAI was attributable to the final demand increase of other sectors, especially sectors 21 (Other Tertiary Industry) and 18 (Construction). This result reflects the necessity of redistributing sectoral responsibilities for ERCE embodied in construction services energy consumption from the perspective of consumer responsibility. Unlike previous studies, the result of this paper clarify the importance of sector 21 (Other Tertiary Industries), instead of manufacturing sectors, in reducing ERCE embodied in construction services.

4.2. ERCE Embodied in CSFAI Caused by Sectoral a Unit of FDEI Increase

Based on the DHI/O-SDA model, the ERCE embodied in the CSFAI consumed by each sector caused by a unit of FDEI increase of each sector were estimated. Table 2 shows the format of the results obtained.

Table 2. Format of DHI/O–SDA model results table.

The Cause of ERCE Embodied in CSFAI	ERCE Embodied in CSFAI Consumed by Sector 1	…	ERCE Embodied in CSFAI Consumed by Sector j	…	ERCE Embodied in CSFAI Consumed by Sector n
Unit FDEI increase of sector 1	ER_{11}	…	ER_{1j}	…	ER_{1n}
⋮	⋮	⋮	⋮	⋮	⋮
Unit FDEI increase of sector i	ER_{i1}	…	ER_{ij}	…	ER_{in}
⋮	⋮	⋮	⋮	⋮	⋮
Unit FDEI increase of sector n	ER_{n1}	…	ER_{nj}	…	ER_{nn}

In the table, ER_{ij} represents the ERCE embodied in the CSFAI consumed by sector j and caused by a unit of FDEI increase of sector i. The results table can be analyzed in the horizontal and vertical directions. In the horizontal direction, there are three indicators: (1) *ER*, which is the ERCE embodied in the CSFAI consumed by each sector caused by a unit of FDEI increase in any specific sector; (2) *TE*, which is the sum of the *ER* in each sector caused by a unit of FDEI increase in any specific sector; and (3) *PEC*, which is the proportion of *ER* in the *TE* caused by a unit of FDEI increase in any specific sector.

ER_{ij} is an element of the results table. The *ER* in each sector represents the ERCE embodied in the CSFAI consumed by each sector when the FDEI in any specific sector increases by one unit (meaning the increase in one sector leads to *ER* in *n* sectors). By analyzing *ER*, we can determine which sector will produce the most ERCE embodied in construction services because of a unit of FDEI increase in any specific sector.

TE is sum of ER_{ij} in a row, which can be estimated by Equations (14). The *TE* caused by sector *i* represents the total influence of a unit of FDEI increase in sector *i* on the total ERCE embodied in the CSFAI. By analyzing *TE*, we evaluated the efficiency of reducing the ERCE embodied in the CSFAI by restricting the FDEI increase of sector *i*:

$$TE_i = \sum_j ER_{ij} \tag{14}$$

PEC is the proportion of ER_{ii} to the *TE* of sector *i*, which can be estimated by Equation (15). PEC represents the driving influence of sector *i*'s unit FDEI increase on the CSFAI in various sectors. The higher the PEC, the less the FDEI increase of the sector results in the CSFAI in other sectors:

$$PEC_i = \frac{ER_{ii}}{\sum_j ER_{ij}} \tag{15}$$

In the vertical direction, the results demonstrate the influence of each sector's unit FDEI increase on the *ER* in one sector (meaning the increase in *n* sectors leads to *ER* in one sector). The results can be used to identify the sector for which the FDEI increase would be most influential in causing *ER* in the target sector.

Appendix B provides the full results of the ERCE embodied in the CSFAI consumed by each sector caused by a unit of final demand increase of a specific sector in China in 2007–2008, 2008–2009, 2009–2010, 2010–2011, and 2011–2012.

4.2.1. Analysis in the Horizontal Direction

ER

For most sectors (20 of 21 in 2007–2010 and 19 of 21 in 2010–2012), a unit FDEI increase in one sector caused the most *ER* in the sector itself. However, sector 18 (Construction) was an exception in that a unit FDEI increase in this sector resulted in the most *ER* in sectors 21 (Other Tertiary Industry), 19 (Transport), 04 (Electric Power), 08 (Nonmetallic), and 06 (Ferrous Metals) because these sectors are important in supporting the production of sector 18 by providing materials, transportation, assistant services, and energy supply.

A unit FDEI increase in most sectors caused a relatively high *ER* in sector 21 (Other Tertiary Industry), which indicates that the increase in the FDEI in various sectors requires a large increase in products from sector 21. The products of sector 21 provide important support for the CSFAI caused by the FDEI increase in various sectors. Therefore, although not directly providing most of the construction services, sector 21 is key to conserving construction services from the perspective of consumer responsibility.

TE

The *TE* caused by each sector's FDEI increase is shown in Figure 3.

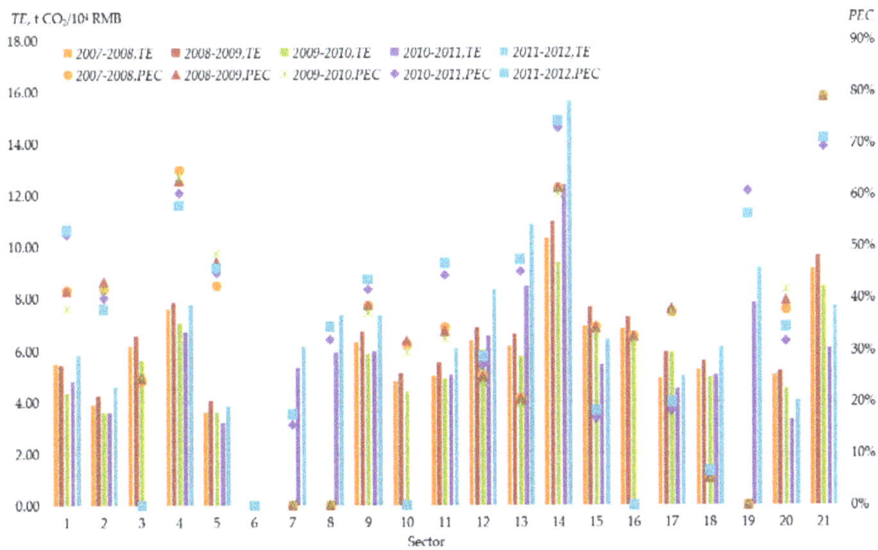

Figure 3. *TE* (left coordinate axis, bar) and the *PEC* (right coordinate axis, point) caused by a unit increase of sectoral FDEI.

In 2007–2012, a unit FDEI increase in sector 14 (Textile) caused the most *TE*. The strong influence of a unit FDEI increase in sector 14 (Textile) on the *TE* is because the total ERCE embodied in the CSFAI caused by the FDEI increase in sector 14 was small (less than 2% of all the ERCE for fixed assets investment in 2007–2012), but the total increase in FDEI of sector 14 was much smaller.

In 2007–2012, the *TEs* caused by a unit increase in FDEI in sector 14 (Textile), sector 21 (Other Tertiary Industry), and sector 04 (Electric Power) remained high and were always more than 6t CO_2. Therefore, an increase in the FDEIs of these three sectors would lead to the fastest increase in the ERCE embodied in the CSFAI in China. However, as the total quantity of the ERCE embodied in the CSFAI caused by sector 14 (Textile) and sector 04 (Electric Power) was small, policy makers might not focus on these two sectors to reduce the ERCE embodied in construction services. Sector 21 (Other Tertiary Industry) was in the top five sectors on the *TE* list in 2007–2012. The total quantity of ERCE embodied in the CSFAI caused by the FDEI increase in sector 21 was also the largest. Thus, to reduce the energy consumption and ERCE embodied in construction services, the most efficient approach would be to restrict the FDEI of sector 21.

PEC

The *PEC* of each sector is shown in Figure 3. In 2007–2012, the *PEC* of some sectors, such as sectors 21 (Other Tertiary Industry), 14 (Textile), and 04 (Electric Power), were more than 50%. Therefore, an increase in the FDEI of these three sectors would cause most of the ERCEs embodied in CSFAI in their own sectors, which has relatively little impact on the ERCEs embodied in the CSFAI of other sectors. When the FDEI of these three sectors increases, other sectors consume only a little of the ERCEs embodied in CSFAI.

4.2.2. Analysis in the Vertical Direction

Two sectors in 2010–2011 were selected as target sectors to illustrate the vertical analysis. Sector 04 (Electric Power) is adopted as a common case, and sector 06 (Ferrous Metals) is adopted as a special case.

The result of sector 04 (Electric Power) is relatively common for most sector, because a unit FDEI increase in sector 04 (Electric Power) caused the most *ER* in this sector. Besides sector 04 (Electric Power), a unit FDEI increase in sector 07 (Non-ferrous Metals), followed by sector 09 (Chemical) and sector 08 (Non-metallic), caused the most *ER* in sector 04 (Electric Power). The production of these sectors requires a huge amount of electricity. To produce more FDEI in these sectors, sector 04 (Electric Power) needs to expand production capacity to provide more electricity, which results in more *ER*. Therefore, to restrict the energy consumption and the ERCE embodied in the CSFAI consumed by sector 04 (Electric Power), reducing the FDEI increase in sector 04 (Electric Power) would be the most efficient strategy, whereas decreasing the FDEI of sector 07 (Non-ferrous Metals), sector 09 (Chemical), and sector 08 (Non-metallic) would be less efficient.

The result for sector 06 (Ferrous Metals) is a special case because the FDEI in sector 06 (Ferrous Metals) decreased in 2010–2011. The most *ER* in sector 06 (Ferrous Metals) was caused by a unit FDEI increase in sector 12 (Machinery), followed by sector 13 (Automobiles) and sector 18 (Construction). Sector 06 (Ferrous Metals) provides the main materials for the production of these three sectors, so it is reasonable for a large increase in *ER* in sector 06 (Ferrous Metals) to be caused by a unit of FDEI increase of these sectors. Therefore, instead of continuously reducing the FDEI in sector 06 (Ferrous Metals), restricting the FDEI in sector 12 (Machinery), followed by sector 13 (Automobiles) and sector 18 (Construction) would be most efficient to reduce the energy consumption and the ERCE embodied in CSFAI consumed by sector 06 (Ferrous Metals). This result could be a reference case for policy makers to eliminate capacity expansion in sector 06 (Ferrous Metals), which is an important task stipulated by the State Council of China [41].

The results of vertical analysis are useful for policy makers to reduce energy consumption and ERCE embodied in construction services consumed by a specific sector. From the vertical analysis of the results, it can be found that the energy consumption and ERCE embodied in construction services consumed by a specific sector is affected by the FDEI increase of several sectors, or even all the sectors, in the economy to varying degrees. Therefore, to reduce energy consumption and ERCE embodied in construction services consumed by a specific sector more efficiently, policy makers need to consider not only the restriction of the FDEI of the specific sector itself, but also the impact of the FDEI increase of each sector on the specific sector, and make a systematic policy. To this end, it is necessary to adopt the vertical analysis proposed in this paper, which can provide quantitative data support.

4.3. *Uncertainties and Discussions*

4.3.1. Verification of the Model Result

To verify the validity of our model and estimation, we compare ERCE embodied in the final demands of various sectors of China in 2007 estimated separately in our paper with those of Fu's work. In this paper, the result of the total ERCE responsibility of China in 2007 was 5.9 Gt. The results of this paper are reasonable in comparison to those of Fu et al. [5], where the total ERCE of China in 2007 was reported as 6.3 Gt. The total ERCE result of these two papers are close, with the main reasons for the difference are the different processing methods for system errors in the energy balance tables and the input–output tables and different carbon emission factors.

4.3.2. Comparisons with the Results from 2012 to 2015

To compare with the new development in China, we update the analysis results of decomposing ERCE embodied in CSFAI to various sectors from the perspective of consumer responsibility from 2012 to 2015, as shown in Figure 4.

After 2012, the main reasons for China's ERCE embodied in CSFAI are the increase in the FDEI of sector 21 (Other Tertiary Industry), the increase in the FDEI of sector 18 (Construction) and technological changes. The proportion of ERCE embodied in CSFAI caused by these three factors in total ERCE embodied in CSFAI increases from 57% in 2011–2012 to 79% in 2014–2015. This shows that restricting the increase of FDEI in sector 18 (Construction) and 21 (Other Tertiary Industry) is more important for reducing ERCE embodied in CSFAI in 2012 to 2015.

Compared with the results of 2011–2012, the changes of ERCE embodied in CSFAI caused by the increase of FDEI in five sectors were more than 100 Mt in 2012 to 2015, which were sector 12 (Machinery), 16 (Other Manufacture), 18 (Construction), 19 (Transport) and 21 (Other Tertiary Industry), respectively. The ERCE embodied in CSFAI caused by the increase of FDEI in sector 12 (Machinery) and 19 (Transport) in 2012 to 2015 is less than it in 2011–2012, whereas the ERCE embodied in CSFAI caused by the increase of FDEI in sector 16 (Other Manufacture), 18 (Construction) and 21 (Other Tertiary Industry) increases in 2012 to 2015. Although the results have changed, the main conclusion of distributing the responsibilities of ERCE embodied in consumption services from the perspective of consumer responsibility in 2012 to 2015 is basically consistent with those in 2007 to 2012, which is that sector 18 (Construction) and 21 (Other Tertiary Industry) bear the highest responsibility for China's ERCE embodied in construction services, and should be given priority attention in reducing ERCE embodied in construction services.

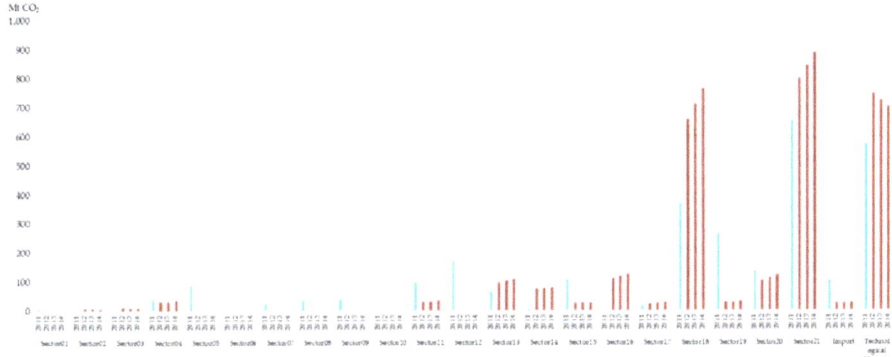

Figure 4. ERCE embodied in CSFAI caused by the final demand increase of each sector and technological changes from 2011 to 2015.

4.3.3. Comparison of DHI/O and DHI/O-SDA Models

The purpose of our paper is to observe the responsibility of ERCE embodied in CSFAI caused by the final demand increase of each sector. However, the construction services will be caused by technological changes and the increase of sectoral final demand at the same time. Therefore, it is necessary to remove the impact of technological changes on construction services. SDA was used to remove the impact of technological changes by distinguishing the output increase caused by the final demand increase of various sectors and the output increase caused by technological changes. Without the SDA, the total output increase caused by technological changes would be apportioned to the total output increase caused by the change in final demand, and the ERCE embodied in the CSFAI caused by final demand increase would contain an uncertain proportion of influence due to technological changes. This coupling of impacts would prevent policymakers from analyzing the impact of the increased final demand of each sector on construction services and the ERCE caused by construction services.

We used China in 2010–2011 as an example to illustrate the difference between with and without SDA in the DHI/O model to estimate the ERCE embodied in the CSFAI, as shown in Figure 5. Here, the results of most sectors are affected by the SDA method. Without SDA, the estimation of ERCE embodied in the CSFAI of sector 08 (Non–metallic), sector 09 (Chemical), sector 18 (Construction), sector 21 (Other Tertiary Industry), and Import are significantly higher, and the impact of technology changes cannot be analyzed.

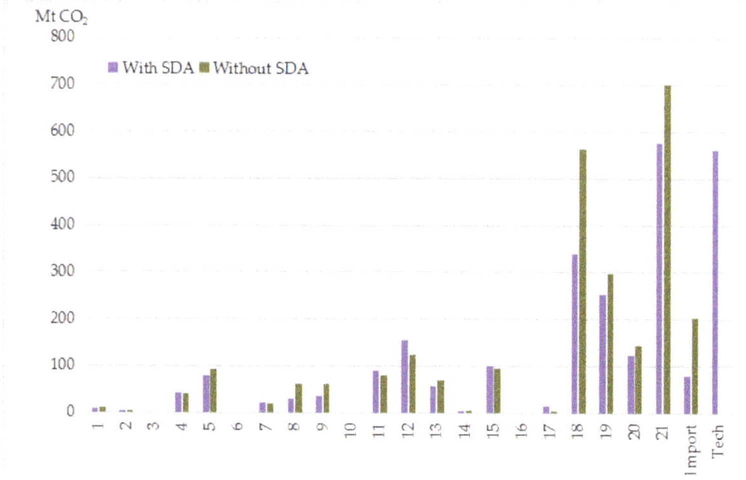

Figure 5. Comparing DHI/O modeling with and without SDA for estimating ERCE embodied in CSFAI.

4.3.4. Sensitivity Analysis of $pcsfai^t$

The $pcsfai^t$ of various sectors was assumed to be the same as that of sector 04 (Electric Power), which is a relatively common sector that directly influences the CSFAI. We analyzed the sensitivity of each sector's $pcsfai^t$ to the ERCE embodied in the CSFAI caused by the sector in 2010–2011, which is the ERCE responsibility of each sector from the consumer side, as shown in Figure 6.

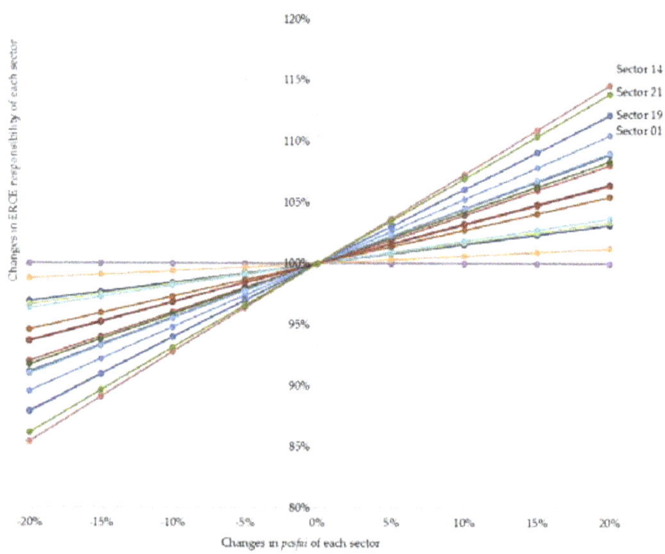

Figure 6. Sensitivity analysis of $pcsfai^t$.

The influence is linear and the ERCE embodied in the CSFAI increases with $pcsfai^t$. The results show that an influence of a 20% change in $pcsfai^t$ in most sectors will cause changes of less than 10% in the ERCE responsibility of the sector. The ERCE embodied in the CSFAI caused by sectors 14 (Textile), 21 (Other Tertiary Industry), 19 (Transport), and 01 (Coal) are most affected by changes in $pcsfai^t$.

Therefore, for some sectors, $pcsfai^t$ has an impact on their ERCE responsibilities, so we should estimate the $pcsfai^t$ of each sector as accurately as possible. For 9 of the 21 sectors, the changes in $pcsfai^t$ have little effect on their ERCE responsibilities (changes less than 5%), so they can be said to be insensitive to changes in $pcsfai^t$.

4.3.5. Statistical Error Analysis

The statistical errors in this paper mainly come from the errors of input–output tables and energy balance tables published by NBS. The errors of input–output tables are mainly caused by the unequal total input and output of each sector in the table, whereas the errors of energy balance tables are mainly caused by the difficulty in accurately measuring losses in energy conversion, utilization and transportation. In this paper, we deal with the errors as one of the final demands in the DHI/O-SDA model. There are two advantages of this treatment. First, it guarantees the balance of input and output in the input-output tables and energy balance tables. Second, the position of the errors in this model is consistent with it in the original input–output tables and energy balance tables, without introducing new assumptions and errors. However, this will inevitably lead to some errors in the intermediate production of some sectors to be calculated as errors in the final demand, which will result in the inaccurate estimation of the direct input matrix A. The calculation based on matrix A will have different degrees of errors. Because there is a strong correlation between the input and output of each sector in the input–output model, statistical errors of any sector may affect the calculation results of all sectors in the model.

4.3.6. Limitations of the DHI/O-SDA Model

In addition to the common limitations of the hybrid input–output model, the DHI/O-SDA model has two main limitations, corresponding to its main assumptions. The first limitation is that some of the fixed assets may not be installed in one time step, as assumed in this study. Some fixed assets are invested to meet the final demand increases of the next several time steps. To simulate this phenomenon, it is necessary to set a construction delay in the fixed assets and to classify the CSFAI into several categories caused by the final demand increase in various time steps. This approach requires using a ratio to determine the quantities in each CSFAI category. With more detailed data and longer time series data, these multiple time step influences on CSFAI could be analyzed by a multiple time delay model, which would give a full view of the energy consumption and ERCE embodied in the CSFAI from time series. The development and testing of multiple time delay models could be the subject of future research work.

The second limitation is that the DHI/O-SDA model does not take into account the changes in capacity utilization rate in various sectors. We assume that the capacity utilization rate of each sector has always been constant, so when the production demand increases, sectors need to consume construction services to meet the higher production demand. However, there is another way for sectors to meet higher production demand, which is to maintain the existing production capacities but increasing capacity utilization rate. To introduce the influence of capacity utilization rate on the quantities of construction services consumed by various sectors in the model, it is necessary to establishing mathematical relations between the capacity utilization rate of each sector and its production demand gap. The model with these relations can better reflect the changes in energy consumption and ERCE embodied in construction services of the economy caused by the increase of sectoral final demand. This will also be another subject direction of improving this DHI/O-SDA model.

5. Conclusions and Policy Implications

A DHI/O-SDA model was proposed in this paper. Using this model, the ERCE of construction services (or CSFAI) due to the final demand increase in various sectors and technological changes can be estimated, and the key sector for reducing the ERCE embodied in construction services can be identified. The ERCE embodied in construction services consumed by each sector caused by a unit of development in each sector can be analyzed in the horizontal and vertical directions. This analysis can reveal several sectoral interrelationships.

China in 2007–2012 was used as a case study to demonstrate our DHI/O-SDA model. The main findings by the DHI/O-SDA model, which can't be revealed by previous models, are: (1) the final demand increase in sector 21 (Other Tertiary Industry) was the main cause of the ERCE embodied in construction services. (2) For most sectors in China, a FDEI increase in one sector causes the greatest ERCE embodied in construction services consumed by the same sector. A FDEI increase in most sectors causes a relatively high ERCE embodied in construction services consumed by sector 21 (Other Tertiary Industry). (3) The vertical analysis of model results for a target sector provide data support for formulating policies for reducing the ERCE embodied in construction services.

The Chinese government has realized the importance of reducing construction services for energy conservation and ERCE reduction. It has published extremely strict policies to restrict capacity expansion in the major industries providing construction services (such as the steel industry and the cement industry) in order to restrict the continued consumption of construction services in these industries. For example, China's State Council has issued two policies that strictly prohibit new production capacity in the steel industry [41] and cement industry [42]. These policies are reasonable in China, because the capacity utilization rate of these industries is not saturated. However, if China continues to build infrastructure and let the tertiary industries such as real estate develop at a rapid pace as in recent years, it is debatable whether it is appropriate to continue to implement these policies. First, after the capacity utilization rate of the steel industry and the cement industry has reached saturation, these two industries must consume construction services to provide more products to meet the increased final demand of other sectors. Second, the effect of simply preventing the steel industry and the cement industry from consuming construction services to reduce energy consumption and ERCE embodied in construction services may not be as effective as controlling the increase in final demand of tertiary industries such as real estate. Our paper can provide quantitative data support for this debate. The DHI/O-SDA model can estimate the amount of construction services that the steel industry and the cement industry must consume as final demand increases in other sectors after capacity utilization rate is saturated, and the energy consumption and carbon emissions embodied in the construction services. The results show that for reducing the ERCE embodied in construction services consumed by one specific sector, we must analyze all the sectors that affect the specific sector and implement a systematic control policy, rather than just restrict the FDEI of the specific industry itself.

Considering the significant regional disparity in energy and industry development across the world, future application of the DHI/O-SDA model in other developing regions would be insightful. The model can be modified to include a multiple time delay function. Another potential future study would involve building mathematical relations between the capacity utilization rate of each sector and its production demand gap in the model. The model can also be used in forecasting energy consumption and the ERCE embodied in construction services caused by various influencing factors.

Author Contributions: Data curation, X.Z.; Methodology, X.Z., Z.L., and L.M.; Writing—original draft, X.Z. and L.M.; Writing—review & editing, X.Z., Z.L., L.M., C.C., and W.N.

Funding: This work was supported by the National Natural Science Foundation of China (Project No. 71690245).

Acknowledgments: The authors gratefully acknowledge financial support from BP Company in the context of the Phase II and III collaboration between BP and Tsinghua University.

Conflicts of Interest: The authors requested to avoid reviewers who are not familiar with China's energy and economic development situation.

Abbreviations

CO_2	Carbon dioxide
CSFAI	Constructive Services-related Fixed Assets Investment
DHI/O	Dynamic Hybrid Input–output
DHI/O-SDA model	Dynamic Hybrid Input–output model combined with Structural Decomposition Analysis
ER	ERCE embodied in CSFAI consumed by each sector caused by a unit of final demand increase in any specific sector
ERCE	Energy-related Carbon Emission
FDEI	Final Demand Excluding Import
IPCC	Intergovernmental Panel on Climate Change
NBS	National Bureau of Statistics
pcsfai	Ratio of CSFAI to the total fixed assets investment
PEC	Proportion of ER in the TE caused by a unit of final demand increase of any specific sector
SDA	Structural Decomposition Analysis
TE	Sum of ER in each sector caused by a unit of final demand increase of any specific sector

Appendix A

The original input–output tables for 2007, 2010, and 2012 have 135, 42, and 139 sectors respectively. However, the sectors in the original input–output table do not match those in the energy balance table. Thus, in this study, the sectors were merged into 21 sectors to match the energy balance table, as shown in Table A1.

Table A1. Sector setting.

Sector (Full Name)	Sector (Abbreviation)	Code in This Paper	Code in I/O Table, 2007	Code in I/O Table, 2010	Code in I/O Table, 2012
Mining and Washing of Coal	Coal	01	006	02	06006
Extraction of Petroleum and Natural Gas	Petroleum and Natural Gas	02	007	03	07007
Processing of Petroleum, Coking, and Processing of Nuclear Fuel	Petroleum Processing	03	037 038	11	25039 25040
Production and Supply of Electric Power, Heat Power, and Gas	Electric Power	04	092 093	23 24	44096 45097
Agriculture, Forestry, Animal Husbandry, and Fishing	Agriculture	05	001 002 003 004	01	01001 02002 03003 04004
Smelting and Pressing of Ferrous Metals	Ferrous Metals	06	057 058 059 060	14	31059 31060 31061
			062		32063
Manufacture of Raw Chemical Materials and Chemical Products	Chemical	09	039 040 041 042 043 044 045 046 047 048 049	12	26041 26042 26043 26044 26045 26046 26047 27048 28049 29050 29051

Table A1. *Cont.*

Sector (Full Name)	Sector (Abbreviation)	Code in This Paper	Code in I/O Table, 2007	Code in I/O Table, 2010	Code in I/O Table, 2012
Non-Energy Mining	Mining of Non-energy	10	008 009 010	04 05	08008 09009 10010
Manufacture of Foods, Drinks, and Tobacco	Foods	11	011 012 013 014 015 016 017 018 019	06	13012 13013 13014 13015 13016 13017 13018 14019 14020
Manufacture of Machinery	Machinery	12	064 065 066 067 068 069 070 071 072	16	34065 34066 34067 34068 34069 34070 35071 35072 35073 35074
Manufacture of Automobiles, Railway, Ship, Aerospace, and Other Equipment	Automobiles	13	073 074 075 076	17	36075 36076 37077 37078 37079
Manufacture of Textile, Apparel, Accessories, Leather, Fur, Feather, and Related Products, and Footwear	Textile	14	025 026 027 028 029 030 031	07 08	17026 17027 17028 17029 17030 18031 19032 19033
Manufacture of Paper; Paper Products; Articles for Culture, Education, and Arts and Crafts; and Printing and Reproduction of Recording Media	Paper	15	034 035 036	10	22036 23037 24038
Wholesale, Retail Trade, and Hotel, Restaurants	Wholesale	20	108 109 110	30 31	51103 61112 62113
			122 123 124 125 126 127 128 129 130 131 132 133 134 135		77126 78127 79128 80129 82130 83131 84132 85133 86134 87135 88136 89137 93138 90139

Appendix B

The results of the ERCE embodied in the CSFAI consumed by each sector caused by a unit of final demand increase of a specific sector of China are provided in Table A2.

Table A2. The ERCE embodied in the CSFAI consumed by each sector caused by a unit of final demand increase in a specific sector in China.

2007–2008																					
Sector	**01**	**02**	**03**	**04**	**05**	**06**	**07**	**08**	**09**	**10**	**11**	**12**	**13**	**14**	**15**	**16**	**17**	**18**	**19**	**20**	**21**
Caused by sector 01	2.28	0.07	0.07	0.47	0.05	0.17	0.06	0.03	0.16	0.05	0.03	0.14	0.03	0.09	0.03	0.23	0.05	0.00	0.52	0.11	0.85
Caused by sector 02	0.05	1.64	0.07	0.34	0.03	0.15	0.04	0.02	0.14	0.04	0.02	0.12	0.03	0.08	0.03	0.16	0.03	0.00	0.27	0.08	0.57
Caused by sector 03	0.31	1.24	1.49	0.60	0.04	0.15	0.05	0.03	0.21	0.04	0.04	0.15	0.03	0.09	0.04	0.20	0.04	0.00	0.48	0.13	0.83
Caused by sector 04	0.44	0.09	0.07	4.94	0.03	0.08	0.05	0.02	0.11	0.03	0.03	0.07	0.03	0.08	0.03	0.26	0.03	0.00	0.28	0.08	0.87
Caused by sector 05	0.04	0.05	0.05	0.24	1.54	0.03	0.02	0.01	0.28	0.02	0.21	0.04	0.02	0.04	0.03	0.07	0.02	0.00	0.25	0.08	0.60
Caused by sector 06	0.00	0.00	0.00	0.00	0.00	0.00	0.00	0.00	0.00	0.00	0.00	0.00	0.00	0.00	0.00	0.00	0.00	0.00	0.00	0.00	0.00
Caused by sector 07	0.00	0.00	0.00	0.00	0.00	0.00	0.00	0.00	0.00	0.00	0.00	0.00	0.00	0.00	0.00	0.00	0.00	0.00	0.00	0.00	0.00
Caused by sector 08	0.00	0.00	0.00	0.00	0.00	0.00	0.00	0.00	0.00	0.00	0.00	0.00	0.00	0.00	0.00	0.00	0.00	0.00	0.00	0.00	0.00
Caused by sector 09	0.12	0.14	0.11	0.81	0.13	0.08	0.06	0.03	2.43	0.08	0.08	0.09	0.03	0.17	0.09	0.19	0.05	0.00	0.49	0.14	0.99
Caused by sector 10	0.07	0.07	0.07	0.51	0.04	0.09	0.06	0.04	0.28	1.48	0.03	0.16	0.04	0.10	0.04	0.22	0.04	0.00	0.55	0.11	0.76
Caused by sector 11	0.05	0.05	0.05	0.32	0.76	0.04	0.02	0.02	0.27	0.02	1.72	0.04	0.02	0.07	0.09	0.10	0.03	0.00	0.41	0.13	0.79
Caused by sector 12	0.09	0.13	0.15	0.57	0.06	0.47	0.20	0.04	0.31	0.14	0.05	1.59	0.05	0.13	0.06	0.53	0.09	0.00	0.53	0.18	1.01
Caused by sector 13	0.08	0.12	0.12	0.54	0.07	0.37	0.16	0.04	0.38	0.11	0.06	0.28	1.26	0.21	0.06	0.43	0.09	0.00	0.50	0.20	1.08
Caused by sector 14	0.08	0.06	0.06	0.58	0.33	0.05	0.03	0.02	0.53	0.03	0.14	0.07	0.02	6.32	0.09	0.15	0.04	0.00	0.45	0.14	1.12
Caused by sector 15	0.09	0.08	0.07	0.67	0.17	0.09	0.07	0.02	0.62	0.05	0.07	0.09	0.04	0.33	2.39	0.27	0.20	0.00	0.48	0.16	0.99
Caused by sector 16	0.09	0.10	0.11	0.61	0.08	0.28	0.32	0.05	0.42	0.13	0.06	0.15	0.04	0.19	0.08	2.23	0.09	0.00	0.50	0.18	1.14
Caused by sector 17	0.05	0.05	0.05	0.37	0.21	0.09	0.04	0.02	0.34	0.03	0.06	0.06	0.02	0.23	0.06	0.16	1.83	0.00	0.38	0.11	0.74
Caused by sector 18	0.09	0.13	0.14	0.49	0.06	0.36	0.10	0.29	0.30	0.14	0.05	0.14	0.04	0.13	0.06	0.35	0.09	0.29	0.84	0.17	1.02
Caused by sector 19	0.00	0.00	0.00	0.00	0.00	0.00	0.00	0.00	0.00	0.00	0.00	0.00	0.00	0.00	0.00	0.00	0.00	0.00	0.00	0.00	0.00
Caused by sector 20	0.04	0.05	0.05	0.24	0.17	0.03	0.02	0.01	0.14	0.02	0.22	0.04	0.03	0.12	0.06	0.11	0.03	0.00	0.53	1.90	1.23
Caused by sector 21	0.04	0.04	0.05	0.25	0.06	0.05	0.04	0.02	0.24	0.02	0.05	0.05	0.03	0.14	0.11	0.20	0.04	0.00	0.33	0.15	7.23
2008–2009																					
ERCE of Sector	**01**	**02**	**03**	**04**	**05**	**06**	**07**	**08**	**09**	**10**	**11**	**12**	**13**	**14**	**15**	**16**	**17**	**18**	**19**	**20**	**21**
Caused by sector 01	2.24	0.07	0.07	0.41	0.06	0.17	0.06	0.03	0.17	0.05	0.03	0.15	0.03	0.09	0.03	0.23	0.05	0.00	0.51	0.11	0.84
Caused by sector 02	0.05	1.83	0.07	0.34	0.03	0.16	0.04	0.02	0.16	0.05	0.03	0.14	0.03	0.08	0.03	0.17	0.03	0.00	0.28	0.09	0.59
Caused by sector 03	0.34	1.38	1.62	0.58	0.05	0.16	0.05	0.03	0.22	0.05	0.05	0.16	0.04	0.10	0.04	0.21	0.04	0.00	0.47	0.13	0.86
Caused by sector 04	0.49	0.11	0.08	4.96	0.04	0.09	0.05	0.02	0.13	0.03	0.03	0.08	0.03	0.08	0.03	0.28	0.03	0.00	0.29	0.09	0.94
Caused by sector 05	0.04	0.05	0.05	0.23	1.92	0.03	0.02	0.01	0.30	0.02	0.23	0.04	0.02	0.05	0.03	0.07	0.02	0.00	0.27	0.09	0.60
Caused by sector 06	0.00	0.00	0.00	0.00	0.00	0.00	0.00	0.00	0.00	0.00	0.00	0.00	0.00	0.00	0.00	0.00	0.00	0.00	0.00	0.00	0.00
Caused by sector 07	0.00	0.00	0.00	0.00	0.00	0.00	0.00	0.00	0.00	0.00	0.00	0.00	0.00	0.00	0.00	0.00	0.00	0.00	0.00	0.00	0.00
Caused by sector 08	0.00	0.00	0.00	0.00	0.00	0.00	0.00	0.00	0.00	0.00	0.00	0.00	0.00	0.00	0.00	0.00	0.00	0.00	0.00	0.00	0.00
Caused by sector 09	0.12	0.15	0.12	0.77	0.17	0.08	0.06	0.03	2.61	0.09	0.09	0.10	0.03	0.18	0.09	0.21	0.05	0.00	0.52	0.16	1.08
Caused by sector 10	0.07	0.07	0.08	0.49	0.05	0.11	0.06	0.04	0.30	1.62	0.04	0.19	0.04	0.10	0.04	0.24	0.04	0.00	0.56	0.13	0.81
Caused by sector 11	0.05	0.05	0.05	0.31	0.97	0.04	0.03	0.02	0.29	0.02	1.86	0.05	0.02	0.08	0.10	0.11	0.03	0.00	0.44	0.15	0.85
Caused by sector 12	0.10	0.15	0.16	0.57	0.08	0.52	0.22	0.05	0.33	0.16	0.06	1.70	0.06	0.14	0.06	0.58	0.08	0.00	0.56	0.20	1.11
Caused by sector 13	0.09	0.13	0.13	0.53	0.09	0.40	0.18	0.04	0.41	0.13	0.06	0.30	1.35	0.22	0.06	0.46	0.08	0.00	0.53	0.22	1.17
Caused by sector 14	0.08	0.07	0.06	0.56	0.44	0.06	0.04	0.02	0.56	0.03	0.16	0.08	0.03	6.76	0.10	0.16	0.04	0.00	0.47	0.15	1.14
Caused by sector 15	0.11	0.09	0.08	0.67	0.22	0.11	0.08	0.03	0.67	0.06	0.08	0.10	0.04	0.36	2.62	0.35	0.19	0.00	0.52	0.18	1.10
Caused by sector 16	0.09	0.11	0.11	0.60	0.11	0.30	0.34	0.06	0.44	0.15	0.07	0.16	0.04	0.20	0.09	2.38	0.08	0.00	0.52	0.20	1.23
Caused by sector 17	0.06	0.06	0.06	0.41	0.30	0.10	0.05	0.03	0.40	0.04	0.07	0.07	0.03	0.27	0.08	0.19	2.26	0.00	0.44	0.13	0.88
Caused by sector 18	0.10	0.14	0.15	0.47	0.08	0.37	0.10	0.32	0.32	0.15	0.06	0.15	0.04	0.14	0.06	0.37	0.09	0.29	0.89	0.19	1.09

Table A2. *Cont.*

	01	02	03	04	05	06	07	08	09	10	11	12	13	14	15	16	17	18	19	20	21
Caused by sector 19	0.00	0.00	0.00	0.00	0.00	0.00	0.00	0.00	0.00	0.00	0.00	0.00	0.00	0.00	0.00	0.00	0.00	0.00	0.00	0.00	0.00
Caused by sector 20	0.04	0.05	0.05	0.23	0.22	0.03	0.02	0.01	0.14	0.02	0.24	0.04	0.03	0.12	0.06	0.10	0.03	0.00	0.51	2.06	1.20
Caused by sector 21	0.04	0.05	0.05	0.24	0.07	0.05	0.04	0.02	0.26	0.03	0.05	0.06	0.03	0.15	0.11	0.21	0.04	0.00	0.35	0.16	7.62
2009–2010																					
Sector	**01**	**02**	**03**	**04**	**05**	**06**	**07**	**08**	**09**	**10**	**11**	**12**	**13**	**14**	**15**	**16**	**17**	**18**	**19**	**20**	**21**
Caused by sector 01	1.66	0.06	0.06	0.34	0.05	0.14	0.05	0.02	0.14	0.04	0.03	0.12	0.03	0.07	0.03	0.19	0.04	0.00	0.50	0.10	0.71
Caused by sector 02	0.04	1.49	0.06	0.30	0.03	0.13	0.04	0.02	0.14	0.04	0.02	0.12	0.02	0.07	0.02	0.14	0.02	0.00	0.28	0.08	0.52
Caused by sector 03	0.27	1.12	1.38	0.51	0.05	0.13	0.04	0.03	0.19	0.04	0.04	0.14	0.03	0.08	0.03	0.17	0.03	0.00	0.46	0.12	0.74
Caused by sector 04	0.39	0.09	0.07	4.50	0.04	0.07	0.04	0.02	0.11	0.03	0.03	0.07	0.03	0.07	0.03	0.23	0.02	0.00	0.30	0.08	0.85
Caused by sector 05	0.03	0.04	0.04	0.19	1.77	0.02	0.01	0.01	0.25	0.01	0.20	0.03	0.01	0.04	0.03	0.06	0.01	0.00	0.27	0.08	0.50
Caused by sector 06	0.00	0.00	0.00	0.00	0.00	0.00	0.00	0.00	0.00	0.00	0.00	0.00	0.00	0.00	0.00	0.00	0.00	0.00	0.00	0.00	0.00
Caused by sector 07	0.00	0.00	0.00	0.00	0.00	0.00	0.00	0.00	0.00	0.00	0.00	0.00	0.00	0.00	0.00	0.00	0.00	0.00	0.00	0.00	0.00
Caused by sector 08	0.00	0.00	0.00	0.00	0.00	0.00	0.00	0.00	0.00	0.00	0.00	0.00	0.00	0.00	0.00	0.00	0.00	0.00	0.00	0.00	0.00
Caused by sector 09	0.10	0.11	0.10	0.67	0.17	0.07	0.06	0.03	2.19	0.07	0.08	0.09	0.03	0.15	0.08	0.18	0.03	0.00	0.53	0.15	0.98
Caused by sector 10	0.06	0.06	0.06	0.42	0.05	0.10	0.06	0.04	0.25	1.30	0.03	0.17	0.04	0.09	0.03	0.21	0.03	0.00	0.56	0.12	0.71
Caused by sector 11	0.04	0.04	0.04	0.27	0.91	0.03	0.02	0.02	0.25	0.02	1.57	0.04	0.02	0.06	0.08	0.10	0.02	0.00	0.45	0.14	0.75
Caused by sector 12	0.08	0.12	0.13	0.51	0.07	0.45	0.19	0.04	0.28	0.14	0.05	1.43	0.05	0.12	0.05	0.49	0.05	0.00	0.58	0.19	1.00
Caused by sector 13	0.07	0.10	0.11	0.48	0.08	0.34	0.15	0.04	0.34	0.11	0.05	0.25	1.14	0.19	0.05	0.39	0.06	0.00	0.54	0.21	1.06
Caused by sector 14	0.06	0.05	0.05	0.49	0.43	0.05	0.03	0.02	0.46	0.03	0.14	0.06	0.02	5.67	0.08	0.14	0.03	0.00	0.48	0.14	0.97
Caused by sector 15	0.09	0.07	0.07	0.60	0.21	0.10	0.08	0.02	0.56	0.05	0.07	0.09	0.04	0.30	2.25	0.35	0.12	0.00	0.55	0.17	1.01
Caused by sector 16	0.07	0.09	0.09	0.53	0.10	0.25	0.28	0.05	0.37	0.13	0.06	0.14	0.03	0.17	0.07	1.99	0.05	0.00	0.53	0.19	1.11
Caused by sector 17	0.06	0.06	0.06	0.41	0.31	0.10	0.04	0.03	0.38	0.04	0.07	0.07	0.03	0.26	0.08	0.19	2.21	0.00	0.51	0.14	0.89
Caused by sector 18	0.08	0.12	0.13	0.41	0.08	0.30	0.09	0.27	0.26	0.12	0.05	0.12	0.03	0.11	0.05	0.31	0.07	0.25	0.92	0.17	0.97
Caused by sector 19	0.00	0.00	0.00	0.00	0.00	0.00	0.00	0.00	0.00	0.00	0.00	0.00	0.00	0.00	0.00	0.00	0.00	0.00	0.00	0.00	0.00
Caused by sector 20	0.03	0.04	0.04	0.20	0.20	0.03	0.02	0.01	0.11	0.01	0.21	0.03	0.02	0.09	0.05	0.08	0.02	0.00	0.47	1.87	0.98
Caused by sector 21	0.03	0.04	0.04	0.21	0.06	0.04	0.03	0.01	0.22	0.02	0.04	0.04	0.03	0.12	0.09	0.17	0.03	0.00	0.36	0.15	6.68
2010–2011																					
ERCE of Sector	**01**	**02**	**03**	**04**	**05**	**06**	**07**	**08**	**09**	**10**	**11**	**12**	**13**	**14**	**15**	**16**	**17**	**18**	**19**	**20**	**21**
Caused by sector 01	2.52	0.05	0.06	0.27	0.04	0.15	0.02	0.03	0.14	0.06	0.04	0.13	0.07	0.10	0.01	0.30	0.01	0.00	0.32	0.05	0.45
Caused by sector 02	0.12	1.44	0.07	0.27	0.03	0.14	0.02	0.03	0.15	0.06	0.03	0.14	0.06	0.10	0.01	0.24	0.01	0.00	0.20	0.04	0.43
Caused by sector 03	0.00	0.00	0.00	0.00	0.00	0.00	0.00	0.00	0.00	0.00	0.00	0.00	0.00	0.00	0.00	0.00	0.00	0.00	0.00	0.00	0.00
Caused by sector 04	0.65	0.09	0.09	4.07	0.03	0.08	0.02	0.02	0.11	0.04	0.04	0.07	0.07	0.10	0.01	0.41	0.01	0.00	0.22	0.05	0.54
Caused by sector 05	0.05	0.03	0.05	0.17	1.44	0.03	0.01	0.01	0.29	0.02	0.29	0.04	0.04	0.06	0.01	0.10	0.00	0.00	0.20	0.05	0.31
Caused by sector 06	0.00	0.00	0.00	0.00	0.00	0.00	0.00	0.00	0.00	0.00	0.00	0.00	0.00	0.00	0.00	0.00	0.00	0.00	0.00	0.00	0.00
Caused by sector 07	0.18	0.08	0.10	0.88	0.05	0.12	0.83	0.07	0.28	0.66	0.06	0.14	0.09	0.15	0.02	0.45	0.03	0.00	0.38	0.08	0.66
Caused by sector 08	0.23	0.10	0.13	0.53	0.06	0.14	0.03	1.90	0.38	0.25	0.07	0.14	0.10	0.19	0.04	0.45	0.02	0.00	0.48	0.09	0.62
Caused by sector 09	0.16	0.11	0.12	0.60	0.14	0.08	0.03	0.05	2.48	0.11	0.13	0.10	0.08	0.26	0.03	0.31	0.01	0.00	0.41	0.09	0.65
Caused by sector 10	0.00	0.00	0.00	0.00	0.00	0.00	0.00	0.00	0.00	0.00	0.00	0.00	0.00	0.00	0.00	0.00	0.00	0.00	0.00	0.00	0.00
Caused by sector 11	0.07	0.04	0.05	0.23	0.73	0.04	0.01	0.03	0.27	0.03	2.24	0.04	0.05	0.10	0.03	0.16	0.01	0.00	0.35	0.09	0.48
Caused by sector 12	0.13	0.11	0.16	0.46	0.06	0.52	0.10	0.07	0.32	0.23	0.07	1.79	0.15	0.18	0.02	0.92	0.02	0.00	0.45	0.11	0.70

Table A2. *Cont.*

Caused by sector 13	0.12	0.10	0.13	0.43	0.07	0.38	0.08	0.07	0.38	0.18	0.08	0.28	3.83	0.30	0.02	0.72	0.02	0.00	0.43	0.12	0.73
Caused by sector 14	0.10	0.05	0.06	0.44	0.34	0.05	0.01	0.03	0.51	0.04	0.20	0.07	0.06	9.07	0.03	0.23	0.01	0.00	0.37	0.10	0.63
Caused by sector 15	0.14	0.07	0.08	0.53	0.17	0.10	0.05	0.04	0.62	0.09	0.10	0.10	0.10	0.54	0.92	0.57	0.04	0.00	0.42	0.10	0.67
Caused by sector 16	0.00	0.00	0.00	0.00	0.00	0.00	0.00	0.00	0.00	0.00	0.00	0.00	0.00	0.00	0.00	0.00	0.00	0.00	0.00	0.00	0.00
Caused by sector 17	0.09	0.06	0.07	0.38	0.25	0.11	0.02	0.04	0.43	0.06	0.11	0.08	0.08	0.41	0.03	0.36	0.82	0.00	0.40	0.08	0.59
Caused by sector 18	0.13	0.11	0.15	0.37	0.06	0.37	0.04	0.45	0.30	0.20	0.07	0.14	0.10	0.18	0.02	0.57	0.03	0.31	0.64	0.09	0.69
Caused by sector 19	0.11	0.20	0.28	0.24	0.07	0.09	0.02	0.03	0.18	0.05	0.07	0.10	0.34	0.17	0.02	0.25	0.01	0.00	4.76	0.08	0.77
Caused by sector 20	0.05	0.03	0.04	0.17	0.15	0.03	0.01	0.01	0.12	0.02	0.27	0.03	0.06	0.13	0.02	0.14	0.01	0.00	0.32	1.04	0.66
Caused by sector 21	0.05	0.04	0.05	0.19	0.05	0.05	0.02	0.02	0.24	0.04	0.07	0.05	0.09	0.19	0.04	0.32	0.01	0.00	0.28	0.08	4.21

2011–2012

ERCE of Sector	01	02	03	04	05	06	07	08	09	10	11	12	13	14	15	16	17	18	19	20	21
Caused by sector 01	3.11	0.05	0.08	0.29	0.04	0.18	0.02	0.04	0.17	0.07	0.05	0.14	0.06	0.12	0.01	0.34	0.02	0.00	0.27	0.06	0.70
Caused by sector 02	0.26	1.73	0.07	0.28	0.03	0.16	0.02	0.03	0.18	0.06	0.05	0.14	0.06	0.11	0.01	0.28	0.01	0.00	0.20	0.06	0.84
Caused by sector 03	0.00	0.00	0.00	0.00	0.00	0.00	0.00	0.00	0.00	0.00	0.00	0.00	0.00	0.00	0.00	0.00	0.00	0.00	0.00	0.00	0.00
Caused by sector 04	0.81	0.11	0.11	4.52	0.03	0.10	0.02	0.03	0.14	0.05	0.05	0.08	0.06	0.10	0.01	0.53	0.01	0.00	0.22	0.06	0.75
Caused by sector 05	0.06	0.04	0.06	0.17	1.77	0.03	0.01	0.01	0.38	0.02	0.38	0.04	0.04	0.07	0.01	0.10	0.00	0.00	0.19	0.06	0.42
Caused by sector 06	0.00	0.00	0.00	0.00	0.00	0.00	0.00	0.00	0.00	0.00	0.00	0.00	0.00	0.00	0.00	0.00	0.00	0.00	0.00	0.00	0.00
Caused by sector 07	0.22	0.09	0.12	0.92	0.07	0.13	1.09	0.09	0.36	0.71	0.08	0.12	0.08	0.19	0.02	0.41	0.08	0.00	0.37	0.09	0.92
Caused by sector 08	0.27	0.12	0.16	0.56	0.07	0.16	0.04	2.55	0.48	0.32	0.09	0.16	0.11	0.26	0.04	0.49	0.02	0.00	0.48	0.11	0.88
Caused by sector 09	0.20	0.13	0.16	0.62	0.19	0.08	0.03	0.05	3.21	0.11	0.19	0.10	0.07	0.37	0.03	0.32	0.01	0.00	0.43	0.13	0.91
Caused by sector 10	0.00	0.00	0.00	0.00	0.00	0.00	0.00	0.00	0.00	0.00	0.00	0.00	0.00	0.00	0.00	0.00	0.00	0.00	0.00	0.00	0.00
Caused by sector 11	0.07	0.05	0.06	0.23	0.87	0.04	0.01	0.03	0.32	0.03	2.86	0.05	0.05	0.12	0.03	0.16	0.01	0.00	0.36	0.13	0.62
Caused by sector 12	0.15	0.13	0.19	0.47	0.07	0.61	0.13	0.08	0.41	0.26	0.10	2.40	0.19	0.23	0.03	1.19	0.03	0.00	0.47	0.15	1.05
Caused by sector 13	0.14	0.11	0.16	0.44	0.08	0.45	0.12	0.09	0.47	0.21	0.10	0.30	5.17	0.37	0.02	0.89	0.03	0.00	0.48	0.17	1.06
Caused by sector 14	0.12	0.06	0.08	0.46	0.39	0.05	0.02	0.03	0.64	0.04	0.24	0.07	0.06	11.62	0.03	0.24	0.01	0.00	0.38	0.18	0.90
Caused by sector 15	0.15	0.07	0.10	0.53	0.21	0.10	0.10	0.04	0.74	0.11	0.13	0.09	0.08	0.80	1.18	0.45	0.07	0.00	0.43	0.14	0.92
Caused by sector 16	0.00	0.00	0.00	0.00	0.00	0.00	0.00	0.00	0.00	0.00	0.00	0.00	0.00	0.00	0.00	0.00	0.00	0.00	0.00	0.00	0.00
Caused by sector 17	0.10	0.06	0.08	0.35	0.28	0.10	0.03	0.05	0.49	0.06	0.13	0.09	0.07	0.48	0.03	0.41	1.00	0.00	0.38	0.11	0.74
Caused by sector 18	0.16	0.12	0.19	0.39	0.08	0.50	0.06	0.56	0.40	0.24	0.09	0.14	0.10	0.23	0.03	0.71	0.04	0.41	0.52	0.12	1.06
Caused by sector 19	0.14	0.24	0.38	0.26	0.08	0.11	0.02	0.03	0.22	0.05	0.10	0.11	0.46	0.18	0.02	0.30	0.01	0.01	5.18	0.11	1.18
Caused by sector 20	0.05	0.03	0.04	0.16	0.15	0.03	0.01	0.01	0.13	0.02	0.29	0.03	0.06	0.13	0.02	0.18	0.01	0.00	0.29	1.38	0.99
Caused by sector 21	0.06	0.04	0.06	0.19	0.07	0.06	0.02	0.03	0.28	0.04	0.09	0.05	0.11	0.24	0.05	0.42	0.01	0.01	0.29	0.11	5.47

References

1. Intergovernmental Panel on Climate Change. Climate Change 2018: Global Warming of 1.5 °C. Available online: https://www.ipcc.ch/sr15/ (accessed on 28 March 2019).
2. Intergovernmental Panel on Climate Change. Climate Change 2014: Mitigation of Climate Change. Available online: https://www.ipcc.ch/report/ar5/wg3/ (accessed on 29 December 2018).
3. Ma, L.; Liu, P.; Fu, F.; Li, Z.; Ni, W. Integrated energy strategy for the sustainable development of China. *Energy* **2010**, *36*, 1143–1154. [CrossRef]
4. Fu, F.; Pan, L.Y.; Ma, L.W.; Li, Z. A simplified method to estimate the energy–saving potentials of frequent construction and demolition process in China. *Energy* **2013**, *49*, 316–322. [CrossRef]
5. Fu, F.; Ma, L.; Li, Z.; Polenske, K.R. The implications of China's investment–driven economy on its energy consumption and carbon emissions. *Energy Convers. Manag.* **2014**, *85*, 573–580. [CrossRef]
6. Acquaye, A.A.; Duffy, A.P. Input–output analysis of Irish construction sector greenhouse gas emissions. *Build. Environ.* **2010**, *45*, 784–791. [CrossRef]
7. Liu, Z.; Geng, Y.; Lindner, S.; Zhao, H.; Fujita, T.; Guan, D. Embodied energy use in China's industrial sectors. *Energy Policy* **2012**, *49*, 751–758. [CrossRef]
8. Huang, L.; Krigsvoll, G.; Johansen, F.; Liu, Y.; Zhang, X. Carbon emission of global construction sector. *Renew. Sustain. Energy Rev.* **2018**, *81*, 1906–1916. [CrossRef]
9. Holz, C.A. New capital estimates for China. *China Econ. Rev.* **2006**, *17*, 142–185. [CrossRef]
10. Bullard, C.W.I.; Herendeen, R.A. Energy Impact of Consumption Decisions. *Proc. IEEE* **1975**, *63*, 484–493. [CrossRef]
11. Leontief, W. The Dynamic Inverse. In *Contributions to Input-Output Analysis*; Carter, A.P., Brody, A., Eds.; North Holland: Amsterdam, The Netherlands, 1970.
12. Miller, R.E.; Blair, P.D. *Input-Output Analysis: Foundations and Extensions*; Cambridge University Press: New York, NY, USA, 2009.
13. Pauliuk, S.; Wood, R.; Hertwich, E.G. Dynamic Models of Fixed Capital Stocks and Their Application in Industrial Ecology. *J. Ind. Ecol.* **2015**, *19*, 104–116. [CrossRef]
14. Rhoten, R.P. Dynamic input-output analysis of the economics of energy. In Proceedings of the Energy '78 IEEE 1978 Region V Annual Conference, Tulsa, OK, USA, 16–18 April 1978.
15. Tamura, H.; Ito, K. Application of dynamic input–output analysis to evaluating trade–offs between environmental pollution and energy consumption. *Syst. Control* **1984**, *28*, 529–535.
16. Song, J.; Yang, W.; Higano, Y.; Wang, X.E. Dynamic integrated assessment of bioenergy technologies for energy production utilizing agricultural residues: An input–output approach. *Appl. Energy* **2015**, *158*, 178–189. [CrossRef]
17. Yu, S.; Zheng, S.; Ba, G.; Wei, Y.M. Can China realise its energy-savings goal by adjusting its industrial structure? *Econ. Syst. Res.* **2016**, *28*, 273–293. [CrossRef]
18. Pan, L.; Liu, P.; Li, Z.; Wang, Y. A dynamic input–output method for energy system modeling and analysis. *Chem. Eng. Res. Des.* **2018**, *131*, 183–192. [CrossRef]
19. Dobos, I.; Tallos, P. A dynamic input-output model with renewable resources. *Cent. Eur. J. Oper. Res.* **2013**, *21*, 295–305. [CrossRef]
20. Cruz, J.B., Jr.; Tan, R.R.; Culaba, A.B.; Ballacillo, J.A. A dynamic input-output model for nascent bioenergy supply chains. *Appl. Energy* **2009**, *861*, S86–S94. [CrossRef]
21. Nie, H.; Kemp, R.; Vivanco, D.F. Véronique Vasseur. Structural decomposition analysis of energy–related CO_2 emissions in China from 1997 to 2010. *Energy Effic.* **2016**, *9*, 1351–1367. [CrossRef]
22. Wang, C.; Wang, F.; Zhang, X.; Zhang, H. Influencing mechanism of energy–related carbon emissions in Xinjiang based on the input–output and structural decomposition analysis. *J. Geogr. Sci.* **2017**, *27*, 365–384. [CrossRef]
23. Wood, R. Structural decomposition analysis of Australia's greenhouse gas emissions. *Energy Policy* **2009**, *37*, 4943–4948. [CrossRef]
24. Chenery, H.; Shishido, S.; Watanabe, T. The pattern of Japanese Growth, 1914–1954. *Ecomometrica* **1962**, *30*, 98–139. [CrossRef]
25. Gould, B.; Kulshreshtha, S. An interindustry analysis of structural change and energy use linkage in the Saskatchewan economy. *Energy Econ.* **1986**, *8*, 186–196. [CrossRef]

26. Rose, A.; Chen, C.Y. Sources of change in energy use in the US economy, 1972–1982. *Resour. Energy* **1991**, *13*, 1–21. [CrossRef]
27. Lee, C.F.; Lin, S.J. Structural decomposition of CO_2 emissions from Taiwan's petrochemical industries. *Energy Policy* **2001**, *29*, 237–244. [CrossRef]
28. Zhang, Y. Structural decomposition analysis of sources of decarbonizing economic development in China; 1992–2006. *Ecol. Econ.* **2009**, *68*, 2399–2405. [CrossRef]
29. Butnar, I.; Llop, M. Structural decomposition analysis and input–output subsystems: Changes in CO_2 emissions of Spanish service sectors (2000–2005). *Ecol. Econ.* **2011**, *70*, 2012–2019. [CrossRef]
30. Yamakawa, A.; Peters, G.P. Structural Decomposition Analysis of Greenhouse Gas Emissions in Norway 1990–2002. *Econ. Syst. Res.* **2011**, *23*, 303–318. [CrossRef]
31. Tian, X.; Chang, M.; Tanikawa, H.; Shi, F.; Imura, H. Structural decomposition analysis of the carbonization process in Beijing: A regional explanation of rapid increasing carbon dioxide emission in China. *Energy Policy* **2013**, *53*, 279–286. [CrossRef]
32. Intergovernmental Panel on Climate Change. *2006 IPCC Guidelines for National Greenhouse Gas Inventories*; IGES: Prefecture, Japan, 2006.
33. State Council of People's Republic of China. Measures for Expanding Domestic Demand and Promoting Economic Growth Deployed at the Standing Meeting of the State Council. 9 November 2008. Available online: http://www.gov.cn/ldhd/2008-11/09/content_1143689.htm (accessed on 28 February 2019).
34. National Bureau of Statistics. The Input-Output Table of CHINA 2007. Available online: http://data.stats.gov.cn/ifnormal.htm?u=/files/html/quickSearch/trcc/trcc01.html&h=740 (accessed on 28 February 2019).
35. National Bureau of Statistics. The Input-Output Table of CHINA 2010. Available online: http://data.stats.gov.cn/ifnormal.htm?u=/files/html/quickSearch/trcc/trcc01.html&h=740 (accessed on 28 February 2019).
36. National Bureau of Statistics. The Input-Output Table of CHINA 2012. Available online: http://data.stats.gov.cn/ifnormal.htm?u=/files/html/quickSearch/trcc/trcc01.html&h=740 (accessed on 28 February 2019).
37. National Bureau of Statistics. *China Energy Statistical Yearbook 2013*; China Statistics Press: Beijing, China, 2014.
38. National Bureau of Statistics. *China Statistical Yearbook 2016*; China Statistics Press: Beijing, China, 2017.
39. Zhang, X.; Li, Z.; Ma, L.; Chong, C.; Ni, W. Forecasting the Energy Embodied in Construction Services Based on a Combination of Static and Dynamic Hybrid Input-Output Models. *Energies* **2019**, *12*, 300. [CrossRef]
40. Hu, J. Hu Jintao's Report at the Eighteenth National Congress of the Communist Party of China. People's Daily. 18 November 2012. Available online: http://politics.people.com.cn/n/2012/1118/c1001-19612670.html (accessed on 28 February 2019).
41. State Council of People's Republic of China. About the Steel Industry to Resolve Excess Production Capacity. 4 February 2016. Available online: http://www.gov.cn/zhengce/content/2016-02/04/content_5039353.htm (accessed on 28 February 2019).
42. State Council of People's Republic of China. Opinions on Enhancing the Protection of Ecological Environment in an All-Round Way and Fighting the Strong Battle of Pollution Prevention and Control. 24 June 2018. Available online: http://www.gov.cn/zhengce/2018-06/24/content_5300953.htm (accessed on 28 February 2019).

energies

MDPI

Article

Analysing the Energy Efficiency of EU Member States: The Potential of Energy Recovery from Waste in the Circular Economy

George Halkos * and Kleoniki Natalia Petrou

Laboratory of Operations Research, Department of Economics, University of Thessaly, 28is Octovriou, 33888 Volos, Greece; kleoniki.petrou12@imperial.ac.uk

* Correspondence: halkos@econ.uth.gr

Received: 29 August 2019; Accepted: 27 September 2019; Published: 28 September 2019

Abstract: This paper examines energy efficiency across 28 selected European Union (EU) Member States and reviews the potential for energy recovery from waste according to the efficiency scores obtained. The efficiencies are assessed through data envelopment analysis (DEA) and the following variables are used, inputs: final energy consumption, labour, capital, population density and outputs: gross domestic product (GDP), nitrogen oxide (NOx) emissions, sulphur oxide (SOx) emissions and greenhouse gas (GHG) emissions for the years 2008, 2010, 2012, 2014 and 2016. Results show that most countries maintain their efficiency scores with only a few marginally improving theirs and at the same time, it is noticed that most are decreasing after 2012. Based on these efficiency scores, this paper recommends moving towards waste-to-energy with two main objectives, namely sufficient and sustainable energy production and effective treatment of municipal solid waste (MSW). This option would enhance the circular economy, whereas prioritization needs to be given to prevention, preparation for reuse, recycling and energy recovery through to disposal. Together with the EU Commission's competition strategy, these options would ensure reliable energy supplies at rational prices and with the least environmental impacts. Moreover the efficiency scores need to be examined along the financial crisis which has been affecting the EU since 2008, showing a decrease in those efficiency scores after 2012 under a more imminent crisis.

Keywords: waste; energy; environmental efficiency; energy recovery; data envelopment analysis; circular economy

1. Introduction

One of the major challenges of the 21st century is the continuous increase of municipal solid waste (MSW) production as well as its management. According to the European Union (EU) Waste Framework Directive (WFD) 2008/98/EC, 'any substance or object which the holder discards or intends or is required to discard' is defined as waste. The main treatment options for MSW include, among others, landfill, incineration, recycling and composting. Both developed and developing countries have been dealing with the issue of sustainable waste management and are investigating ways to meet national and international standards in order to reduce their overall environmental impact. The main issues that have been pestering developed countries are potential ways to decrease the amount of waste going to landfill and increase the recycling and recovery of materials. The Waste Hierarchy (Figure 1) has been affecting countries' management options as it gives priority to preventing waste, but even if and when it is created, it should be prepared for reuse, recycling and energy recovering and only disposed to landfill if no other option is possible [1].

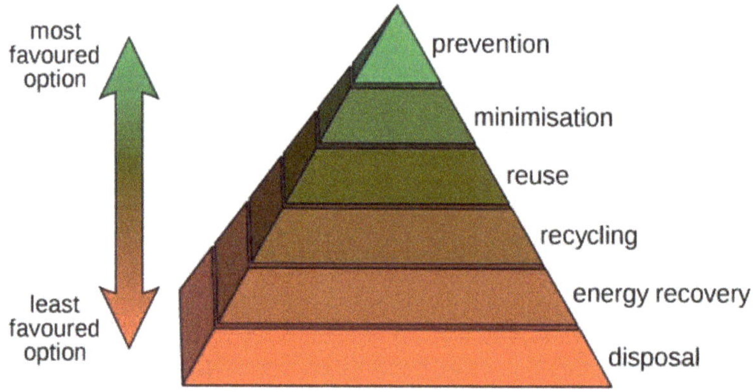

Figure 1. Waste hierarchy [1].

To date many EU Member States have failed to implement waste prevention practices and therefore the regulations that have been set out by WFD [2]. In general Southern and Eastern Europe countries are shown to have the largest implementation gaps regarding their waste management systems [2]. Figure 2 shows EU Member States that have been performing above (blue), below (purple) and average (green) regarding their waste management.

Figure 2. Waste management performance across Europe [2].

A significant part of the Europe 2020 growth strategy has been sustainable growth towards a 'smart, sustainable and inclusive economy' under the notion of the circular economy, while achieving lower greenhouse gas emissions by 20% compared to levels of 1990, generating 20% of its energy from renewable sources and to increase energy efficiency by 20% [3]. These measures could bring net savings to EU Member States, while increasing resource productivity by 30% by 2030, enhancing Gross Domestic Product (GDP) by nearly 1% and creating 2 M additional jobs while also reducing EU carbon emissions by 450 Mt by 2030 [4]. The framework of measures for the promotion of energy efficiency is set out by Directive 2012/27/EU of the European Parliament and of the Council of 25 October 2012 on energy efficiency addressing the achievement of the 20% target on energy efficiency in 2020.

In addition to those, the 2030 climate and energy framework covers EU-wide targets and policy objectives for the period 2021 to 2030, with the main targets being: at least 40% cuts in greenhouse gas (GHG) emissions (from 1990 levels), at least 32% share for renewable energy and at least 32.5% improvement in energy efficiency [5]. Moreover the 2050 EU long-term strategy stresses the opportunities that a climate neutral Europe may bring as well as challenges that may appear, without revising the 2030 targets nor launching new policies [6]. Overall this strategy is meant to provide a framework for the EU to achieve the Paris Agreement objectives and tackle climate change by limiting global warming to below 2 °C and attempting to limit it to 1.5 °C [6].

Generally it is noticed that the global economy is highly reliant on fossil fuels such as oil, gas and coal, resulting in higher GHG emissions [7,8]. Due to the volatile price of oil and the environmental degradation occurring because of fossil fuels' use, a turn towards renewable energy sources has been noticed [9]. Along those lines the public has become more sensitive to environmental issues, therefore most countries will be forced to make real changes in their energy mix [10].

Energy efficiency improvement can provide many benefits apart from cost efficiency such as energy savings, air pollution control and GHG emission reduction as well as energy security and health benefits [11,12]. It is essential to combine technological options and implementation approaches to improve energy recovery efficiency of the urban and industrial system and achieve low-carbon cities [13]. In those regards, the development of advanced computational techniques has enabled the evaluation of energy efficiency [14].

Such a tool is data envelopment analysis (DEA) which has been accepted throughout the academic community as a useful benchmarking technique [14]. DEA is a non-parametric linear programming method used to measure the efficiency of selected decision making units (DMUs) [15]. Initially it was intended to be applied in microeconomic studies, but comes handy in macroeconomic analysis too [16].

In the present paper DEA was used at a macroeconomic level, to evaluate energy efficiency in 28 selected EU Member States with the aim to identify the current levels of efficiency as well as to assess the potential of using MSW to regain energy and ensure reliable supplies for all at reasonable prices with the least potential impacts taking the financial crisis into account too. The existing literature shows that there is a major gap in current research as researchers have not attempted to evaluate energy efficiency among EU Member States in order to understand what this means and its potential implications for the MSW sector especially under the circular economy concept. This study therefore aims to also provide EU energy efficiency levels that could act as an incentive to move more towards energy recovery from waste and realise a circular economy in full.

Apart from this Introduction, the rest of the paper is structured as presented below. Section 2 provides the background research on this topic by reviewing the main elements of energy recovery from waste (Section 2.1) as well as the relevant existing DEA studies (Section 2.2) with Section 3 showing the proposed methodology along with the data used. Section 4 presents the empirical findings while Section 5 analyses the results and their implications. Finally the last section (Section 6) concludes the paper.

2. Background

This section provides some main points to introduce the topic of energy recovery from waste and the various options with which this can be made (Section 2.1). Then Section 2.2 reviews some of the studies that have used DEA in evaluating energy efficiency, leading to the methodology part in the following section, Section 3.

2.1. Energy Recovery from Waste

As already mentioned, the burning of waste for recovering energy is called incineration and it happens under a high temperature, therefore it is also called thermal treatment [17]. MSW can act as a source of energy through waste incineration; for instance in Denmark waste incineration covers approximately 5% of the electricity demand and 20% of the district heating demand [8]. With the use of incineration wastes' form is reduced from 95 to 96%, depending how many materials can be recovered as well as their composition; therefore incineration does not achieve the omission of landfill completely but reduces the amount of waste disposed that way [17]. Figure 3 presents the main inputs and outputs from incineration.

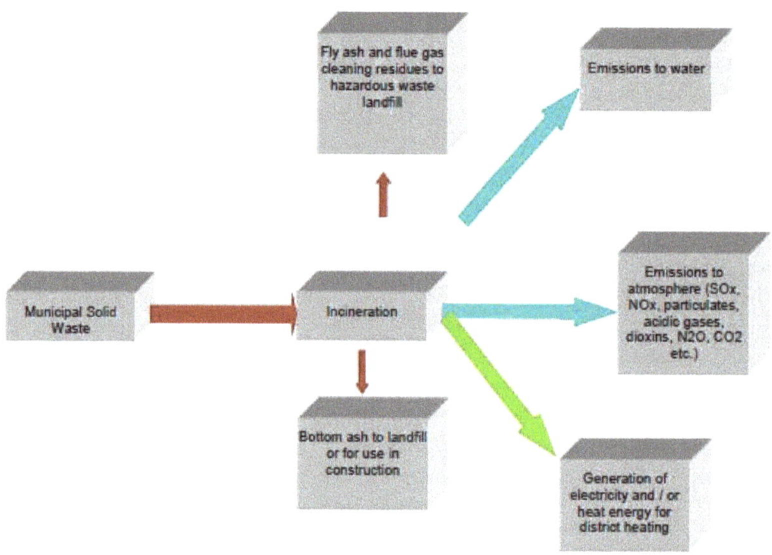

Figure 3. Schematic representation of incineration inputs and outputs [18].

In 2009 there were 449 Incineration plants across 20 Western and Central European countries with a total throughput of around 69.4 Mt [19]. In 2016 there were 512 plants in Europe alone, providing a total incineration capacity of 93 Mt [20]. In many countries such as Germany and Japan, incinerators are widely used to treat both MSW and industrial waste [21]. Incineration has been raising a lot of controversy regarding its potential use. Generally public disagreement can affect political willingness to support incineration, which has been the case especially for Spain and Greece [22].

Some other relatively new technologies include pyrolysis and gasification but these have not yet been fully employed in the EU [18]. Pyrolysis is the thermal decomposition of materials in the absence of oxygen [23]. The pyrolysis of biomass results in the production of char, liquid and gaseous products (Figure 4) [24]. It can be divided into three main parts: conventional pyrolysis, fast pyrolysis and flash pyrolysis [25]. More recently research has focused on fast pyrolysis in which case waste is decomposed quickly under high temperatures and produces bio-oil. The main features of a fast pyrolysis process are [26]:

- very high heating and heat transfer rates
- carefully controlled temperature of around 500 °C
- rapid cooling of the pyrolysis vapours.

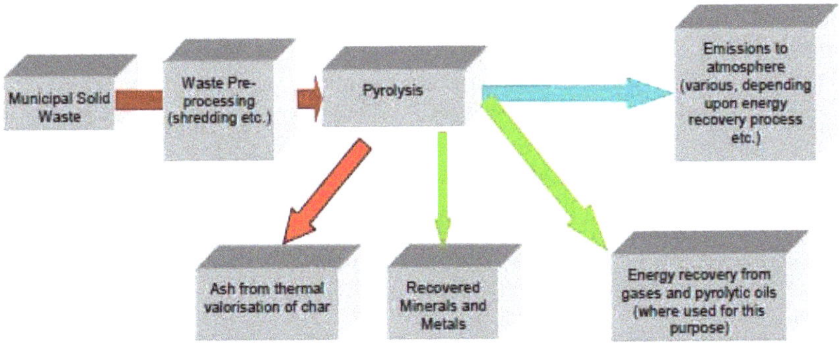

Figure 4. Schematic representation of single pyrolysis process inputs and outputs [18].

Bio-oil that is produced through pyrolysis can replace fuel oil or diesel, for instance in boilers, furnaces, engines and turbines for producing electricity [23]. Even though the production of crude bio-oils has been researched extensively, little progress has been made to produce additives or transportation fuel extenders from these oils, therefore this is an area that has to be further examined [27].

Gasification is actually a process between pyrolysis and incineration because it comprises of the partial oxidation of waste [19]. Gasification involves heating carbon rich waste in sub-stochiometric conditions, whereas the majority of carbon is transformed into a gaseous material leaving an inert residue from the breakdown of organic molecules [18]. In gasification (Figure 5) carbon based wastes are heated in the absence of oxygen to produce a solid, low in carbon and energy from syngas which is a fuel gas mixture consisting of hydrogen and carbon monoxide [19], and can therefore be considered as a thermochemical process. Gasification is highly efficient and has low environmental emission rates therefore it is a quite desirable technology [28]. It is a viable alternative to incineration specifically for thermal treatment of homogeneous carbon-based waste and for pre-treated heterogeneous waste [29].

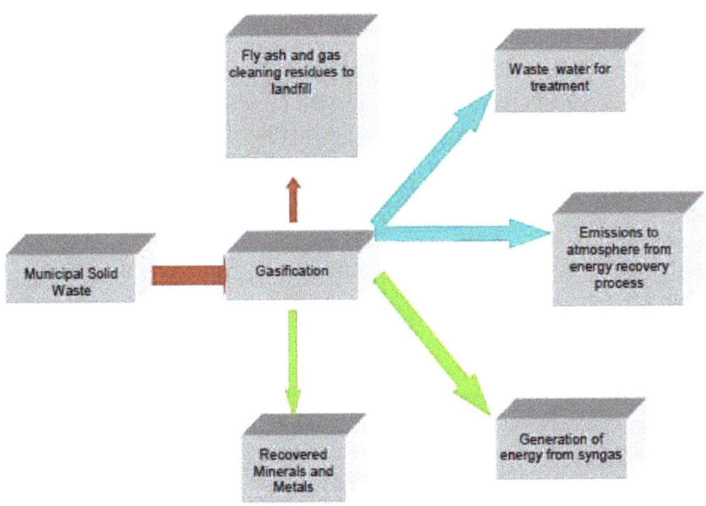

Figure 5. Schematic representation of gasification inputs and outputs [18].

In addition to the methods described above, a further treatment method is anaerobic digestion (AD) which includes the bacterial decomposition of organic material in almost anaerobic conditions whose by-products include biogas and digestate [18]. There are two main types of anaerobic digestion called thermophilic and mesophilic – the primary difference between them is the temperatures used in the process; thermophilic processes reach temperatures of up to 60 °C whereas the mesophilic ones normally run at about 35–40 °C [30].

The high degree of flexibility associated with AD is considered one of the most important advantages of the method, since it can treat several types of waste, ranging from wet to dry and from clean organics to grey waste [18]. Therefore it's a quite desirable option, and for instance in the UK alone there were about 378 AD combined heat and electricity (CHP) plants in 2015 [31]. AD (Figure 6) can in comparison to composting better treat waste with a higher moisture content and can occur usually between 60% and 99% moisture content [18]. Hence kitchen waste and other putrescible wastes which are high in moisture can be an excellent feedstock for AD, whereas woody wastes including a higher proportion of lignocellulosic materials are better suited to composting [32].

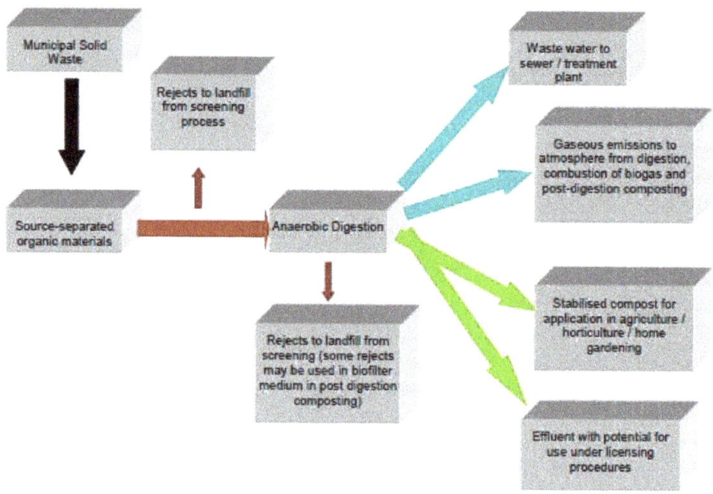

Figure 6. Schematic representation of AD inputs and outputs [18].

The process of AD provides a source of renewable energy, since waste is broken down to produce biogas (a mixture of methane and carbon dioxide), which can be used to produce energy. The biogas can be used threefold: to generate electricity, to power on-site equipment and any excess electricity can be exported to the national grid [18]. Possible uses include its potential to provide heat, electricity or both. Alternatively, the biogas can be 'upgraded' to pure methane, often called biomethane, by removing other gases. One cubic metre of biogas at 60% methane content converts to 6.7 kWh energy [33].

Therefore incentives are provided by the EU to encourage small to medium enterprises and farms to employ AD to gain economic benefits from their organic waste; for instance the UK Renewable Heat Incentive (RHI) scheme provides quarterly payments over twenty years for non-domestic thermal energy production using renewable resources [31].

Finally another important waste-to-energy technology is mechanical biological treatment (MBT) through which the so-called refuse derived fuel (RDF) or solid recovered fuel (SRF) can be produced. RDF generally includes sewage sludge, waste wood, calorific fractions of household and commercial waste, shredder lightweight fractions, scrap tyres, food byproducts [34]. MBT is a process designed to optimise the use of resources by recovering materials for one or more purposes and stabilising the organic fraction of residual waste [18].

Some of the benefits of MBT include the fact that materials and energy can be recovered, space requirements are reduced and gas and leachate emissions from landfill are reduced at the same time [18]. MBT systems basically comprise of two simple ideas: either to separate the waste and then treat or to treat the waste and then separate [19]. Aerobic biological unit processes are used to 'stabilise' the organic fraction, to reduce its biodegradability and therefore its ability to generate methane, whereas anaerobic biological unit processes can help produce biogas from the organic portion of MSW [35]. Figure 7 presents a schematic representation of the MBT inputs and outputs. In those regards RDF must fulfill general quality requirements in order to be safely and efficiently used such as [34]:

- well defined calorific value
- low chlorine content
- quality controlled composition (few impurities)
- defined grain size
- defined bulk density
- availability of sufficient quantities with required specifications.

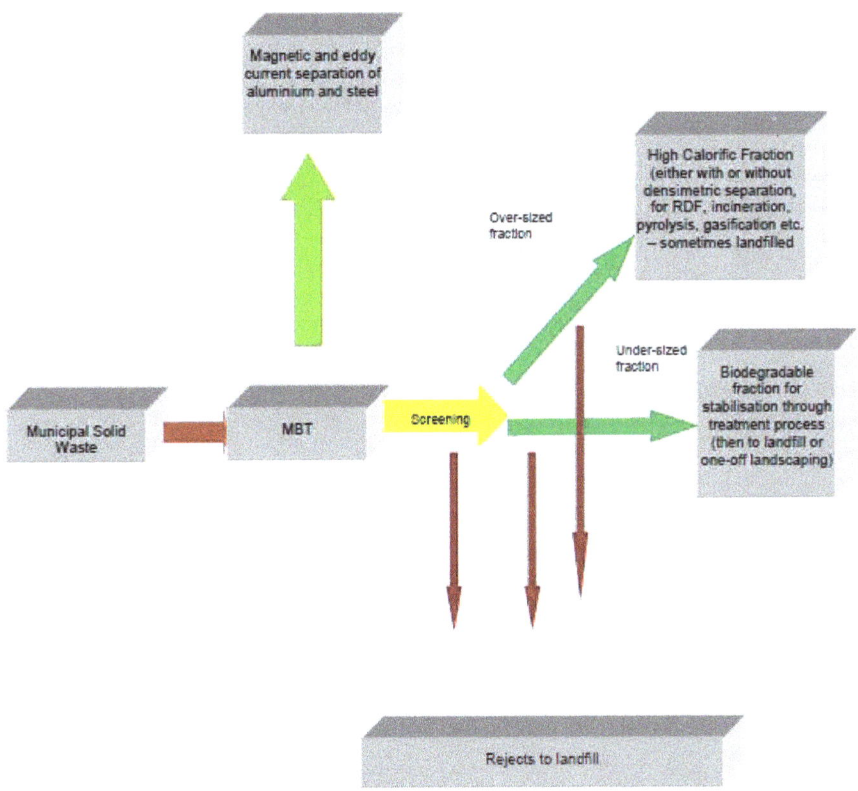

Figure 7. Schematic representation of MBT inputs and outputs [18].

In relation to the aforementioned information, it should be noted that there is an increasing interest in the development and application of heat recovery systems worldwide, driven by government regulatory requirements both from an environmental and an economic perspective [36]. For instance in the EU, 70% of the total energy use in the industrial sector is for thermal processes and about 1/3 of this energy is waste, which could be recovered and used instead [37].

The waste heat recovery market is projected to continue to rise and thus far the EU is the dominant player accounting for 38% of the global market [36]. Of course every industry shows different potential for waste heat which can be seen in Table 1.

Table 1. Waste heat potential per industry (%) [37].

Industry	Waste Heat Potential (%)
Iron and steel	11.4
Chemical and petrochemical	11.0
Non-ferrous metal	9.59
Non-metallic minerals	11.4
Food and tobacco	8.64
Paper pulp and print	10.56
Wood and wood products	6.00
Textile and leather	11.04
Other	10.38

Moreover efficient energy recovery means that access to heat distribution infrastructure which utilises recovered excess heat is essential [38]. Closing this section it should be noted that the EU Commission suggests that more investments should be made to AD processes than incineration, in order to ensure that increases in recycling and reuse do not find any obstacles [39].

2.2. Use of DEA in Energy Efficiency Studies

Many studies have focused on the field of energy and environmental efficiency with the use of DEA. Efficiency is the ratio of output to input; a state of absolute efficiency is achieved when the best possible output per input is realised and it is not possible to amend this without changing technology or any other factors in the production process [35]. Some researchers have composed a list of the main studies working on this topic [40,41]. In more detail, Mardani et al. [40] identified a total of 144 papers between 2006 and 2015. The specific focus of those studies can be seen in Table 2.

Table 2. Distribution papers based on application areas (Adapted from [40]).

Application Fields	Number of Papers	Percentage (%)
Environmental efficiency	23	15.97
Economic and eco-efficiency	14	9.72
Energy efficiency	35	24.31
Renewable and sustainable energy	23	15.97
Water efficiency	4	2.78
Energy performance	8	5.56
Energy saving	6	4.17
Integrated energy efficiency	6	4.17
Other application areas	25	17.36
Total	144	100

Sueyoshi et al. [41] present DEA applications from 1980 to 2010 (693 studies) and a considerable increase in research has been noticed after 2000. The first research work on energy efficiency was by Färe et al. [42]. Further studies focused both on developed countries such as Canada [43], USA [44], selected Organisation for Economic Co-operation and Development (OECD) ones [45,46] and developing countries like Korea [47] and India [48].

These studies have focused on different aspects of energy efficiency. For instance Zhou et al. [45] by using DEA measured the carbon emissions' performance of eight regions worldwide in 2002, while they examined the environmental efficiency of 26 OECD countries from 1995 to 1997 [45]. Halkos and Tzeremes [7] examined energy consumption on countries' economic efficiency levels and DEA in that case presents economic efficiency variations among the examined countries. Additionally the effects of renewable energy on the technical efficiency of 45 economies during 2001–2002 was studied by Chen and Hu [49] showing that increasing the use of renewable energy improves an economy's technical efficiency.

Chen et al. [21] evaluated the performance-based efficiencies of 19 largescale municipal incinerators in Taiwan with different operational conditions for 2002–2005, leading to optimal management strategies for promoting the quality of solid waste incineration. Moreover the renewable energy sector in Greece is examined through DEA for 78 firms for 2006–2008 showing that the majority of the firms operating in the Greek renewable sector are based on the production of wind energy [10].

Hu and Wang [50] measured the energy efficiency of 29 regions in China and propose a total factor energy efficiency evaluation method. The technical efficiency of energy utilities in China and Taiwan was also studied by Yeh et al. [51]. The same approach but with the incorporation of environmental efficiency as well was followed by Bian and Yang [52]. Furthermore Zhou and Ang [53] measured energy efficiency using both energy and non-energy inputs. Wang et al. [54] created a mixed efficiency model which includes both economic and environmental efficiency attempting to proportionally increase desirable outputs and decrease undesirable outputs.

Wang et al. [55] evaluated energy and environmental efficiency of 29 regions in China with an improved DEA model. Yang et al. [56] modeled carbon emissions from travel in Beijing using microsimulation modelling and investigating the effects of the major transport policies. Finally Song et al. [57] developed an improved method by which to evaluate resource and environmental efficiency with the evaluation of resource inputs into the objective function and focus on resource inputs, undesirable outputs and desirable outputs simultaneously.

3. Research Methods, Data and Production Frameworks

3.1. The Proposed Methodology—An Overview of DEA

As mentioned already DEA is used to assess the efficiency of selected DMUs, whereas each unit is compared with all others [12] and aims to identify the ones that are operating inefficiently [35]. That way both good and bad outputs can be taken into account [15]. Efficient DMUs achieve a rating of 1 (or 100%) and these constitute the efficiency frontier showing the non-efficient DMUs as well [12,58].

In DEA analysis it is not necessary to assume that there is any specific relationship between inputs and outputs [59]. DEA models are either input-oriented minimizing inputs or output-oriented models maximizing outputs without the use of more inputs [60]. The relevant formulations of those two models are as follows [14]:

Input-oriented Min θ_0

Subject to:

$$\theta_0 x_{i0} - \sum_{k=0}^{n} x_{ik}\lambda_k \geq 0. \forall i \tag{1}$$

$$-y_{j0} + \sum_{k=0}^{n} y_{jk}\lambda_k \geq 0. \forall j$$

$$\lambda_k \geq 0. \forall k$$

Output oriented Max $(1/\theta_0)$

Subject to:

$$x_{i0} - \sum_{k=0}^{n} x_{ik}\lambda_k \geq 0. \forall i \tag{2}$$

$$-\frac{y_{j0}}{\theta_0} + \sum_{k=0}^{n} y_{jk}\lambda_k \geq 0. \forall j$$

$$\lambda_k \geq 0. \forall k$$

where θ_0 is DMU 0's efficiency score, is DMU k's contribution on the targets of DMU 0, y_{j0} is output j quantity for DMU 0, x_{i0} is output i quantity for DMU 0 and n is the quantity of DMUs used on the model. Moreover the decision varialbes are θ and λ. Farrell's [61] input measure operationalization of efficiency was introduced via linear programming estimators by Charnes et al. [62]. Therefore for a given DMU operating at a point it can be defined as:

$$\hat{\Psi}_{DEA} = \Big\{(x,y) \in R_{+}^{p,q} \big| y \leq \sum_{i=1}^{n} \gamma_i Y_i; x \geq \sum_{i=1}^{n} \gamma_i X_i, \text{for } (\gamma_1,\ldots,\gamma_n) \\ \text{s.t.} \sum_{i=1}^{n} \gamma_i = 1; \gamma_i \geq 0, i = 1,\ldots,n\Big\}. \quad (3)$$

where x and y are the input and output vectors.

DEA has been widely used in research mainly because one can use multiple inputs and outputs without assigning weights on those and at the same time efficiencies are calculated based on the best operating DMU and not on average performance levels [63]. On the contrary a disadvantage of DEA is that is that it produces a separate linear programme for each DMU thus creating a computational mess when there are a lot of DMUs taken into account [60].

3.2. Bias Correction and Returns of Scale in DEA

Bootstrap is used in most DEA studies as the DEA estimators have been proven to be biased by construction so it is necessary to correct and estimate the relevant bias [64–66]. In that case a simulation of the data generating process (DGP) is applied whereas the estimator copies the sampling distribution of the original estimator [67]. At the same time the sensitivity of the efficiency scores relative to the sampling variations of the estimated frontier is defined as well [64].

The bootstrap bias estimate for the original DEA estimator θ DEA (x, y) can be calculated as:

$$B\hat{I}AS_B(\hat{\theta}_{DEA}\ (x,y) = B^{-1}\sum_{b=1}^{B} \hat{\theta}^{*}_{DEA,\ b\ (x,y)} - \hat{\theta}_{DEA\ (x,y)} \quad (4)$$

where B stands for bootstrap replications performed.

The biased corrected estimator of (x, y) can be calculated as:

$$\hat{\theta}_{DEA\ (x,y)} = \hat{\theta}_{DEA\ (x,y)} - B\hat{I}AS_B\ (\hat{\theta}_{DEA\ (x,y)}) = 2\,\hat{\theta}_{DEA\ (x,y)} - B^{-1}\sum_{b=1}^{B} \hat{\theta}^{*}_{DEA,\ b\ (x,y)} \quad (5)$$

This procedure also provides confidence limits on the efficiencies in order to present the true efficient frontier within the specified interval [68]. Therefore the $(1-\alpha) \times 100$—percent bootstrap confidence intervals can be obtained for $\theta(x, y)$ as:

$$\frac{1}{\hat{\delta}_{DEA\ (X,Y)-\ nc^{*}_{1-a/2}}} \leq \theta\ (x,y) \leq \frac{1}{\hat{\delta}_{DEA\ (X,Y)-\ nc^{*}_{a/2}}} \quad (6)$$

These calculations have also been applied in this research as will be presented in Section 4.

Another important element that needs to be considered in the DEA analysis is if constant returns to scale (CRS) or variable returns to scale (VRS) deem suitable for each specific case. Under CRS as originally designed by Charnes et al. [62] (CCR model) a full proportionality between all inputs and outputs is assumed [69] which could be the case when firms operate at the optimal level [70]. It is possible to disregard this information by using VRS. As originally designed by Banker et al. [71] (BCC model) VRS accounts for the use of technical and scale efficiencies in DEA. This method includes both increasing and decreasing returns to scale. Therefore following Simar's and Wilson's [64] bootstrap approach we compare between CRS and VRS according to these hypotheses:

$H_0 : \Psi^{\theta}$ is CRS
$H_1 : \Psi^{\theta}$ is VRS
The test statistic mean of the ratios of the efficiency scores is then provided by:

$$T\left(X_n\right) = \frac{1}{n}\sum_{i=1}^{n}\frac{\hat{\theta}_{CRS,\ n}\left(X_I,\ Y_i\right)}{\hat{\theta}_{VRS,\ n}\left(X_I,\ Y_i\right)} \tag{7}$$

Then the *p-value* of the null-hypothesis can be obtained:

$$p - value = prob\ \left(T\ \left(X_n\right) \leq T_{obs}\ \right]\ H_0\ is\ true) \tag{8}$$

where *Tobs* is the value of T computed on the original observed sample *Xn* and *B* is the number of bootstrap reputations. Then the *p-value* can be approximated by the proportion of bootstrap values of T^{*b} less the original observed value of *Tobs* such as:

$$p - value\ \approx \sum_{b=1}^{B}\frac{\mathrm{I}\left(T^{*b} \leq T_{obs}\right)}{B} \tag{9}$$

Based on these equations, calculations have been performed on Stata and it is shown that for the data used (Section 3.3) and the designed frameworks (described in further detail in Section 3.4), CRS is more appropriate following the CCR model [62] as the results obtained are higher than 0.05 thus accepting the null hypothesis (B = 999). The specific results are shown in Table 3.

Table 3. Stata results on testing CRS vs. VRS in this study's two models for all examined years.

Frameworks	2008	2010	2012	2014	2016
M1	0.6507	0.8809	0.2252	0.5075	0.4795
M2	0.6016	0.8138	0.3393	0.5736	0.5816

In the case of the CRS or CCR models, the efficiency frontier is a straight line crossing the point of origin and the best performers (efficient DMUs) [72]. Figure 8 presents the graphical representation of the efficient and inefficient DMUs along the frontier, in which case DMU_2 is the best performer and is used as a reference for all other DMUs. In those regards further improvement of efficiency scores for inefficient DMUs can be achieved through the implementation of good practices of the efficient ones [73].

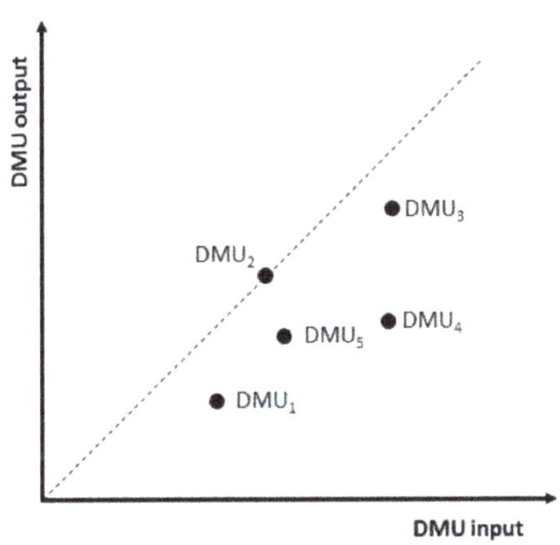

Figure 8. Graphical representation of the efficiency frontier of CCR model [72].

3.3. Data Used

For this paper's analysis the MaxDEA for Data Envelopment Analysis programme (MaxDEA Basic 6.6 – 2015 edition) is used. Table 4 presents the descriptive statistics of the inputs and outputs used in the different DEA frameworks and for all the years and for all the examined countries.

Table 4. Descriptive statistics for all DEA models.

	Final Energy Consumption (Mt Equivalent)	GDP (M Euro)	Labor (Thousand Persons)	Capital (M Euro)	Population Density (Persons per km^2)	SOx Emissions (t)	NOx Emissions (t)	GHG Emissions from Energy (Thousand t of CO_2 Equivalent)
2008								
Mean	42.1	1,781,373.4	7986.4	104,801.4	169.6	130,041.6	74,443.1	142,163.4
St. dev	56.0	5,096,821.3	10,180.0	148,216.2	243.2	176,282.2	100,320.9	197,222.1
Min	0.5	6128.7	158.6	1203.1	17.5	12.0	783.0	2833.4
Max	217.6	27,193,630.0	38,541.5	520,809.0	1295.5	628,644.0	382,978.0	820,242.4
2010								
Mean	41.5	1,787,110.4	7774.2	91,911.8	171.4	96,084.9	66,404.0	135,553.4
St. dev	55.5	5,103,808.5	10,076.9	136,804.0	246.3	132,887.3	94,477.9	189,697.4
Min	0.5	6599.5	162.6	1411.6	17.6	11.0	863.0	2598.1
Max	219.7	27,224,599.0	38,737.8	501,449.0	1311.7	545,404.0	334,748.0	802,121.3
2012								
Mean	39.6	1,878,639.0	7548.8	94,847.8	172.7	84,384.5	65,251.6	128,695.0
St. dev	53.3	5,394,392.0	9951.5	145,091.9	249.9	119,172.7	97,298.8	183,709.8
Min	0.5	7168.4	170.7	1299.8	17.8	10.0	779.0	2818.9
Max	212.1	28,781,064.0	38,320.6	554,746.0	1329.2	485,523.0	366,449.0	785,284.2
2014								
Mean	38.0	2,058,682.8	7622.2	97,341.0	175.1	62,735.0	55,019.0	119,113.9
St. dev	51.3	6,103,555.2	10,096.1	150,749.6	258.2	94,256.0	85,032.2	173,002.8
Min	0.5	8505.4	186.8	1465.4	18.0	15.0	728.0	2470.1
Max	208.9	32,583,424.0	38,907.7	587,549.0	1375.2	425,649.0	300,824.0	762,351.1
2016								
Mean	39.6	2,235,599.6	7819.9	106,622.6	178.8	43,531.1	46,821.9	119,581.1
St. dev	53.1	6,647,143.6	10,365.3	160,493.8	271.3	67,884.9	72,540.6	173,095.0
Min	0.6	10,343.0	204.6	2435.6	18.1	17.0	612.0	1426.9
Max	216.4	35,474,186.0	40,165.1	634,029.0	1,450.2	296,757.0	295,747.0	771,900.6

In this analysis the variables used include: final energy consumption, GDP, labour, capital, population density, nitrogen oxide (NOx) emissions (from energy), sulphur oxide (SOx) emissions (from energy) and GHG emissions (from energy) with data obtained from Eurostat. In total 28 EU Member States are examined for the years 2008, 2010, 2012, 2014 and 2016. The following units are used for each factor of this analysis:

- Final energy consumption: Mt equivalent
- GDP: current prices (M Euro)
- Labor: number of persons (thousand persons)
- Capital: gross fixed capital formation (current prices, M Euro)
- Population density: persons per km^2
- SOx emissions: t (from energy production and distribution)
- NOx emissions: t (from energy production and distribution)
- GHG emissions: thousand t of CO_2 equivalent (from energy production and distribution)

Based on these data, Figure 9 presents the trend of energy consumption levels, SOx, NOx and GHG emissions for all examined years on an average EU basis for the 28 countries taken into account. It is noticed that all indicators have dropped since 2008 especially SOx and NOx emissions, while energy consumption and GHG emissions are on the rise again after 2014.

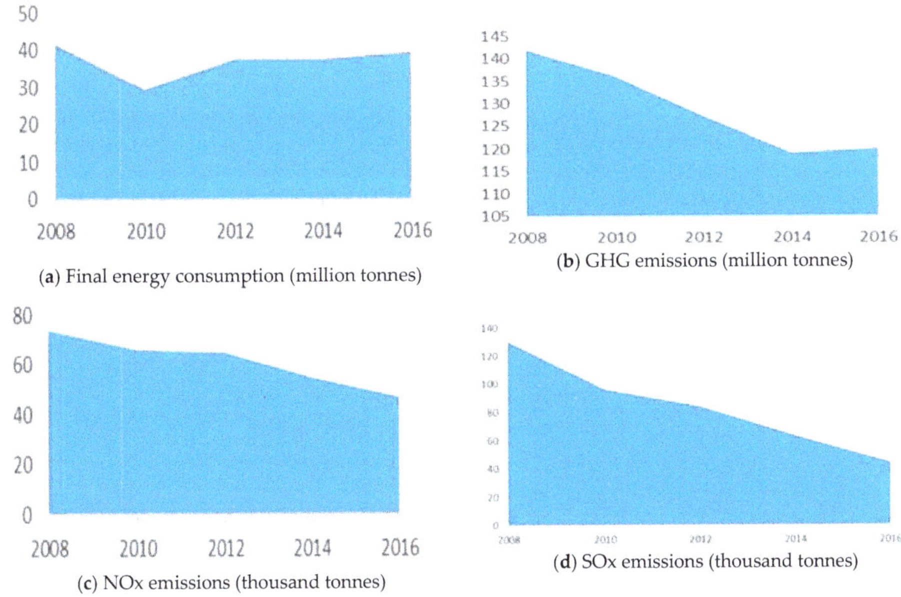

(a) Final energy consumption (million tonnes) (b) GHG emissions (million tonnes) (c) NOx emissions (thousand tonnes) (d) SOx emissions (thousand tonnes)

Figure 9. Trend of the main components of the present analysis.

At the same time it is noticed that the share of energy from renewable sources is also on the rise in the EU Member States as shown in Figure 10, showing also how far those countries are from achieving their 2020 target. So far Sweden, Finland, Denmark, Estonia, Croatia, Lithuania, Romania, Bulgaria, Italy, Czech Republic and Hungary have managed to accomplish this.

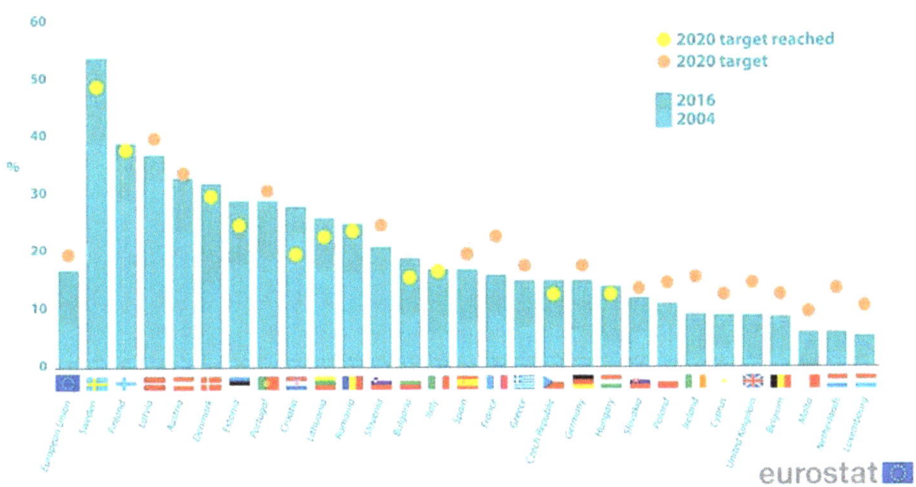

Figure 10. Share of energy from renewable sources 2004–2016 (in % of gross final energy consumption) (Eurostat data).

3.4. Environmental Production Frameworks

Following studies such as those of Wang et al. [55] and Chien and Hu [74], where capital, labor, population density (M2 framework) and energy consumption are used as inputs and GDP (desirable output), carbon dioxide and sulphur dioxide (undesirable outputs), this paper's analysis produces two production frameworks as presented in Figure 11 (M1 framework) and Figure 12 (M2 framework). Population density is a factor that has not been used in previous research regarding energy efficiency but it is a strong inequality measure which affects regional and interregional policies and in turn regional and interregional socioeconomic development [50]. In both frameworks a radial model is used, which is output oriented (as presented in Section 3.1).

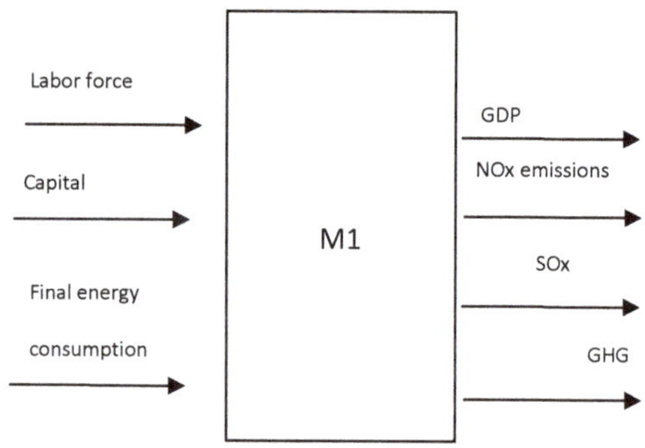

Figure 11. Description of M1 environmental production framework.

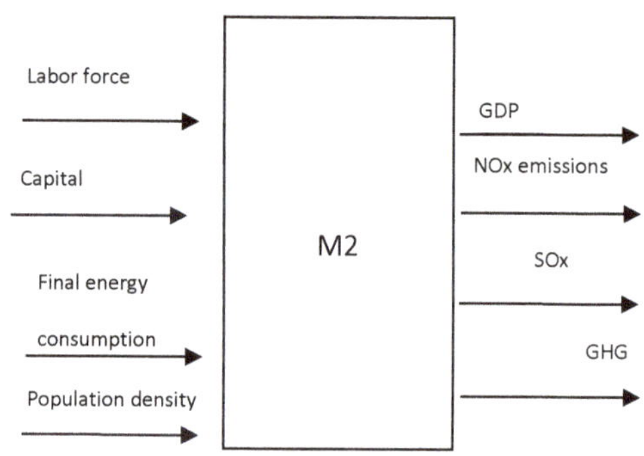

Figure 12. Description of environmental production framework (M2 indicator).

Overall to evaluate the energy efficiency of the studied EU Member States, DEA was used to examine all countries' parameters, classify and quantify the variables, model the problem and determine the best performing DMUs. These were then analysed and recommendations are provided.

4. Result

Under the M1 framework the highest performers are: Hungary, Luxembourg, Sweden; whereas the lowest performers are: Estonia, Bulgaria, Greece and Slovenia. For framework M2 the picture is quite similar. Table 5 shows the efficiency scores of all examined countries for the whole time period studied. Also Table 6 presents the average scores (year-wise) per country per modelling framework.

Table 5. Results of M1 and M2 frameworks for the EU countries for 2008, 2010, 2012, 2014 and 2016.

	M1					M2				
Country	2008	2010	2012	2014	2016	2008	2010	2012	2014	2016
Austria	0.528	0.523	0.529	0.531	0.527	0.528	0.523	0.529	0.531	0.527
Belgium	0.507	0.507	0.508	0.507	0.506	0.507	0.507	0.508	0.507	0.506
Bulgaria	0.501	0.501	0.501	0.501	0.501	0.501	0.501	0.501	0.501	0.501
Croatia	0.515	0.516	0.514	0.513	0.513	0.515	0.516	0.514	0.513	0.513
Cyprus	0.502	0.502	0.502	0.502	0.502	0.502	0.502	0.502	0.502	0.502
CzechRepublic	0.534	0.532	0.529	0.528	0.530	0.534	0.532	0.529	0.528	0.530
Denmark	0.567	0.579	0.588	0.604	0.572	0.567	0.579	0.588	0.604	0.572
Estonia	0.501	0.501	0.501	0.501	0.501	0.501	0.501	0.501	0.501	0.501
Finland	0.503	0.503	0.503	0.503	0.503	0.503	0.503	0.503	0.503	0.503
France	0.510	0.510	0.510	0.512	0.510	0.510	0.510	0.510	0.512	0.510
Germany	0.504	0.504	0.503	0.503	0.503	0.504	0.504	0.503	0.503	0.503
Greece	0.502	0.502	0.502	0.502	0.502	0.502	0.502	0.502	0.502	0.502
Hungary	1.000	1.000	1.000	1.000	1.000	1.000	1.000	1.000	1.000	1.000
Ireland	0.504	0.506	0.506	0.506	0.507	0.504	0.506	0.506	0.506	0.507
Italy	0.508	0.508	0.507	0.507	0.506	0.508	0.508	0.507	0.507	0.506
Latvia	0.511	0.506	0.507	0.507	0.503	0.511	0.506	0.507	0.507	0.503
Lithuania	0.502	0.502	0.502	0.502	0.502	0.502	0.502	0.502	0.502	0.502
Luxembourg	1.000	1.000	1.000	1.000	0.824	1.000	1.000	1.000	1.000	0.824
Malta	0.502	0.502	0.502	0.502	0.505	0.502	0.502	0.502	0.502	0.505
The Netherlands	0.508	0.508	0.507	0.506	0.505	0.508	0.508	0.507	0.506	0.505
Poland	0.504	0.504	0.504	0.503	0.504	0.504	0.504	0.504	0.503	0.504
Portugal	0.503	0.503	0.503	0.502	0.502	0.503	0.503	0.503	0.502	0.502
Romania	0.505	0.506	0.505	0.505	0.506	0.505	0.506	0.505	0.505	0.506
Slovakia	0.502	0.502	0.502	0.502	0.502	0.502	0.502	0.502	0.502	0.502
Slovenia	0.502	0.502	0.502	0.502	0.502	0.502	0.502	0.502	0.502	0.502
Spain	0.503	0.504	0.503	0.503	0.503	0.503	0.504	0.503	0.503	0.503
Sweden	0.649	0.596	0.621	0.665	0.606	0.650	0.608	0.622	0.665	0.606
United Kingdom	0.503	0.503	0.503	0.503	0.503	0.503	0.503	0.503	0.503	0.503

Table 6. Average scores per country and per modelling frameworks.

	M1	M2
Country	Average	Average
Austria	0.528	0.528
Belgium	0.507	0.507
Bulgaria	0.501	0.501
Croatia	0.514	0.514
Cyprus	0.502	0.502
CzechRepublic	0.530	0.530
Denmark	0.582	0.582
Estonia	0.501	0.501
Finland	0.503	0.503
France	0.510	0.510
Germany	0.503	0.503
Greece	0.502	0.502
Hungary	1.000	1.000
Ireland	0.506	0.506
Italy	0.507	0.507
Latvia	0.507	0.507
Lithuania	0.502	0.502
Luxembourg	0.965	0.965
Malta	0.503	0.503
The Netherlands	0.507	0.507
Poland	0.504	0.504
Portugal	0.503	0.503
Romania	0.506	0.506
Slovakia	0.502	0.502
Slovenia	0.502	0.502
Spain	0.503	0.503
Sweden	0.628	0.630
United Kingdom	0.503	0.503

The obtained results are biased and therefore following the bootstrap technique presented in Section 3, the bias corrected results need to be applied in our analysis. Tables A1 and A2 (Appendix A) present the efficiency scores of the 28 countries, the bias corrected efficiency scores and the 95-percent confidence intervals: lower and upper bound obtained by B = 999 bootstrap replications using the algorithm described in Section 3.2.

According to the bias corrected efficiency measures the countries with the higher environmental efficiency scores (i.e., >0.497) over the years are reported to be:

- Framework M1: Bulgaria, Cyprus, Estonia, Greece, Lithuania, Malta and Slovenia.
- Framework M2: Bulgaria, Cyprus, Estonia, Greece, Lithuania and Slovenia.

The two modelling techniques used in this analysis cannot be compared to each other since they use different inputs and outputs. A lack of common environmental policies among EU Member States can be seen in their energy efficiency levels regarding energy consumption and the relevant emissions. With regards to changes over the years and as can be seen in Figure 13, most countries seem to maintain their efficiency scores with only Czech Republic, Finland, Ireland, Malta, Romania and Slovenia marginally improving theirs. At the same time, it can be noticed that most countries have higher environmental efficiency scores over 2010 and 2012 with a decrease after that.

Figure 13. Bias corrected efficiency scores for all countries for all examined years.

5. Discussion

The efficiency scores obtained and presented in the previous section show that EU-wise environmental efficiency levels regarding energy consumption and emissions tend to be quite low. The world's tension level of energy supply is worsened over the years and efforts are being made to replace traditional fossil fuels with more sustainable options achieving a good balance between economic development and environmental protection [75]. Energy from waste is the largest source of renewable energy today in the EU and is expected to hold this place until 2030, reaching a share of 60–70% [40].

The 'International Energy Efficiency Scorecard' published in 2014 by the American Council for an Energy-Efficient Economy stresses that countries can maintain their resources, address global warming, stabilize their economies and reduce the costs of their economic outputs by using energy more efficiently [76]. This can be seen graphically also in Figure 9 where a decrease in emissions' level is generally noticed. The results obtained from the current analysis are also in connection with the EU's targets for energy and climate as presented in Figure 14.

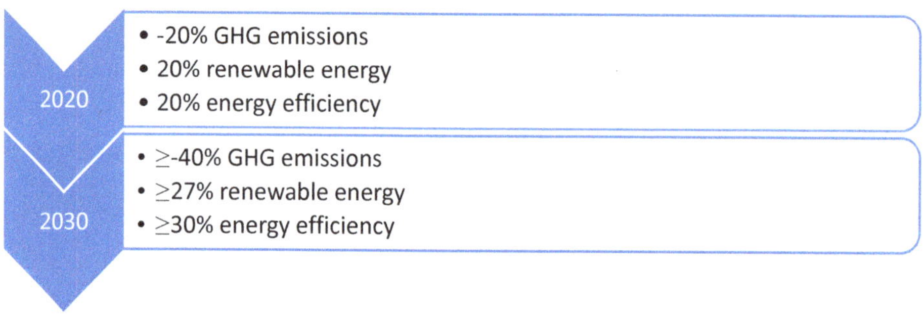

Figure 14. EU's framework for energy and climate for 2020 and 2030 (Adapted by [77]).

In connection to that, nations have been moving towards waste-to-energy with two main objectives, namely sufficient and sustainable energy production and effective treatment of MSW by reducing its volume by about 87% [78]. Both these two factors need to be taken into account when considering this option [79]. A major issue to make sure this option is viable, both from an economic and an environmental perspective, is to take into consideration the resource characteristics, such as their location, amount and quality [80]. The results of this study presenting energy efficiency should be considered to avoid unnecessary entropy production but also to make processes more cost effective and ecofriendly [81]. The main benefits from waste-to-energy include [82]:

- It transforms waste from a problem into a resource.
- Energy generated contributes to primary energy savings from other energy sources.
- It can reduce greenhouse-gas emissions when it replaces more carbon-intensive energy sources.
- Waste to landfill is reduced heavily.
- Waste treatment time is extremely short compared with landfills.
- It also enables treatment of hazardous waste.

At the same time, the main associated risk is that those systems become highly dependent on and justify societies' increasingly uneconomical consumption levels, while also having unintended negative effects (such as higher levels of energy and material use throughout a society, increasing upstream environmental impacts) [82]. Moreover it is essential to create a network of the waste by-products, electricity and heat between multiple sectors throughout the world [83,84]. Figure 15 presents a map of waste-to-energy plants in Europe for 2017, in which capacity is seen to be overall stable compared to 2016, with only the UK increasing its capacity.

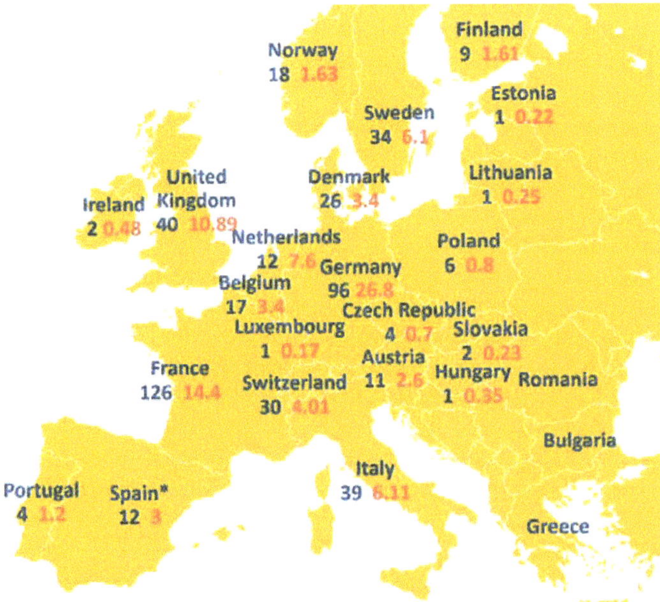

Figure 15. Waste-to-energy in Europe in 2017 (with red: waste thermally treated (in Mt) and with blue: plants operating in Europe) [85].

The necessary treatment that is to be used depends highly on the nature and volume of the waste stream with the main factor taken into account being its energy content (calorific value) and as a rule of thumb waste-to-energy option should be considered when the incoming waste has an average calorific value of at least 7 MJ/kg [86]. Table 7 presents the average net calorific values for most common MSW waste streams.

Table 7. Approximate net calorific value (MJ/kg) [87].

Fraction	Value
Paper	16
Organic material	4
Plastics	35
Glass	0
Metals	0
Textiles	19
Other material	11

Overall the European Commission 2017 (Ref. [38]) recommends the main technologies that could be used [88]:

- co-incineration in combustion plants: with gasification of SRF and co-incineration of the resulting syngas in the combustion plant.
- co-incineration in cement kilns.
- incineration in dedicated facilities:
    - the use of super heaters and heat pumps
    - the utilisation of the energy contained in flue gas
    - distributing chilled water through district cooling networks.

- Bio-methane for further distribution and utilisation.

In this regard Scarlat et al. [20] perform a suitability analysis as to where waste-to-energy plants are best to be built, which can be seen graphically on Figure 16. The potential plants (shown in green) are interrelated with the results of the current analysis, as according to their analysis, there is great potential to build plants for instance in the Czech Republic, Croatia, France, Hungary, Italy, Spain and UK. For those countries the current analysis found that energy efficiency scores are overall quite low in comparison to other countries. Also Greece and Bulgaria show a great potential for building waste-to-energy plants which makes sense according to this analysis as for these countries efficiency scores are quite low as well.

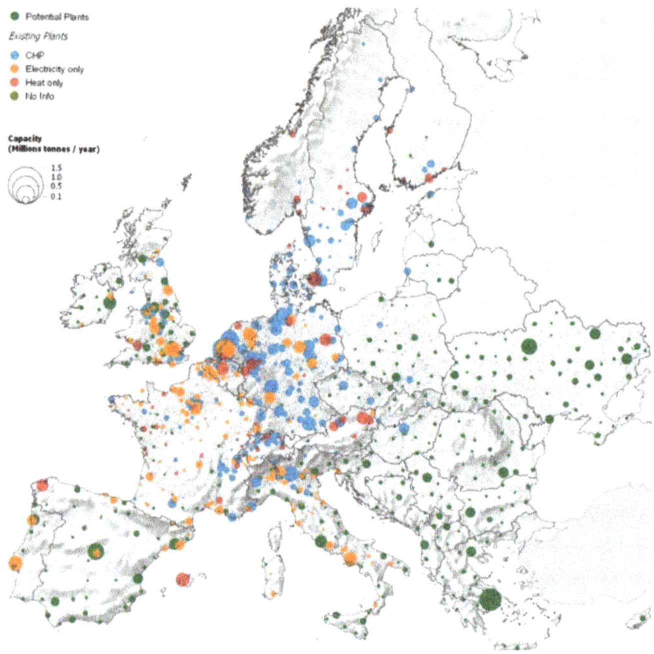

Figure 16. Suitability map for waste-to-energy plant location [20].

Energy efficiency levels across the 28 EU examined countries are quite low overall with only a few differentiated countries. As it stands, waste management is a crucial part in the context of the circular economy whereas prioritization needs to give to prevention, reuse, recycling and energy recovery and as a last resort disposal to landfills [89]. Therefore the circular economy requires a better understanding of existing waste infrastructure, including location and capacity [90]. The circular economy aims to accomplish the optimum production through the 3R principle—reduce, reuse and recycle—while minimizing resource utilization, pollution emissions and waste discarded [91].

To deliver the circular economy governments need to collaborate with various partners to combine scientific research, policies and regulations, thus adopting a long-term policy framework [90]. Along those lines, the EU Commission's Circular Economy Package drives the treatment options that have been used by EU Member States. This package's aim is the acceleration of Europe's transition towards circular economy as well as the waste reduction targets across EU Member States [92]. Therefore it is essential to preserve the worth of products, materials and resources in the economy as much as possible and minimize waste generated [4]. Hence waste-to-energy addresses the problems of energy demand, waste management and GHG emissions at the same time, achieving a circular economy system [93].

By 2020 196 billion kWh of sustainable energy could be produced through waste-to-energy plants which makes an equivalent of the energy produced by 6-9 nuclear stations or 25 coal power plants [94].

At the same time one of the EU Commission's priorities is also a European Energy Union which ensures reliable energy supplies at rational prices for businesses and consumers and with the least environmental impacts [95]. This union would enhance the economy and attract investments thus creating new jobs opportunities [77]. Competition policy in the EU is essential for the internal market with the first liberalisation directives established in 1996 (electricity) and 1998 (gas) and the second liberalisation directives adopted in 2003 [95]. This competition policy aims mainly to ensure that companies compete fairly, providing more choices to consumers and helping reduce prices and improve quality [96,97].

Despite these regulations, markets seem to be largely national and with relatively few cross-border trade, therefore the EU Commission has paid great attention into controlling potential mergers (such as the proposed merger between EDP and GDP in Portugal), into setting up rules for mergers and in controlling state aid to energy companies across the EU [95]. In more detail, it is essential to have an EU competition policy, mainly to achieve [96]:

- Low prices for all: more people can afford to buy products and businesses are encouraged to produce.
- Better quality: competition encourages businesses to improve the quality of goods and services they sell and to attract more customers and expand their market share.
- More choice: businesses will try to make their products unique.
- Innovation: in their product concepts, design, production techniques, services, etc.
- Better competitors in global markets: competition would enhance European companies' strength outside the EU and enable them to hold their own against global competitors.

Also waste-to-energy could relieve the EU from foreign imports, for instance in 2012 it imported 4 million TJ of natural gas from Russia, whereas waste-to-energy could substitute 19% of Russian gas imports [94]. Unfair competition will only hinder the clean energy transition as far as Member States continue to provide fossil fuel subsidies, such as direct subsidies to uneconomical coal mines, capacity mechanisms for emission intensive power plants, tax relief for company cars or diesel fuel and similar measures [77]. More detailed research conducted in China by Zhang et al. [97] shows that raw material price subsidies increase profits both for recycling and biofuel companies, but investment subsidies only produce greater profits for recycling companies. One important and unexpected issue that needs to be taken into account and has undoubtedly affected energy efficiency in EU Member States is the financial crisis from which the EU has been suffering severely after 2008. This can also be noticed in the efficiency scores obtained through the present analysis, whereas efficiencies have decreased after 2012 when the crisis became more imminent.

As for the future steps, the EU plans for a climate-neutral Europe by 2050 through investments to realistic technological solutions, the empowering of citizens and aligning action in key areas such as industrial policy, finance, or research [6]. In those regards studies suggest that the potential for using heat from waste could be an equivalent to 200 billion kWh per year by 2050 [90]. Therefore it is essential to already have consultations with young people, citizens affected by the energy transition, inventors, social partners and civil society, mayors and other politicians to show the potential of realizing this energy transition [78].

6. Conclusions and Policy Implications

The current paper examines energy efficiency across 28 selected EU Member States and reviews the potential for energy recovery from waste according to the efficiency scores obtained for the examined Member States. The efficiencies are assessed through DEA under CRS and the following variables are examined: final energy consumption, GDP, labour, capital, population density, NOx emissions (from energy), SOx emissions (from energy) and GHG emissions (from energy) from Eurostat data and for the years 2008, 2010, 2012, 2014 and 2016. The two models that are designed use two outputs one desirable (GDP) and one undesirable (aerial gas emissions – GHG, SOx and NOx) with different inputs in each case.

The bias corrected efficiency scores show that overall Bulgaria, Cyprus, Estonia, Greece, Lithuania and Slovenia are efficient under both frameworks. Also most countries seem to maintain their efficiency scores with only the Czech Republic, Finland, Ireland, Malta, Romania and Slovenia marginally improving theirs. At the same time, it can be noticed that most countries have higher environmental efficiency scores over 2010 and 2012 with a decrease after that.

These efficiency scores show that EU-wise environmental efficiency levels regarding energy consumption and emissions tend to be quite low overall, therefore it is suggestible to move towards waste-to-energy with two main objectives, namely sufficient and sustainable energy production and effective treatment of MSW. This option would enhance the circular economy, whereas prioritization needs to give to prevention, preparation for reuse, recycling and energy recovery through to disposal, such as landfilling. Waste to energy addresses the problems of energy demand, waste management and GHG emissions simultaneously.

Together with the EU Commission's competition strategy, these options would ensure reliable energy supplies at rational prices for businesses and consumers and with the least environmental impacts. Along with these and taking into account the current analysis' results, it is essential to account for the financial crisis which affects EU since 2008. Namely the efficiency scores show a decrease after 2012 when the crisis became more imminent (Figure 13). Regarding future steps towards a climate neutral Europe, investments into technology along with the empowering of citizens and industry need to be considered.

The models of the present research could be enriched with additional control variables which could incorporate specific characteristics of EU countries, such as their technological level regarding waste management especially, their institutional background and their education level to name a few. Moreover once data become available it would be useful to expand this research with more recent data to better reflect today's situation.

Author Contributions: For research articles with several authors, a short paragraph specifying their individual contributions must be provided. The following statements should be used conceptualization, G.H. and K.N.P.; methodology, G.H.; software, G.H.; formal analysis, K.N.P.; investigation, K.N.P.; resources, K.N.P.; data curation, K.N.P.; writing—original draft preparation, K.N.P.; writing—review and editing, G.H. and K.N.P.; supervision, G.H.; project administration, G.H.; funding acquisition, K.N.P.

Funding: This research was funded by the General Secretariat for Research and Technology and the Hellenic Foundation for Research and Innovation (HFRI).

Acknowledgments: This work has been supported by the General Secretariat for Research and Technology and the Hellenic Foundation for Research and Innovation (HFRI). Thanks are due to the three anonymous reviewers and the Editors for their helpful and constructive comments. Any remaining errors are solely the authors' responsibility.

Conflicts of Interest: The authors declare no conflict of interest.

Appendix A

Table A1. Bias corrected efficiency scores of countries' by modelling framework. Framework M1.

Country	Score Original	Bias Corrected	Bias	Std	Lower	Upper	2008
Austria	0.528	0.461	0.067	0.027	0.402	0.498	
Belgium	0.507	0.490	0.017	0.006	0.475	0.499	
Bulgaria	0.501	0.498	0.004	0.001	0.494	0.499	
Croatia	0.515	0.475	0.040	0.012	0.439	0.493	
Cyprus	0.502	0.496	0.006	0.002	0.491	0.499	
Czech Republic	0.534	0.446	0.087	0.026	0.367	0.484	
Denmark	0.567	0.396	0.171	0.052	0.241	0.474	

Table A1. *Cont.*

Country	Score Original	Bias Corrected	Bias	Std	Lower	Upper	2008
Estonia	0.501	0.499	0.002	0.001	0.496	0.500	
Finland	0.503	0.495	0.009	0.003	0.486	0.499	
France	0.510	0.486	0.023	0.008	0.465	0.498	
Germany	0.504	0.494	0.010	0.003	0.485	0.499	
Greece	0.502	0.496	0.006	0.002	0.491	0.499	
Hungary	1.000	−0.304	1.304	0.385	−1.490	0.248	
Ireland	0.504	0.494	0.010	0.003	0.484	0.498	
Italy	0.508	0.487	0.021	0.006	0.468	0.496	
Latvia	0.511	0.485	0.025	0.010	0.463	0.499	
Lithuania	0.502	0.496	0.006	0.002	0.491	0.499	
Luxembourg	1.000	−0.287	1.287	0.331	−1.037	0.255	
Malta	0.502	0.497	0.005	0.002	0.492	0.499	
The Netherlands	0.508	0.489	0.019	0.007	0.472	0.498	
Poland	0.504	0.494	0.010	0.003	0.485	0.498	
Portugal	0.503	0.495	0.008	0.002	0.488	0.499	
Romania	0.505	0.492	0.013	0.004	0.480	0.498	
Slovakia	0.502	0.496	0.006	0.002	0.491	0.499	
Slovenia	0.502	0.497	0.005	0.002	0.492	0.499	
Spain	0.503	0.495	0.009	0.003	0.487	0.499	
Sweden	0.649	0.294	0.355	0.142	−0.025	0.490	
United Kingdom	0.503	0.495	0.007	0.002	0.489	0.499	
							2010
Austria	0.523	0.466	0.057	0.020	0.415	0.495	
Belgium	0.507	0.489	0.018	0.006	0.472	0.498	
Bulgaria	0.501	0.498	0.004	0.001	0.494	0.499	
Croatia	0.516	0.475	0.042	0.013	0.437	0.494	
Cyprus	0.502	0.496	0.006	0.002	0.491	0.499	
Czech Republic	0.532	0.449	0.082	0.024	0.375	0.484	
Denmark	0.579	0.376	0.203	0.061	0.192	0.466	
Estonia	0.501	0.499	0.002	0.001	0.497	0.500	
Finland	0.503	0.496	0.007	0.002	0.489	0.499	
France	0.510	0.485	0.026	0.009	0.461	0.497	
Germany	0.504	0.494	0.009	0.003	0.486	0.499	
Greece	0.502	0.497	0.006	0.002	0.492	0.499	
Hungary	1.000	−0.313	1.313	0.376	−1.508	0.229	
Ireland	0.506	0.491	0.015	0.005	0.478	0.498	
Italy	0.508	0.486	0.022	0.006	0.467	0.496	
Latvia	0.506	0.492	0.014	0.005	0.479	0.499	
Lithuania	0.502	0.497	0.005	0.001	0.492	0.499	
Luxembourg	1.000	−0.302	1.302	0.338	−1.195	0.226	
Malta	0.502	0.496	0.006	0.002	0.491	0.499	
The Netherlands	0.508	0.488	0.020	0.007	0.470	0.498	
Poland	0.504	0.494	0.010	0.003	0.485	0.498	
Portugal	0.503	0.495	0.008	0.003	0.487	0.499	
Romania	0.506	0.491	0.014	0.004	0.478	0.498	
Slovakia	0.502	0.497	0.006	0.002	0.491	0.499	
Slovenia	0.502	0.497	0.005	0.002	0.492	0.499	
Spain	0.504	0.493	0.011	0.003	0.483	0.499	
Sweden	0.596	0.359	0.237	0.084	0.144	0.479	
United Kingdom	0.503	0.496	0.007	0.002	0.489	0.499	
							2012
Austria	0.529	0.459	0.070	0.027	0.397	0.497	
Belgium	0.508	0.488	0.020	0.007	0.469	0.498	
Bulgaria	0.501	0.498	0.003	0.001	0.495	0.499	
Croatia	0.514	0.478	0.037	0.011	0.444	0.494	
Cyprus	0.502	0.497	0.006	0.002	0.491	0.499	
Czech Republic	0.529	0.454	0.076	0.022	0.385	0.486	

Table A1. *Cont.*

Country	Score Original	Bias Corrected	Bias	Std	Lower	Upper	2008
Denmark	0.588	0.375	0.213	0.078	0.184	0.488	
Estonia	0.501	0.499	0.002	0.001	0.497	0.500	
Finland	0.503	0.495	0.008	0.002	0.488	0.499	
France	0.510	0.485	0.025	0.008	0.462	0.497	
Germany	0.503	0.495	0.008	0.003	0.488	0.499	
Greece	0.502	0.497	0.004	0.001	0.494	0.499	
Hungary	1.000	−0.306	1.306	0.383	−1.494	0.244	
Ireland	0.506	0.492	0.014	0.005	0.479	0.498	
Italy	0.507	0.489	0.018	0.006	0.472	0.498	
Latvia	0.507	0.491	0.016	0.006	0.477	0.499	
Lithuania	0.502	0.497	0.005	0.002	0.492	0.499	
Luxembourg	1.000	−0.279	1.279	0.347	−1.133	0.273	
Malta	0.502	0.497	0.005	0.001	0.492	0.499	
The Netherlands	0.507	0.489	0.018	0.006	0.472	0.498	
Poland	0.504	0.494	0.010	0.003	0.485	0.498	
Portugal	0.503	0.496	0.007	0.002	0.489	0.499	
Romania	0.505	0.492	0.013	0.004	0.480	0.498	
Slovakia	0.502	0.497	0.006	0.002	0.491	0.499	
Slovenia	0.502	0.497	0.004	0.001	0.493	0.499	
Spain	0.503	0.495	0.008	0.002	0.488	0.499	
Sweden	0.621	0.328	0.293	0.112	0.064	0.485	
United Kingdom	0.503	0.496	0.007	0.002	0.490	0.499	
							2014
Austria	0.531	0.460	0.071	0.030	0.396	0.499	
Belgium	0.507	0.489	0.018	0.006	0.472	0.497	
Bulgaria	0.501	0.498	0.003	0.001	0.495	0.499	
Croatia	0.513	0.480	0.033	0.010	0.450	0.494	
Cyprus	0.502	0.497	0.005	0.001	0.493	0.499	
Czech Republic	0.528	0.456	0.072	0.022	0.391	0.487	
Denmark	0.604	0.358	0.246	0.097	0.138	0.492	
Estonia	0.501	0.499	0.002	0.001	0.497	0.500	
Finland	0.503	0.495	0.008	0.002	0.489	0.499	
France	0.512	0.482	0.030	0.009	0.454	0.496	
Germany	0.503	0.496	0.007	0.002	0.490	0.499	
Greece	0.502	0.498	0.004	0.001	0.494	0.499	
Hungary	1.000	−0.297	1.297	0.396	−1.475	0.270	
Ireland	0.506	0.491	0.016	0.005	0.476	0.498	
Italy	0.507	0.489	0.019	0.006	0.471	0.497	
Latvia	0.507	0.491	0.015	0.007	0.477	0.500	
Lithuania	0.502	0.496	0.006	0.002	0.491	0.499	
Luxembourg	1.000	−0.117	1.117	0.458	−1.106	0.465	
Malta	0.502	0.497	0.006	0.002	0.491	0.499	
The Netherlands	0.506	0.491	0.016	0.005	0.476	0.498	
Poland	0.503	0.495	0.009	0.003	0.486	0.498	
Portugal	0.502	0.496	0.006	0.002	0.490	0.499	
Romania	0.505	0.492	0.014	0.004	0.479	0.498	
Slovakia	0.502	0.496	0.006	0.002	0.491	0.499	
Slovenia	0.502	0.497	0.005	0.001	0.493	0.499	
Spain	0.503	0.496	0.007	0.002	0.489	0.499	
Sweden	0.665	0.281	0.384	0.162	−0.064	0.496	
United Kingdom	0.503	0.496	0.007	0.002	0.489	0.499	
							2016
Austria	0.527	0.456	0.071	0.022	0.402	0.488	
Belgium	0.506	0.487	0.018	0.004	0.478	0.496	
Bulgaria	0.501	0.498	0.003	0.001	0.495	0.499	
Croatia	0.513	0.480	0.033	0.009	0.450	0.492	
Cyprus	0.502	0.497	0.005	0.001	0.493	0.499	

Table A1. *Cont.*

Country	Score Original	Bias Corrected	Bias	Std	Lower	Upper	2008
Czech Republic	0.530	0.458	0.072	0.021	0.382	0.480	
Denmark	0.572	0.326	0.246	0.056	0.236	0.469	
Estonia	0.501	0.499	0.002	0.001	0.497	0.500	
Finland	0.503	0.496	0.008	0.002	0.488	0.498	
France	0.510	0.479	0.030	0.007	0.462	0.493	
Germany	0.503	0.496	0.007	0.002	0.490	0.498	
Greece	0.502	0.498	0.004	0.001	0.494	0.499	
Hungary	1.000	−0.297	1.297	0.350	−1.482	0.148	
Ireland	0.507	0.492	0.016	0.005	0.473	0.496	
Italy	0.506	0.487	0.019	0.004	0.476	0.496	
Latvia	0.503	0.488	0.015	0.003	0.488	0.499	
Lithuania	0.502	0.496	0.006	0.002	0.492	0.499	
Luxembourg	0.824	−0.293	1.117	0.258	−0.607	0.376	
Malta	0.505	0.499	0.006	0.003	0.482	0.497	
The Netherlands	0.505	0.489	0.016	0.003	0.481	0.497	
Poland	0.504	0.495	0.009	0.003	0.486	0.498	
Portugal	0.502	0.496	0.006	0.002	0.490	0.499	
Romania	0.506	0.493	0.014	0.005	0.475	0.496	
Slovakia	0.502	0.497	0.006	0.002	0.491	0.498	
Slovenia	0.502	0.498	0.005	0.002	0.491	0.499	
Spain	0.503	0.496	0.007	0.002	0.489	0.498	
Sweden	0.606	0.222	0.384	0.086	0.110	0.457	
United Kingdom	0.503	0.496	0.007	0.002	0.488	0.498	

Table A2. Bias corrected efficiency scores of countries' by modelling framework. Framework M2.

Country	Score Original	Bias Corrected	Bias	Std	Lower	Upper	2008
Austria	0.528	0.461	0.068	0.027	0.400	0.498	
Belgium	0.507	0.490	0.017	0.006	0.475	0.499	
Bulgaria	0.501	0.498	0.004	0.001	0.494	0.499	
Croatia	0.515	0.475	0.040	0.012	0.439	0.493	
Cyprus	0.502	0.496	0.006	0.002	0.491	0.499	
Czech Republic	0.534	0.446	0.087	0.026	0.367	0.484	
Denmark	0.567	0.396	0.171	0.052	0.241	0.474	
Estonia	0.501	0.499	0.002	0.001	0.496	0.500	
Finland	0.503	0.495	0.009	0.003	0.486	0.499	
France	0.510	0.486	0.023	0.008	0.465	0.498	
Germany	0.504	0.494	0.010	0.003	0.485	0.499	
Greece	0.502	0.496	0.006	0.002	0.491	0.499	
Hungary	1.000	−0.304	1.304	0.385	−1.490	0.249	
Ireland	0.504	0.494	0.010	0.003	0.484	0.498	
Italy	0.508	0.487	0.021	0.006	0.468	0.496	
Latvia	0.511	0.485	0.025	0.010	0.463	0.499	
Lithuania	0.502	0.496	0.006	0.002	0.491	0.499	
Luxembourg	1.000	−0.287	1.287	0.330	−1.037	0.255	
Malta	0.502	0.497	0.005	0.002	0.492	0.499	
The Netherlands	0.508	0.489	0.019	0.007	0.472	0.498	
Poland	0.504	0.494	0.010	0.003	0.485	0.498	
Portugal	0.503	0.495	0.008	0.002	0.488	0.499	
Romania	0.505	0.492	0.013	0.004	0.480	0.498	
Slovakia	0.502	0.496	0.006	0.002	0.491	0.499	
Slovenia	0.502	0.497	0.005	0.002	0.492	0.499	
Spain	0.503	0.495	0.009	0.003	0.487	0.499	
Sweden	0.650	0.218	0.431	0.173	−0.178	0.387	
United Kingdom	0.503	0.495	0.007	0.002	0.489	0.499	
							2010
Austria	0.523	0.466	0.058	0.020	0.414	0.495	

Table A2. *Cont.*

Country	Score Original	Bias Corrected	Bias	Std	Lower	Upper	2008
Belgium	0.507	0.489	0.019	0.006	0.472	0.498	
Bulgaria	0.501	0.498	0.004	0.001	0.494	0.499	
Croatia	0.516	0.475	0.042	0.013	0.436	0.494	
Cyprus	0.502	0.496	0.006	0.002	0.491	0.499	
Czech Republic	0.532	0.449	0.082	0.024	0.375	0.484	
Denmark	0.579	0.376	0.203	0.062	0.192	0.467	
Estonia	0.501	0.499	0.002	0.001	0.497	0.500	
Finland	0.503	0.496	0.007	0.002	0.489	0.499	
France	0.510	0.485	0.026	0.008	0.461	0.497	
Germany	0.504	0.494	0.009	0.003	0.486	0.499	
Greece	0.502	0.497	0.006	0.002	0.492	0.499	
Hungary	1.000	−0.311	1.311	0.378	−1.504	0.233	
Ireland	0.506	0.491	0.015	0.005	0.478	0.498	
Italy	0.508	0.487	0.022	0.006	0.467	0.496	
Latvia	0.506	0.491	0.014	0.005	0.479	0.499	
Lithuania	0.502	0.497	0.005	0.001	0.492	0.499	
Luxembourg	1.000	−0.306	1.306	0.337	−1.203	0.219	
Malta	0.502	0.496	0.006	0.002	0.491	0.499	
The Netherlands	0.508	0.488	0.020	0.007	0.470	0.498	
Poland	0.504	0.494	0.010	0.003	0.485	0.498	
Portugal	0.503	0.495	0.008	0.003	0.487	0.499	
Romania	0.506	0.491	0.014	0.004	0.478	0.498	
Slovakia	0.502	0.496	0.006	0.002	0.491	0.499	
Slovenia	0.502	0.497	0.005	0.002	0.492	0.499	
Spain	0.504	0.493	0.011	0.003	0.483	0.499	
Sweden	0.608	0.139	0.468	0.176	−0.262	0.304	
United Kingdom	0.503	0.496	0.007	0.002	0.489	0.499	
							2012
Austria	0.529	0.459	0.071	0.027	0.395	0.497	
Belgium	0.508	0.488	0.020	0.007	0.469	0.498	
Bulgaria	0.501	0.498	0.003	0.001	0.495	0.499	
Croatia	0.514	0.478	0.037	0.011	0.444	0.494	
Cyprus	0.502	0.497	0.006	0.002	0.491	0.499	
Czech Republic	0.529	0.454	0.076	0.022	0.385	0.486	
Denmark	0.588	0.375	0.213	0.079	0.183	0.488	
Estonia	0.501	0.499	0.002	0.001	0.497	0.500	
Finland	0.503	0.495	0.008	0.002	0.488	0.499	
France	0.510	0.485	0.025	0.008	0.462	0.497	
Germany	0.503	0.495	0.008	0.003	0.488	0.499	
Greece	0.502	0.497	0.004	0.001	0.494	0.499	
Hungary	1.000	−0.306	1.306	0.383	−1.494	0.244	
Ireland	0.506	0.492	0.014	0.005	0.479	0.498	
Italy	0.507	0.489	0.018	0.006	0.472	0.498	
Latvia	0.507	0.491	0.016	0.006	0.477	0.499	
Lithuania	0.502	0.497	0.005	0.002	0.492	0.499	
Luxembourg	1.000	−0.279	1.279	0.347	−1.133	0.273	
Malta	0.502	0.497	0.005	0.001	0.492	0.499	
The Netherlands	0.507	0.489	0.018	0.006	0.472	0.498	
Poland	0.504	0.494	0.010	0.003	0.485	0.498	
Portugal	0.503	0.496	0.007	0.002	0.489	0.499	
Romania	0.505	0.492	0.013	0.004	0.480	0.498	
Slovakia	0.502	0.497	0.006	0.002	0.491	0.499	
Slovenia	0.502	0.497	0.004	0.001	0.493	0.499	
Spain	0.503	0.495	0.008	0.002	0.488	0.499	
Sweden	0.622	0.190	0.431	0.170	−0.212	0.349	
United Kingdom	0.503	0.496	0.007	0.002	0.490	0.499	
							2014
Austria	0.531	0.459	0.072	0.031	0.395	0.500	

Table A2. *Cont.*

Country	Score Original	Bias Corrected	Bias	Std	Lower	Upper	2008
Belgium	0.507	0.489	0.018	0.006	0.472	0.497	
Bulgaria	0.501	0.498	0.003	0.001	0.495	0.499	
Croatia	0.513	0.480	0.033	0.010	0.450	0.494	
Cyprus	0.502	0.497	0.005	0.001	0.493	0.499	
Czech Republic	0.528	0.456	0.072	0.022	0.391	0.487	
Denmark	0.604	0.358	0.246	0.098	0.136	0.491	
Estonia	0.501	0.499	0.002	0.001	0.497	0.500	
Finland	0.503	0.495	0.008	0.002	0.489	0.499	
France	0.512	0.482	0.030	0.009	0.454	0.496	
Germany	0.503	0.496	0.007	0.002	0.490	0.499	
Greece	0.502	0.498	0.004	0.001	0.494	0.499	
Hungary	1.000	−0.297	1.297	0.396	−1.475	0.270	
Ireland	0.506	0.491	0.016	0.005	0.476	0.498	
Italy	0.507	0.489	0.019	0.006	0.471	0.497	
Latvia	0.507	0.491	0.015	0.007	0.477	0.500	
Lithuania	0.502	0.496	0.006	0.002	0.491	0.499	
Luxembourg	1.000	−0.117	1.117	0.458	−1.106	0.465	
Malta	0.502	0.497	0.006	0.002	0.491	0.499	
The Netherlands	0.506	0.491	0.016	0.005	0.476	0.498	
Poland	0.503	0.495	0.009	0.003	0.486	0.498	
Portugal	0.502	0.496	0.006	0.002	0.490	0.499	
Romania	0.505	0.492	0.014	0.004	0.479	0.498	
Slovakia	0.502	0.496	0.006	0.002	0.491	0.499	
Slovenia	0.502	0.497	0.005	0.001	0.493	0.499	
Spain	0.503	0.496	0.007	0.002	0.489	0.499	
Sweden	0.665	0.277	0.389	0.162	-0.073	0.489	
United Kingdom	0.503	0.496	0.007	0.002	0.489	0.499	
							2016
Austria	0.527	0.460	0.067	0.022	0.401	0.489	
Belgium	0.506	0.491	0.015	0.004	0.478	0.496	
Bulgaria	0.501	0.498	0.004	0.001	0.495	0.499	
Croatia	0.513	0.479	0.034	0.009	0.450	0.492	
Cyprus	0.502	0.497	0.005	0.001	0.493	0.499	
Czech Republic	0.530	0.451	0.079	0.021	0.382	0.480	
Denmark	0.572	0.392	0.180	0.056	0.235	0.469	
Estonia	0.501	0.499	0.002	0.001	0.497	0.500	
Finland	0.503	0.495	0.008	0.002	0.488	0.498	
France	0.510	0.484	0.025	0.007	0.462	0.493	
Germany	0.503	0.496	0.007	0.002	0.490	0.498	
Greece	0.502	0.497	0.004	0.001	0.494	0.499	
Hungary	1.000	-0.322	1.322	0.350	−1.482	0.148	
Ireland	0.507	0.489	0.018	0.005	0.473	0.496	
Italy	0.506	0.490	0.016	0.004	0.476	0.496	
Latvia	0.503	0.495	0.008	0.003	0.488	0.499	
Lithuania	0.502	0.497	0.006	0.002	0.492	0.499	
Luxembourg	0.824	0.070	0.754	0.258	−0.607	0.376	
Malta	0.505	0.493	0.012	0.003	0.482	0.497	
The Netherlands	0.505	0.492	0.013	0.003	0.481	0.497	
Poland	0.504	0.494	0.009	0.003	0.486	0.498	
Portugal	0.502	0.496	0.006	0.002	0.490	0.499	
Romania	0.506	0.490	0.017	0.005	0.475	0.496	
Slovakia	0.502	0.496	0.006	0.002	0.491	0.498	
Slovenia	0.502	0.496	0.006	0.002	0.491	0.499	
Spain	0.503	0.495	0.007	0.002	0.489	0.498	
Sweden	0.606	0.211	0.395	0.140	−0.149	0.360	
United Kingdom	0.503	0.495	0.008	0.002	0.488	0.498	

References

1. Defra. *Guidance on Applying the Waste Hierarchy*; Department for Environment, Food and Rural Affairs: London, UK, 2011.
2. FhG-IBP. *Waste 2 Go: D 2.2 Waste Profiling*; FhG-IBP (Fraunhofer-Gesellschaft zur Förderung der angewandten Forschung e.V.): Munich, Germany, 2014; Available online: http://www.waste2go.eu/downlo%20ad/1/D2.2_Waste%20profiling.pdf (accessed on 25 June 2019).
3. European Commission. *EUROPE 2020: A European Strategy for Smart, Sustainable and Inclusive Growth*; European Commission, Communication from the Commission: Brussels, Belgium, 2010; Available online: http://ec.europa.eu/eu2020/pdf/COMPLET%20EN%20BARROSO%20%20%20007%20-%20Europe%202020%20-%20EN%20version.pdf (accessed on 4 September 2019).
4. European Commission. *The Road to a Circular Economy*; European Commission, Environment: Brussels, Belgium, 2016; Available online: http://ec.europa.eu/environment/news/efe/articles/2014/08/%20article_20140806_01_en.htm (accessed on 5 July 2019).
5. European Commission. *2030 Climate & Energy Framework*; European Commission: Brussels, Belgium, 2019; Available online: https://ec.europa.eu/clima/policies/strategies/2030_en#tab-0-0 (accessed on 5 September 2019).
6. European Commission. *A Clean Planet for All—A European Strategic Long-Term Vision for a Prosperous, Modern, Competitive and Climate Neutral Economy*; European Commission: Brussels, Belgium, 2018.
7. Halkos, G.E.; Tzeremes, N.G. Renewable energy consumption and economic efficiency: Evidence from European countries. *Renew. Energy* **2013**, *5*, 041803. [CrossRef]
8. Fruergaard, T.; Astrup, T. Optimal utilization of waste-to-energy in an LCA perspective. *Waste Manag.* **2011**, *31*, 572–582. [CrossRef] [PubMed]
9. Apergis, N.; Payne, J.E. Renewable energy consumption and economic growth: Evidence from a panel of OECD countries. *Energy Policy* **2010**, *38*, 656–660. [CrossRef]
10. Halkos, G.E.; Tzeremes, N.G. Analyzing the Greek renewable energy sector: A Data Envelopment Analysis approach. *Renew. Sustain. Energy Rev.* **2012**, *16*, 2884–2893. [CrossRef]
11. Zhou, Y.; Ma, M.; Kong, F.; Wang, K.; Bi, J. Capturing the co-benefits of energy efficiency in China—A perspective from the water-energy nexus. *Resour. Conserv. Recycl.* **2018**, *132*, 93–101. [CrossRef]
12. Silva, R.D.S.; Oliveira, R.C.; Tostes, M.E.L. Analysis of the Brazilian Energy Efficiency Program for Electricity Distribution Systems. *Energies* **2017**, *10*, 1391. [CrossRef]
13. Ohnishi, S.; Fujii, M.; Ohata, M.; Rokuta, I.; Fujita, T. Efficient energy recovery through a combination of waste-to-energy systems for a low-carbon city. *Resour. Conserv. Recycl.* **2018**, *128*, 394–405. [CrossRef]
14. De Alencar Bezerra, S.; Jackson dos Santos, F.; Rogerio Pinheiro, P.; Rocha Barbosa, F. Dynamic Evaluation of the Energy Efficiency of Environments in Brazilian University Classrooms Using DEA. *Sustainability* **2017**, *9*, 2373. [CrossRef]
15. Boussofiane, A.; Dyson, R.G.; Thanassoulis, E. Applied data envelopment analysis. *Eur. J. Oper. Res.* **1991**, *52*, 1–15. [CrossRef]
16. Honma, S.; Hu, J.L. Efficient waste and pollution abatements for regions in Japan. *Int. J. Sustain. Dev. World Ecol.* **2009**, *16*, 270–285. [CrossRef]
17. WMR. *Incineration*; Waste Management Resources: Haverfordwest, UK, 2009.
18. Eunomia. *Economic Analysis of Options for Managing Biodegradable Municipal Waste*; Final Report to the European Commission; Eunomia: Bristol, UK, 2011.
19. Defra. *Incineration of Municipal Solid Waste*; Department for Environment, Food and Rural Affairs: London, UK, 2013.
20. Scarlat, N.; Fahl, F.; Dallemand, J.F. Status and Opportunities for Energy Recovery from Municipal Solid Waste in Europe. *Waste Biomass Valorization* **2018**, *10*, 2425–2444. [CrossRef]
21. Chen, H.W.; Chang, N.B.; Chen, J.C.; Tsai, S.J. Environmental performance evaluation of large-scale municipal solid waste incinerators using data envelopmfent analysis. *Waste Manag.* **2010**, *30*, 1371–1381. [CrossRef] [PubMed]
22. De Beer, J.; Cihlar, J.; Hensing, I.; Zabeti, M. *Status and Prospects of Co-Processing of Waste in EU Cement Plants*; Ecofys Publication: Utrecht, The Netherlands, 2017.

23. Bridgwater, A.V. Review of fast pyrolysis of biomass and product upgrading. *Biomass Bioenergy* **2012**, *38*, 68–94. [CrossRef]
24. Maschio, G.; Koufopanos, C.; Lucches, A. Pyrolysis, a Promising Route for Biomass Utilization. *Bioresour. Technol.* **1992**, *42*, 219–231. [CrossRef]
25. Derimbas, A.; Arin, G. An Overview of Biomass Pyrolysis. *Energy Sources* **2002**, *24*, 471–482.
26. Bridgwater, A.V.; Peacocke, G.V. Fast pyrolysis processes for biomass. *Renew. Sustain. Energy Rev.* **2000**, *4*, 1–73. [CrossRef]
27. Garcia-Perez, M.; Shen, J.; Wang, S.X.; Li, C.Z. Production and fuel properties of fast pyrolysis oil/bio-diesel blends. *Fuel Process. Technol.* **2010**, *91*, 296–305. [CrossRef]
28. Higman, C. Gasification. In *Combustion Engineering Issues for Solid Fuel Systems*; Academic Press: Cambridge, MA, USA, 2008; Chapter 11.
29. Belgiorno, V.; De Feo, G.; Della Rocca, C.; Napoli, R.M.A. Energy from gasification of solid wastes. *Waste Manag.* **2003**, *23*, 1–15. [CrossRef]
30. WRAP. *Anaerobic Digestion*; The Waste and Resources Action Programme: Banbury, UK, 2016; Available online: http://www.wrap.org.uk/content/anaerobic-digestion-1 (accessed on 5 June 2019).
31. Verman, S. Anaerobic Digestion of Biodegradable Organics in Municipal Solid Wastes. Master's Thesis, Columbia University, New York, NY, USA, 2002.
32. Oreggioni, G.D.; Gowreesunker, B.L.; Tassou, S.A.; Bianchi, G.; Reilly, M.; Kirby, M.E.; Toop, T.A.; Theodorou, M.K. Potential for Energy Production from Farm Wastes Using Anaerobic Digestion in the UK: An Economic Comparison of Different Size Plants. *Energies* **2017**, *10*, 1396. [CrossRef]
33. Defra; DECC; NNFCC. The Official Information Portal on Anaerobic Digestion. Available online: http://www.biogas-info.co.uk/ (accessed on 10 June 2019).
34. Sarc, R.; Lorber, K.E. Production, quality and quality assurance of Refuse Derived Fuels (RDFs). *Waste Manag.* **2013**, *33*, 1825–1834. [CrossRef]
35. UNEP. *Global Waste Management Outlook*; United Nations Environment Programme: Nairobi, Kenya, 2015; Available online: http://www.unep.org/ietc/Portals/136/Publications/Waste%20%20Management/GWMO%20report/GWMO%20full%20report.pdf (accessed on 15 June 2019).
36. Sherman, H.D.; Zhu, J. Data Envelopment Analysis explained. In *Service Productivity Management, Improving Service Performance Using Data Envelopment Analysis (DEA)*; Springer: New York, NY, USA, 2006; Chapter 2.
37. Panayiotou, G.P.; Bianchi, G.; Georgiou, G.; Arestia, L.; Argyrou, M.; Agathokleous, R.; Tsamos, K.M.; Tassou, S.A.; Florides, G.; Kalogirou, S.; et al. Preliminary assessment of waste heat potential in major European industries. *Energy Procedia* **2017**, *123*, 335–345. [CrossRef]
38. European Commission. *Communication from the Commission to the European Parliament, the Council, the European Economic and Social Committee and the Committee of the Regions, The Role of Waste-To-Energy in the Circular Economy*; European Commission: Brussels, Belgium, 2017; Available online: https://ec.europa.eu/environment/waste/waste-to-energy.pdf (accessed on 13 July 2019).
39. Agathokleous, R.; Bianchi, G.; Panayiotou, G.; Arestia, L.; Argyrou, M.C.; Georgiou, G.S.; Tassou, S.A.; Jouhara, H.; Kalogirou, S.A.; Florides, G.A.; et al. Waste Heat Recovery in the EU industry and proposed new technologies. *Energy Procedia* **2019**, *161*, 489–496. [CrossRef]
40. Persson, U.; Munster, M. Current and future prospects for heat recovery from waste in European district heating systems: A literature and data review. *Energy* **2016**, *110*, 116–128. [CrossRef]
41. Mardani, A.; Zavadskas, E.K.; Streimikiene, D.; Jusoh, A.; Khoshnoudi, M. A comprehensive review of data envelopment analysis (DEA) approach in energy efficiency. *Renew. Sustain. Energy Rev.* **2017**, *70*, 1298–1322. [CrossRef]
42. Sueyoshi, T.; Yuana, Y.; Goto, M. A literature study for DEA applied to energy and environment. *Energy Econ.* **2017**, *62*, 104–124. [CrossRef]
43. Färe, R.; Grosskop, F.S.; Logan, J. The relative efficiency of Illinois electric utilities. *Renew. Energy* **1983**, *5*, 349–367. [CrossRef]
44. Hailu, A.; Veeman, T.S. Non-parametric productivity analysis with undesirable outputs: An application to the Canadian pulp and paper industry. *Am. J. Agric. Econ.* **2011**, *83*, 605–616. [CrossRef]
45. Mukherjee, K. Energy use efficiency in US manufacturing: A nonparametric analysis. *Energy Econ.* **2008**, *30*, 76–96. [CrossRef]

46. Zhou, P.; Poh, K.L.; Ang, B.W. A non-radial DEA approach to measuring environmental performance. *Eur. J. Oper. Res.* **2007**, *178*, 1–9. [CrossRef]
47. Halkos, G.E.; Tzeremes, N.G. Exploring the existence of Kuznets curve in countries' environmental efficiency using DEA window analysis. *Ecol. Econ.* **2009**, *68*, 2168–2176. [CrossRef]
48. Lee, J.D.; Park, J.B.; Kim, T.Y. Estimation of the shadow prices of pollutants with production/environment inefficiency taken into account: A nonparametric directional distance function approach. *J. Environ. Manag.* **2002**, *64*, 365–375. [CrossRef]
49. Mukherjee, K. Measuring energy efficiency in the context of an emerging economy: The case of Indian manufacturing. *Eur. J. Oper. Res.* **2010**, *201*, 933–941. [CrossRef]
50. Chen, T.; Hu, J.L. Renewable energy and macroeconomic efficiency of OECD and non-OECD economies. *Energy Policy* **2007**, *35*, 3606–3615. [CrossRef]
51. Hu, J.L.; Wang, S.C. Total-factor energy efficiency of regions in China. *Energy Policy* **2006**, *34*, 3206–3217. [CrossRef]
52. Yeh, T.L.; Chen, T.Y.; Lai, P.Y. A comparative study of energy utilization efficiency between Taiwan and China. *Energy Policy* **2010**, *38*, 2386–2394. [CrossRef]
53. Bian, Y.W.; Fang, Y. Resource and environment efficiency analysis of provinces in China: A DEA approach based on Shannon's entropy. *Energy Policy* **2010**, *38*, 1909–1917. [CrossRef]
54. Zhou, P.; Ang, B.W. Linear programming models for measuring economy-wide energy efficiency performance. *Energy Policy* **2008**, *36*, 2911–2916. [CrossRef]
55. Wang, K.; Yuc, S.; Zhang, W. China's regional energy and environmental efficiency: A DEA window analysis based dynamic evaluation. *Math. Comput. Model.* **2013**, *58*, 1117–1127. [CrossRef]
56. Wang, Q.; Zhou, P.; Zhou, D. Efficiency measurement with carbon dioxide emissions: The case of China. *Appl. Energy* **2012**, *90*, 161–166. [CrossRef]
57. Yang, Y.; Wang, C.; Liu, W.; Zhou, P. Microsimulation of low carbon urban transport policies in Beijing. *Energy Policy* **2017**, *107*, 561–572. [CrossRef]
58. Song, M.; Peng, J.; Wang, J.; Dong, L. Better resource management: An improved resource and environmental efficiency evaluation approach that considers undesirable outputs. *Resour. Conserv. Recycl.* **2018**, *128*, 197–205. [CrossRef]
59. Dostalova, K. Efficiency Evaluation of Municipal Solid Waste Management in the Czech Republic Using DEA Method. Master's Thesis, Masaryk University, Faculty of Economics and Administration, Brno, Czech Republic, 2014.
60. Seiford, L.M.; Thrall, R.M. Recent developments in DEA: The mathematical programming approach to frontier analysis. *J. Econom.* **1990**, *46*, 7–38. [CrossRef]
61. Halkos, G.; Petrou, K.N. Assessing 28 EU member states' environmental efficiency in national waste generation with DEA. *J. Clean. Prod.* **2018**, *208*, 509–521. [CrossRef]
62. Farrell, M.J. The Measurement of Productive Efficiency. *J. R. Stat. Soc.* **1957**, *120*, 253–290. [CrossRef]
63. Charnes, A.; Cooper, W.W.; Rhodes, E. Measuring the efficiency of decision making units. *Eur. J. Oper. Res.* **1978**, *2*, 429–444. [CrossRef]
64. Vyas, G.S.; Jha, K.N. Benchmarking green building attributes to achieve cost effectiveness using a data envelopment analysis. *Sustain. Cities Soc.* **2017**, *28*, 127–134. [CrossRef]
65. Simar, L.; Wilson, P.W. Sensitivity analysis of efficiency scores: How to bootstrap in non parametric frontier models. *Manag. Sci.* **1998**, *44*, 49–61. [CrossRef]
66. Simar, L.; Wilson, P.W. A general methodology for bootstrapping in nonparametric frontier models. *J. Appl. Stat.* **2000**, *27*, 779–802. [CrossRef]
67. Simar, L.; Wilson, P.W. Non parametric tests of return to scale. *Eur. J. Oper. Res.* **2002**, *139*, 115–132. [CrossRef]
68. Efron, B. Bootstrap methods: Another look at the jackknife. *Ann. Stat.* **1979**, *7*, 1–26. [CrossRef]
69. Dyson, R.G.; Shale, E.A. Data envelopment analysis, operational research and Uncertainty. *J. Oper. Res. Soc.* **2010**, *61*, 25–34. [CrossRef]
70. Podinovski, V.V. Bridging the Gap between the Constant and Variable Returns-to-Scale Models: Selective Proportionality in Data Envelopment Analysis. *J. Oper. Res. Soc.* **2004**, *55*, 265–276. [CrossRef]
71. Coelli, T.J.; Rao, D.S.P.; O'Donell, C.J.; Battese, G.E. *An Introduction to Efficiency and Productivity Analysis*; Springer: New York, NY, USA, 2005.

72. Banker, R.D.; Charnes, A.; Cooper, W.W. Some models for estimating technical and scale inefficiencies in data envelopment analysis. *Manag. Sci.* **1984**, *30*, 1078–1092. [CrossRef]
73. Laso, J.; Hoehn, D.; Margallo, M.; García-Herrero, I.; Batlle-Bayer, L.; Bala, A.; Fullana-i-Palmer, P.; Vázquez-Rowe, I.; Irabien, A.; Aldaco, R. Assessing Energy and Environmental Efficiency of the Spanish Agri-Food System Using the LCA/DEA Methodology. *Energies* **2018**, *11*, 3395. [CrossRef]
74. Vlontzos, G.; Niavis, S.; Pardalos, P. Testing for Environmental Kuznets Curve in the EU Agricultural Sector through an Eco-(in)Efficiency Index. *Energies* **2017**, *10*, 1992. [CrossRef]
75. Portnov, B.A.; Erell, E. Interregional inequalities in Israel, 1948–1995: Divergence or convergence? *Socio-Econ. Plan. Sci.* **2004**, *38*, 255–289. [CrossRef]
76. Song, M.; Yang, L.; Wu, J.; Lv, D. Energy saving in China: Analysis on the energy efficiency via bootstrap-DEA approach. *Energy Policy* **2013**, *57*, 1–6. [CrossRef]
77. Suzuki, S.; Nijkamp, P. An evaluation of energy-environment-economic efficiency for EU, APEC and ASEAN countries: Design of a Target-Oriented DFM model with fixed factors in Data Envelopment Analysis. *Energy Policy* **2016**, *88*, 100–112. [CrossRef]
78. European Commission. *Communication from the Commission to the European Parliament, The Council, The European Economic and Social Committee, The Committee of the Regions and the European Investment Bank, COM(2017) 688 Final. Third Report on the State of the Energy Union*; European Commission: Brussels, Belgium, 2017.
79. Miranda, M.L.; Hale, B. Waste not, want not: The private and social costs of waste-to-energy production. *Energy Policy* **1997**, *25*, 587–600. [CrossRef]
80. Miranda, M.L.; Hale, B. Paradise recovered: Energy production and waste management in island environments. *Energy Policy* **2005**, *33*, 1691–1702. [CrossRef]
81. Milbrandt, A.; Seiple, T.; Heimiller, D.; Skaggs, R.; Coleman, A. Wet waste-to-energy resources in the United States. *Resour. Conserv. Recycl.* **2018**, *137*, 32–47. [CrossRef]
82. Krajacic, G.; Duic, N.; Vujanovic, M.; Kilkis, S.; Rosen, M.A.; Al-Nimr, M.A. Sustainable development of energy, water and environment systems for future energy technologies and concepts. *Energy Convers. Manag.* **2016**, *125*, 1–14. [CrossRef]
83. WWF. *Denmark Waste to Energy*; WWF: Gland, Switzerland, 2012; Available online: http://wwf.panda.org/?204596/Denmark-waste-to-energy (accessed on 17 July 2019).
84. Geng, Y.; Fujita, T.; Park, H.S.; Chiu, A.S.F.; Huisingh, D. Recent progress on innovative eco-industrial development. *J. Clean. Prod.* **2016**, *114*, 1–10. [CrossRef]
85. CEWEP. *Waste to Energy Plants in Europe in 2017*; CEWEP: Dusseldorf, Germany, 2016; Available online: http://www.cewep.eu/waste-to-energy-plants-in-europe-in-2017/ (accessed on 20 September 2019).
86. World Energy Council. *World Energy Resources Waste to Energy*; World Energy Council: London, UK, 2016.
87. ISWA. *Guidelines: Waste-To-Energy in Low and Middle Income Countries*; International Solid Waste Association: Wien, Austria, 2013.
88. Malinauskaite, J.; Jouhara, H.; Czajczyńska, D.; Stanchev, P.; Katsou, E.; Rostkowski, P.; Thorne, R.J.; Colón, J.; Ponsá, S.; Al-Mansour, F.; et al. Municipal solid waste management and waste-to-energy in the context of a circular economy and energy recycling in Europe. *Energy* **2017**, *141*, 2013–2044. [CrossRef]
89. European Commission. *EU Waste Legislation*; European Commission, Environment: Brussels, Belgium, 2015; Available online: https://ec.europa.eu/environment/circular-economy/index_en.htm (accessed on 16 July 2019).
90. Velenturf, A.P.M.; Purnell, P.; Tregent, M.; Ferguson, J.; Holmes, A. Co-Producing a Vision and Approach for the Transition towards a Circular Economy: Perspectives from Government Partners. *Sustainability* **2018**, *10*, 1401. [CrossRef]
91. Wu, H.G.; Shi, Y.; Xia, Q.; Zhu, W.D. Effectiveness of the policy of circular economy in China: A DEA-based analysis for the period of 11th five-year-plan. *Resour. Conserv. Recycl.* **2014**, *83*, 163–175. [CrossRef]
92. European Commission. *Directive 2008/98/EC on Waste (Waste Framework Directive)*; European Commission, Environment: Brussels, Belgium, 2015; Available online: http://ec.europa.eu/%20environment/waste/framework/framework_%20directive.htm (accessed on 18 July 2019).
93. Trindade, A.B.; Palacio, J.C.E.; Gonzales, A.M.; Orozco, D.J.R.; Lora, E.E.S.; Renó, M.L.G.; del Olmo, O.A. Advanced exergy analysis and environmental assessment of the steam cycle of an incineration system of municipal solid waste with energy recovery. *Energy Convers. Manag.* **2018**, *157*, 194–214. [CrossRef]

94. Kleppmann, F. *Current Developments in European Waste-To-Energy*; Interessengemeinschaft der Thermischen Abfallbehandlungsanlagen Deutschland e.V.: Düsseldorf, Germany, 2013.
95. European Commission. *Energy and Environment—Overview*; European Commission: Brussels, Belgium, 2012; Available online: http://ec.europa.eu/competition/sectors/energy/overview_en.html (accessed on 18 July 2019).
96. European Commission. *Overview: Making Markets Work Better*; European Commission: Brussels, Belgium, 2015; Available online: http://ec.europa.eu/competition/general/overview_en.html (accessed on 25 July 2019).
97. Zhang, H.; Li, L.; Zhou, P.; Hou, J.; Qiu, Y. Subsidy modes, waste cooking oil and biofuel: Policy effectiveness and sustainable supply chains in China. *Energy Policy* **2014**, *65*, 270–274. [CrossRef]

MDPI
St. Alban-Anlage 66
4052 Basel
Switzerland
Tel. +41 61 683 77 34
Fax +41 61 302 89 18
www.mdpi.com

Energies Editorial Office
E-mail: energies@mdpi.com
www.mdpi.com/journal/energies

www.ingramcontent.com/pod-product-compliance
Lightning Source LLC
Chambersburg PA
CBHW061832210926
51850CB00006B/873